Undergraduate Lecture Notes in Physics

Series Editors

Neil Ashby, University of Colorado, Boulder, CO, USA

William Brantley, Department of Physics, Furman University, Greenville, SC, USA

Michael Fowler, Department of Physics, University of Virginia, Charlottesville, VA, USA

Morten Hjorth-Jensen, Department of Physics, University of Oslo, Oslo, Norway

Michael Inglis, Department of Physical Sciences, SUNY Suffolk County Community College, Selden, NY, USA

Barry Luokkala🆔, Department of Physics, Carnegie Mellon University, Pittsburgh, PA, USA

Undergraduate Lecture Notes in Physics (ULNP) publishes authoritative texts covering topics throughout pure and applied physics. Each title in the series is suitable as a basis for undergraduate instruction, typically containing practice problems, worked examples, chapter summaries, and suggestions for further reading.

ULNP titles must provide at least one of the following:

- An exceptionally clear and concise treatment of a standard undergraduate subject.
- A solid undergraduate-level introduction to a graduate, advanced, or non-standard subject.
- A novel perspective or an unusual approach to teaching a subject.

ULNP especially encourages new, original, and idiosyncratic approaches to physics teaching at the undergraduate level.

The purpose of ULNP is to provide intriguing, absorbing books that will continue to be the reader's preferred reference throughout their academic career.

More information about this series at https://link.springer.com/bookseries/8917

Reinhard Hentschke

Thermodynamics

For Physicists, Chemists and Materials
Scientists

Second Edition

 Springer

Reinhard Hentschke
School of Mathematics and Natural Sciences
Bergische Universität
Wuppertal, Germany

ISSN 2192-4791 ISSN 2192-4805 (electronic)
Undergraduate Lecture Notes in Physics
ISBN 978-3-030-93878-9 ISBN 978-3-030-93879-6 (eBook)
https://doi.org/10.1007/978-3-030-93879-6

This Springer imprint is published by the registered company Springer Nature Switzerland AG
The registered company address is: Gewerbestrasse 11, 6330 Cham, Switzerland

Preface

Many of us associate thermodynamics with blotchy photographs of men in old-fashioned garments posing in front of ponderous steam engines. In fact, thermodynamics was developed mainly as a framework for understanding the relation between heat and work and how to convert heat into mechanical work efficiently. Nevertheless, the premises or laws from which thermodynamics is developed are so general that they provide insight far beyond steam engine engineering. Today, new sources of useful energy, energy storage, transport, and conversion, requiring development of novel technology, are of increasing importance. This development strongly affects many key industries. Thus, it seems that thermodynamics will have to be given more prominence particularly in the physics curriculum—something that is attempted in this book.

Pure thermodynamics is developed, without special reference to the atomic or molecular structure of matter, on the basis of bulk quantities like internal energy, heat, and different types of work, temperature, and entropy. The understanding of the latter two is directly rooted in the laws of thermodynamics—in particular the second law. They relate the above quantities and others derived from them. New quantities are defined in terms of differential relations describing material properties like heat capacity, thermal expansion, compressibility, or different types of conductance. The final result is a consistent set of equations and inequalities. Progress beyond this point requires additional information. This information usually consists in empirical findings like the ideal gas law or its improvements, most notably the van der Waals theory, the laws of Henry, Raoult, and others. Its ultimate power, power in the sense that it explains macroscopic phenomena through microscopic theory, thermodynamics attains as part of Statistical Mechanics or more generally Many-body Theory.

The structure of this text is kept simple in order to make the succession of steps as transparent as possible. Chap. 1 (*Two Fundamental Laws of Nature*) explains how the first and the second law of thermodynamics can be cast into a useful mathematical form. It also explains different types of work as well as concepts like temperature and entropy. The final result is the differential entropy change expressed through differential changes in internal energy and the various types of work. This is a fundamental relation throughout equilibrium as well as non-equilibrium thermodynamics. Chap. 2 (*Thermodynamic Functions*), aside from introducing most of the functions used in thermodynamics, in particular internal energy, enthalpy, Helmholtz, and Gibbs free energy, contains examples allowing to practice the development and application of numerous differential relations between thermodynamic functions. The discussion includes important concepts like the relation of the aforementioned free energies to the second law, extensiveness, and intensiveness as well as homogeneity. In Chap. 3 (*Equilibrium and Stability*) the maximum entropy principle is explored systematically. The phase concept is developed together with a framework for the description of stability of phases and phase transitions. The chemical potential is highlighted as a central quantity and its usefulness is demonstrated with a number of applications. Chap. 4 (*Simple Phase Diagrams*) focuses on the calculation of simple phase diagrams based on the concept of interacting molecules. Here the description is still phenomenological. Equations, rules, and principles developed thus far are combined with van der Waals' picture of molecular interaction. As a result, a qualitative theory for simple gases and liquids emerges. This is extended to gas and liquid mixtures as well as to macromolecular solutions, melts, and mixtures based on ideas due to Flory and others. The subsequent chapter (*Microscopic Interactions*) explains how the exact theory of microscopic interactions can be combined with thermodynamics. The development is based on Gibbs' ensemble picture. Different ensembles are introduced and their specific uses are discussed. However, it also becomes clear that exactness usually is not a realistic goal due to the enormous complexity. In Chap. 6 (*Thermodynamics and Molecular Simulation*) it is shown how necessary and crude approximations sometimes can be avoided with the help of computers. Computer algorithms may even allow tackling problems eluding analytical approaches. This chapter is therefore devoted to an introduction of the Metropolis Monte Carlo method and its application in different ensembles. Thus far, the focus has been equilibrium thermodynamics. The last chapter (*Non-equilibrium Thermodynamics*) introduces concepts in non-equilibrium thermodynamics. The starting point is linear irreversible transport described in terms of small fluctuations close to the equilibrium state. Onsager's reciprocity relations are obtained and their significance is illustrated in various examples. Entropy production far from equilibrium is discussed based on the balance equation approach and the concept of local equilibrium. The formation of dissipative structures is discussed focusing on chemical reactions. This chapter also includes a brief discussion of evolution in relation to non-equilibrium thermodynamics. There are several appendices. Appendix A: Thermodynamics does not require much math. Most of the necessary machinery is compiled in this short appendix. The reason that thermodynamics is often perceived difficult is not because of its difficult mathematics. It is because of the physical understanding and meticulous care required when mathematical operations are carried out under constraints imposed by process conditions. Appendix B: The appendix contains a listing of a Grand-Canonical Monte Carlo algorithm in Mathematica. The interested reader may use this program to re-create results presented in the text in the context of equilibrium adsorption. Appendix C: This appendix compiles constants, units, and references to useful tables. Appendix D: References are included in the text and as a separate list in this appendix. Of course, there are other texts on Thermodynamics or Statistical Thermodynamics, which are nice and valuable sources of information—even if or because some of them have been around for a long time. A selected list is contained in a footnote on page 18. Another listing can be found in the preface to Hill (1986).

The first edition of Thermodynamics is structured according to concepts, allowing a compact but nonetheless comprehensive presentation. This I wanted to preserve. However, since the book's first appearance in 2014, various applications of thermodynamics, not or not sufficiently mentioned in the first edition, from a number of modern fields in physics and materials science have caught my attention. In this second edition, I have added those which I consider the most important and most fitting. A stylistic device I have used for this purpose, which was introduced in the first edition, is the "example box". Most of the examples are intended to apply and practice the application of thermodynamic concepts introduced up to this point. The second edition contains a number of more elaborate examples which themselves are self-contained subjects founded nevertheless on thermodynamic concepts.

Experiments carried out during the past three decades have considerably advanced our understanding of the universe. Thermodynamics is indispensable for the interpretation of these experiments. Instead of the two mere remarks in Section 2 in the first edition, the new edition now contains an extensive example discussing the expanding universe as well as its thermal history and future based on the components of the cosmic energy density and their equations of state. This complements the previous description of the temperature of black holes and the temperature at recombination in the context of the Saha equation. Another field shifting more and more into our focus is atmospheric physics. The first edition features short examples addressing the (dry air) temperature profile of the troposphere and the cloud base. In the second edition this is extended significantly adding discussions of the Earth's equilibrium temperature, droplet growth inside clouds, and the effect of moisture on the aforementioned temperature profile. These examples are distributed throughout the text following the discussions of the attendant thermodynamic concepts. A third field which is given more prominence are modern multi-component materials. The morphological structure and thus the performance of these materials is strongly influenced by the free energies of the component's internal interfaces. There are also the more "obvious" interfaces or surfaces when we apply coatings to protect or to change the appearance of surfaces. Unfortunately, the 1st edition lacks a thorough discussion of surface tension in terms of theory and measurement. The new edition includes a rather broad exposition of the underlying theoretical concepts like the Young-Laplace equation, Young's equation, or the OWRK-theory including applications and experimental techniques. Another "materials topic", already present in the first edition and now enhanced in the new edition, is the application of thermodynamics to polymers. The previous discussion of rubber elasticity has been completely rewritten and extended. Additionally, I have joined the discussion of conformation entropy of macromolecules in the context of rubber elasticity to the discussion of the free enthalpy of macromolecules in the context of polymer melts and mixtures. The result is the Flory-Rehner equation describing the swelling equilibrium of polymer networks. It is also worth mentioning that a fair number of mistakes are now corrected. Most of them were minor but some were not. In particular, Fig. 2.16 (capillary rise) and Fig. 3.20 (fraction of ionized hydrogen vs temperature) are replaced by their revised versions. Finally, I want to express my gratitude to Dr. Jan Plagge for his critical reading of the new material and to Alexander Weiss for pointing out a number of mistakes in the first edition.

Wuppertal, Germany   Reinhard Hentschke

Contents

Chapter 1
Two Fundamental Laws of Nature

1.1 Types of Work

1.1.1 Mechanical Work

A gas confined to a cylinder absorbs a certain amount of heat, δq. The process is depicted in Fig. 1.1. According to experimental experience this leads to an expansion of the gas. The expanding gas moves a piston to increase its volume by an amount $\delta V = V_b - V_a$. For simplicity we assume that the motion of the piston is frictionless and that its mass is negligible compared to the mass, m, of the weight pushing down on the piston. We do not yet have a clear understanding of what heat is, but we consider it a form of energy which to some extend can be converted into mechanical work, w.[1] In our case this is the work needed to lift the mass, m, by a height, δs, against the gravitational force $m\vec{g}$. From mechanics we know

$$\delta w_{done\ by\ gas} = \int_a^b d\vec{s} \cdot \vec{f}_{gas} = -\int_a^b d\vec{s} \cdot m\vec{g}$$

$$= P_{ex} \int_{V_a}^{V_b} dV = P_{ex}\delta V.$$

Here $P_{ex} = mg/A$ is the external pressure exerted on the gas due to the force mg acting on the cross-sectional area, A ($\delta V = A\delta s$).

---

[1] Originally it was thought that heat is a sort of fluid and heat transfer is transfer of this fluid. In addition, it was assumed that the overall amount of this fluid is conserved. Today we understand that heat is a form of dynamical energy due to the disordered motion of microscopic particles and that heat can be changed into other forms of energy. This is what we need to know at this point. The microscopic level will be addressed in Chap. 5.

© Springer Nature Switzerland AG 2022
R. Hentschke, *Thermodynamics*, Undergraduate Lecture Notes in Physics,
https://doi.org/10.1007/978-3-030-93879-6_1

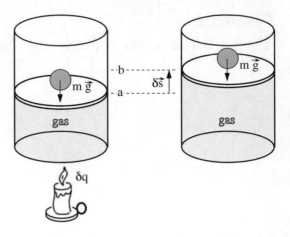

Fig. 1.1 A gas confined to a cylinder absorbs a certain amount of heat, δq

The process just described leads to a change in the total energy content of the gas, δE. The gas receives a positive amount of heat, δq. However, during the expansion it also does work and thereby reduces its total energy content, in the following called internal energy, by $-P_{ex}\delta V$. The combined result is

$$\delta E = \delta q - P_{ex}\delta V.$$

Notice that after the expansion has come to an end we have $P_{ex} = P$, where P is the gas pressure inside the cylinder. In particular we know that P is a function of the volume, V, occupied by the gas, i.e. $P = P(V)$. In the following we assume that the change in gas pressure during a small volume change δV is a second order effect which can be neglected. Therefore for small volume changes we have

$$\boxed{\delta E = \delta q - P\delta V.} \tag{1.1}$$

This is the first law of thermodynamics for this special process. It uses energy conservation to distinguish the different contributions to the total change in internal energy of a system (here the gas) during a thermodynamic process (here absorption of heat plus volume expansion).

We just have introduced two important concepts frequently used in thermodynamics—process and system. The latter requires the ability to define a boundary between "inside" and "outside". Both, the inside and the outside, may be considered systems individually. Systems usually are distinguished according to their degree of openness. Isolated system means that this system exchanges nothing with its exterior. An open system on the other hand may exchange everything there is to exchange, like heat or matter. A closed system holds back matter but allows heat exchange, e.g. the above gas filled cylinder. Systems are sometimes divided into subsystems. Subsystems, however, are still systems. After having defined or (better) prepared a system we may observe what happens to it or we may actively do something to it. This "what happens to it" or "doing something" means that the system undergoes a process (of change). A special type of system is the reservoir. A reservoir usually is in thermal contact with our system of interest. Thermal contact means that heat may be transferred between the reservoir and our system of interest. However, the reservoir is so large that there is no measurable change in any of its physical properties due to the exchange.

Now we proceed replacing the above gas by an elastic medium. Those readers who are not sufficiently familiar with the theory of elastic bodies may skip ahead to "Electric work" (p. 7).

Mechanical Work Involving Elastic Media

We consider an elastic body composed of volume elements dV depicted in Fig. 1.2. The total force acting on the elastic body may be calculated according to

$$\int_V dV f_\alpha \tag{1.2}$$

for every component α ($= 1, 2, 3$ or x, y, z). Here $\vec{f}$ is a force density, i.e. force per volume. Assuming that the f_α are purely elastic forces acting between the boundaries of the aforementioned volume elements inside V, i.e. excluding for instance gravitational forces or other external fields acting on volume elements inside the elastic body, we may define the internal stress tensor, σ, via

$$f_\alpha = \sum_{\beta=1}^{3} \frac{\partial \sigma_{\alpha\beta}}{\partial x_\beta} \equiv \frac{\partial \sigma_{\alpha\beta}}{\partial x_\beta}. \tag{1.3}$$

Fig. 1.2 Elastic body composed of volume elements dV

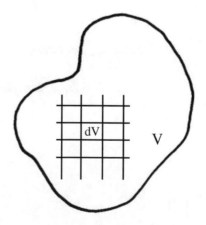

Fig. 1.3 The relation between indices, force components, and the faces of the cubic volume element

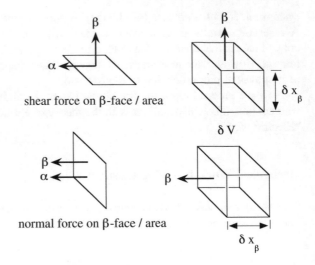

shear force on β-face / area

normal force on β-face / area

Here we apply the summation convention, i.e. if the same index appears twice on the same side of an equation then summation over this index is implicitly assumed (unless explicitly stated otherwise). The relation between indices, force components, and the faces of the cubic volume element is depicted in Fig. 1.3. Upper and lower sketches illustrate the shear and the normal contribution to the force component f_α acting on the volume element in α-direction. Notice that f_α can be written as the sum over two shear stress and one normal stress contribution. The latter are stress differences between adjacent faces of the cubic volume element. Note also that the unit of $\sigma_{\alpha\beta}$ is force per area.

We want to calculate the work δw done by the f_α during attendant small displacements δu_α, i.e.

$$\delta w = \int dV f_\alpha \delta u_\alpha \overset{(1.3)}{=} \int dV \frac{\partial \sigma_{\alpha\beta}}{\partial x_\beta} \delta u_\alpha.$$

The integral may be rewritten using Green's theorem in space:

$$\delta w = \oint \sigma_{\alpha\beta} \delta u_\alpha dA_\beta - \int dV \sigma_{\alpha\beta} \frac{\partial \delta u_\alpha}{\partial x_\beta}.$$

We neglect the surface contribution[2] and use the symmetry property of the stress tensor[3] to obtain

$$\delta w = - \int dV \sigma_{\alpha\beta} \frac{\partial \delta u_\alpha}{\partial x_\beta} = - \frac{1}{2} \int dV \sigma_{\alpha\beta} \delta \underbrace{\left(\frac{\partial u_\alpha}{\partial x_\beta} + \frac{\partial u_\beta}{\partial x_\alpha} \right)}_{\cong 2u_{\alpha\beta}}.$$

The quantity $u_{\alpha\beta}$ is the strain tensor (here for small displacements). The final result is

$$\delta w = - \int dV \sigma_{\alpha\beta} \delta u_{\alpha\beta}. \tag{1.4}$$

We want to work this out in three simple cases. First we consider a homogeneous dilatation of a cubic volume $V = L_x L_y L_z$. We also assume that the shear components of the stress tensor vanish, i.e. $\sigma_{\alpha\beta} = 0$ for $\alpha \neq \beta$. In such a system the normal components of the stress tensor should all be the same, i.e. $\sigma \equiv \sigma_{xx} = \sigma_{yy} = \sigma_{zz}$. We thus have

$$\sigma_{\alpha\beta} \delta u_{\alpha\beta} = \sigma_{xx} \delta u_{xx} + \sigma_{yy} \delta u_{yy} + \sigma_{zz} \delta u_{zz} = \sigma(\delta u_{xx} + \delta u_{yy} + \delta u_{zz}). \tag{1.5}$$

Homogeneous deformation means

$$\frac{\partial u_\alpha}{\partial x_\alpha} = \frac{\delta L_\alpha}{L_\alpha}. \tag{1.6}$$

And because $u_{\alpha\alpha} = \partial u_\alpha / \partial x_\alpha$ (no summation convention here) we obtain

$$\sigma_{\alpha\beta} \delta u_{\alpha\beta} = \sigma \left(\frac{\delta L_x}{L_x} + \frac{\delta L_y}{L_y} + \frac{\delta L_z}{L_z} \right) = \sigma \frac{\delta V}{V}. \tag{1.7}$$

---

[2] For a discussion see Landau et al. (1986).

[3] To show the symmetry of the stress tensor, i.e. $\sigma_{\alpha\beta} = \sigma_{\beta\alpha}$, we compute the torque exerted by the f_α in a particular volume element integrated over the entire body:

$$\int dV(f_\alpha x_\beta - f_\beta x_\alpha) = \int dV \left(\frac{\partial \sigma_{\alpha\gamma}}{\partial x_\gamma} x_\beta - \frac{\partial \sigma_{\beta\gamma}}{\partial x_\gamma} x_\alpha \right)$$
$$= \int dV \frac{\partial(\sigma_{\alpha\gamma} x_\beta - \sigma_{\beta\gamma} x_\alpha)}{\partial x_\gamma} - \int dV(\sigma_{\alpha\gamma} \delta_{\beta\gamma} - \sigma_{\beta\gamma} \delta_{\alpha\gamma})$$
$$= \oint (\sigma_{\alpha\gamma} x_\beta - \sigma_{\beta\gamma} x_\alpha) dA_\gamma - \int dV(\sigma_{\alpha\beta} - \sigma_{\beta\alpha}).$$

The volume integral must vanish in order for the net torque to be entirely due to forces applied to the surface of the body.

Integration over the full volume then yields

$$\delta w = -\sigma \delta V, \tag{1.8}$$

i.e. we recover the above gas case with $P = -\sigma$.

In a second example we consider the homogeneous dilatation of a thin elastic sheet. The sheet's volume is $V = Ah = L_x L_y h$, where the thickness, h, is small and constant. Now we have

$$\sigma_{\alpha\beta}\delta u_{\alpha\beta} = \sigma\left(\frac{\delta L_x}{L_x} + \frac{\delta L_y}{L_y}\right) = \sigma\frac{\delta A}{A} \tag{1.9}$$

and therefore

$$\delta w = -\sigma h \delta A \equiv -\gamma \delta A. \tag{1.10}$$

The quantity γ is the surface tension.

An obvious third example is the homogeneous dilatation of a thin elastic column $V = h^2 L_z$. Here h^2 is the column cross sectional area and L_z is its length. This time we have

$$\sigma_{\alpha\beta}\delta u_{\alpha\beta} = \sigma\left(\frac{\delta L_z}{L_z}\right) \tag{1.11}$$

and thus

$$\delta w = -\sigma A \delta L_z \equiv -T\delta L_z, \tag{1.12}$$

where T is the tension.

Example—Expanding Gas We consider the special case of the first law expressed in Eq. (1.1). If we include the surface tension contribution to the internal energy of the expanding gas, then the resulting equation is

$$\delta E = \delta q - P\delta V + \gamma\delta A. \tag{1.13}$$

We remark that the usual context in which one talks about surface tension refers to interfaces. This may be the interface between two liquids or the surface of a liquid film relative to air, e.g. a soap bubble. In the latter case there are actually two surfaces. In such cases we define $\gamma = f_T/(2l)$, which reflects the presence of two surfaces.

Example—Fusing Bubbles An application of surface tension is depicted in Fig. 1.4. The figure depicts two soap bubbles touching and fusing. We ask whether the small bubble empties its gas content into the large one or vice

Fig. 1.4 An application of
surface tension

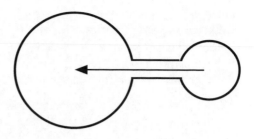

versa. We may answer this question by considering the work done by one
isolated bubble during a small volume change:

$$\delta w_{done\ by\ gas\ in\ bubble} = P_{ex}\delta V + \gamma\delta A.$$

Notice that the sign of the surface tension contribution has changed
compared to Eq. (1.10). This is because in Eq. (1.10) we compute the work
done by the membrane. But here the gas is doing work on the membrane,
which changes the sign of this work contribution. The same work, i.e.
$\delta w_{done\ by\ gas\ in\ bubble}$, can be written in terms of the pressure, P, inside the
bubble,

$$\delta w_{done\ by\ gas\ in\ bubble} = P\delta V.$$

Combining the two equations and using $\delta V = 4\pi r^2 \delta r$ and $\delta A = 8\pi r\delta r$,
where r is the bubble radius, yields

$$P = P_{ex} + \frac{2\gamma}{r}.$$

We conclude that the gas inside the smaller bubble has the higher pressure
and therefore the smaller bubble empties itself into the larger bubble.

1.1.2 Electric Work

We now consider work involving electric and magnetic variables[4],[5] starting with an
example.

---

[4] Here we use Gaussian units. The conversion to SI-units is tabulated in Appendix C.

[5] Three early but very basic papers in this context are: Guggenheim (1936a, b); Koenig (1937).

Example—Charge Transfer Across a Potential Drop A charge δq in an electric field experiences the force $\vec{F} = \delta q \vec{E}$ (Do not confuse this δq with the previously introduced heat change!). Consequently the work done by the charge-field system if the charge moves from point a to point b in space is

$$\delta w_q = \int_a^b d\vec{s} \cdot \vec{F} = -\delta q \int_a^b d\vec{s} \cdot \vec{\nabla}\phi. \tag{1.14}$$

Here ϕ is the potential, i.e. $\vec{E} = -\vec{\nabla}\phi$, and thus

$$\delta w_q = -\delta q \phi_{ba}, \tag{1.15}$$

where $\phi_{ba} = \phi(b) - \phi(a)$ is the potential difference between b and a. The corresponding internal energy of the charge-field system changes by

$$\delta E_q = -\delta w_q = \delta q \phi_{ba}. \tag{1.16}$$

This equation may be restated for a charge current $I = \delta q / \delta t$, where δt is a certain time interval:

$$\delta E_I = I \phi_{ba} \delta t. \tag{1.17}$$

In the presence of the resistance, R, the quantity $\delta q_{Joule} = RI^2 \delta t$ is the Joule heat generated by the current (James Prescott Joule, British physicist, *Salford (near Manchester) 24.12.1818, †Sale (County Cheshire) 11.10.1889; made important contributions to our understanding of heat in relation to mechanical work (Joule heat) and internal energy (Joule-Thomson effect).).

Now we consider the following equations appropriate for continuous dielectric media:

$$\vec{\nabla} \times \vec{H} = \frac{1}{c} \frac{\partial}{\partial t} \vec{D} + \frac{4\pi}{c} \vec{j} \tag{1.18}$$

and

$$\vec{\nabla} \times \vec{E} = -\frac{1}{c} \frac{\partial}{\partial t} \vec{B}. \tag{1.19}$$

The second equation simply follows by the usual spatial averaging procedure applied to the corresponding vacuum Maxwell's equation.[6] Here $\vec{E}(\vec{r})$ is the average electric field in a volume element at point $\vec{r}$. This volume element is large compared to atomic dimensions. In the same sense $\vec{D}$ is the displacement field given by $\vec{D} = \vec{E} + 4\pi\vec{P}$. $\vec{P}$ is the macroscopic polarization, i.e. the local electrical dipole moment per volume. Analogously $\vec{B} = \vec{H} + 4\pi\vec{M}$ is the average magnetic field (magnetic induction), and $\vec{M}$ is the macroscopic magnetization, i.e. the local magnetic dipole moment per volume. The first equation is less obvious and requires a more detailed discussion.

We consider a current density $\vec{j}_e$ inside a medium due to an extra ("injected") charge density ρ_e. The two quantities fulfill the continuity equation

$$\frac{\partial \rho_e}{\partial t} + \vec{\nabla} \cdot \vec{j}_e = 0.$$

We also have

$$\vec{\nabla} \cdot \vec{D} = 4\pi\rho_e.$$

Differentiation of this Maxwell equation with respect to time and inserting the result into the previous equation yields

$$\vec{\nabla} \cdot \left(\frac{\partial \vec{D}}{\partial t} + 4\pi\vec{j}_e \right) = 0.$$

The expression in brackets is a vector, which may be expressed as the curl of another vector $c'\vec{H}'$, i.e.

$$\vec{\nabla} \times \vec{H}' = \frac{1}{c'} \frac{\partial \vec{D}}{\partial t} + \frac{4\pi}{c'} \vec{j}_e.$$

Comparison of this with Ampere's law in vacuum suggests indeed $\vec{H}' = \vec{H}$ and $c' = c$. We thus arrive at Eq. (1.18). An in depths discussion can be found in Lifshitz et al. (2004).

We proceed by multiplying Eq. (1.18) with $c\vec{E}/(4\pi)$ and Eq. (1.19) with $-c\vec{H}/(4\pi)$. Adding the two equations yields

$$\frac{c}{4\pi} \vec{E} \cdot \left(\vec{\nabla} \times \vec{H} \right) - \frac{c}{4\pi} \vec{H} \cdot \left(\vec{\nabla} \times \vec{E} \right) = \frac{1}{4\pi} \vec{E} \cdot \frac{\partial \vec{D}}{\partial t} + \frac{1}{4\pi} \vec{H} \cdot \frac{\partial \vec{B}}{\partial t} + \vec{j} \cdot \vec{E}.$$

---

[6] James Clerk Maxwell, British physicist, *Edinburgh 13.6.1831, †Cambridge 5.11.1879; particularly known for his unified theory of electromagnetism (Maxwell equations).

With the help of the vector identity $\vec{\nabla} \cdot (\vec{a} \times \vec{b}) = \vec{b} \cdot (\vec{\nabla} \times \vec{a}) - \vec{a} \cdot (\vec{\nabla} \times \vec{b})$ this is transformed into

$$\frac{c}{4\pi} \vec{\nabla} \cdot (\vec{H} \times \vec{E}) = \frac{1}{4\pi} \vec{E} \cdot \frac{\partial \vec{D}}{\partial t} + \frac{1}{4\pi} \vec{H} \cdot \frac{\partial \vec{B}}{\partial t} + \vec{j} \cdot \vec{E}.$$

Now we integrate both sides over the volume V and use Green's theorem in space (also called divergence theorem), i.e. $\int_V dV \vec{\nabla} \cdot (\vec{H} \times \vec{E}) = \int_A d\vec{A} \cdot (\vec{H} \times \vec{E})$, where $\vec{A}$ is a surface element on the surface A of the volume oriented towards the outside of V. If we choose the volume so that the fields vanish on its surface, then $\int_A d\vec{A} \cdot (\vec{H} \times \vec{E}) = 0$ (The configuration of the system is fixed during all of this.). Thus our final result is

$$\int dV \left(\vec{E} \cdot \frac{\delta \vec{D}}{4\pi} + \vec{H} \cdot \frac{\delta \vec{B}}{4\pi} + \vec{j} \cdot \vec{E}\delta t \right) = 0. \tag{1.20}$$

The third term in Eq. (1.20) is the work done by the $\vec{E}$-field during the time δt. To see this we imagine a cylindrical volume element whose axis is parallel to $\vec{j}$ depicted in Fig. 1.5. Then $\delta V = A \delta s$ and $\delta V \vec{j} = (q/\delta t)\delta \vec{s}$, where q is the charge passing through the area A during the time δt. Thus $\delta V \vec{j} \cdot \vec{E}\delta t = q\vec{E} \cdot \delta \vec{s}$, where $q\vec{E}$ is the force acting on the charge q doing work (cf. the above example).

We conclude that we may express the work done by the system, $\delta w = \int dV \vec{j} \cdot \vec{E} \, \delta t$, by the other two terms in Eq. (1.20) describing the attendant change of the electromagnetic energy content of the system. For a process during which the system exchanges heat and is doing electrical work we now have

$$\boxed{\delta E = \delta q + \int dV \left(\vec{E} \cdot \frac{\delta \vec{D}}{4\pi} + \vec{H} \cdot \frac{\delta \vec{B}}{4\pi} \right).} \tag{1.21}$$

The quantities $\vec{E}$, $\vec{D}$, $\vec{B}$, and $\vec{H}$ are more difficult to deal with than fields in vacuum. Nevertheless, for the moment we postpone a more detailed discussion and return to Eq. (1.21) on p. 66.

Fig. 1.5 A cylindrical volume element whose axis is parallel to $\vec{j}$

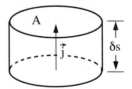

1.1.3 Chemical Work

As a final example consider an open system—one we can add material to. Generally, work must be done to increase the amount of material in a system. The work done depends on the state of the system. If we add δn moles of material,[7] we write the work done on the system as

$$\delta w_{done\ on\ system} = \mu \delta n. \tag{1.22}$$

The quantity μ is called the chemical potential (per mole added). In a more general situation a system may contain different species. We shall say that these are different components i. Now the above equation becomes

$$\delta w_{done\ on\ system} = \sum_i \mu_i \delta n_i. \tag{1.23}$$

Here μ_i is the chemical potential of component i. Thus for a process involving exchange of heat as well as chemical work we have

$$\boxed{\delta E = \delta q + \sum_i \mu_i \delta n_i.} \tag{1.24}$$

1.1.4 The First Law

The first law is expressing conservation of energy. The specific terms appearing in the first law do depend on the types of work occurring in the process of interest. The following box contains a number of examples.

> **Example—Statements of the First Law for Different Processes**
>
> (i) $\delta E = \delta q - P\delta V + \gamma \delta A + \sum_i \mu_i \delta n_i$

---

[7] One mole ($n = 1$) is an amount of substance of a system which contains as many elementary units as there are atoms of carbon in 12 g of the pure nuclide carbon-12. The elementary unit may be an atom, molecule, ion, electron, photon, or a specified group of such units.

This describes a process during which heat is exchanged by the system and its exterior. Mechanical work in the form of volume work and surface work is done in addition. The composition of the system changes as well.

$$(ii) \qquad \delta E = \delta q - \int dV \vec{j} \cdot \vec{E} \delta t$$

Here the process of interest involves heat exchange and electrical work.

$$(iii) \qquad \delta E = \delta q - P\delta V + \int dV \frac{\vec{E} \cdot \delta \vec{D} + \vec{H} \cdot \delta \vec{B}}{4\pi}$$

This example is for a process during which heat is exchanged and both volume and electrical work is done.

More generally the first law is expressed via

$$\delta E = \delta q - \delta w. \tag{1.25}$$

However, there is an alternative sign convention used in some of the literature, i.e.

$$\delta E = \delta q + \delta w. \tag{1.26}$$

The sign preceding δw depends on the meaning of the latter. In Eq. (1.25) δw always is the *work done by the system* for which we write down the change in the system's internal energy, δE, during a process involving both heat transfer and work. In Eq. (1.26) on the other hand δw is understood as *work done on the system*. In the following we shall use the sign convention as expressed in Eq. (1.25)!

Another point worth mentioning is the usage of the symbols δ, Δ, and d. δ denotes a small change (afterwards–before) during a process. Δ basically has the same meaning, except that the change is not necessarily small. Even though d indicates a small change just like δ, it has an additional meaning—indicating exact differentials. This is something we shall discuss in much detail latter in the text. But for the benefit of those who compare the form of Eq. (1.25) to different texts, we must add a provisional explanation.

In principle every process has a beginning and an end. Beginning and end, as we shall learn, are defined in terms of specific values of certain variables (e.g. values of P and V). These two sets of variable values can be connected by different processes or paths in the space in which the variables "live". If a quantity changes during a process and this change only depends on the two endpoints of the path rather than on the path as a whole, then the quantity possesses an exact differential and vice versa. In the case of mechanical work, for instance, we can imagine pushing a cart from point A to point B. There may be two alternative routes—one involving a lot of friction and a "smooth" one causing less friction. In the former case one may find Eq. (1.25) stated as

$$dE = \delta q - \delta w. \tag{1.27}$$

This form explicitly distinguishes between the exact differential dE and the quantities δq and δw, which are not exact differentials. In the case of δw this is in accord with our cart-pushing example, because the work does depend on the path we choose. For the two other quantities we shall show their respective property latter in this text (cf. p. 327ff), when we deal with the mathematics of exact differentials.

However, already at this point we remark that the expressions we have derived in our examples for the various types of work will reappear with d instead of δ. This is because we focus on what we shall call reversible work. Friction, occurring in the cart-pushing example or possibly in Fig. 1.1 when the gas moves the piston, is neglected as well as other types of loss. The following is an example illustrating what we mean by reversible vs. irreversible work.

Example—Reversible and Irreversible Work In an isotropic elastic body the following equation holds (Landau et al. 1986):

$$\sigma_{\alpha\beta} \stackrel{\alpha\neq\beta}{=} 2\mu u_{\alpha\beta} = \mu\left(\frac{\partial u_\alpha}{\partial x_\beta} + \frac{\partial u_\beta}{\partial x_\alpha}\right). \tag{1.28}$$

On the right is the stress tensor and on the left the product of 2μ with the strain tensor (for small strain). The quantity μ is the shear modulus (not to be confused with the chemical potential). This equation is related to the two upper sketches in Fig. 1.3. If in the depicted situation (shear force acting on β-face is applied in α-direction) there is little or ideally no strain in β-direction (this is like shearing a deck of cards), then the above equation may be written as

$$\sigma_\mu \equiv \sigma_{\alpha\beta} = \mu\frac{\partial u_\alpha}{\partial x_\beta} \equiv \mu u_\mu. \tag{1.29}$$

Real shear is accompanied by friction. Experience suggests that friction often can be described by an equation akin to the above:

$$\sigma_\eta = \eta \dot{u}_\eta. \tag{1.30}$$

The quantity η is a friction coefficient and $\dot{u}_\eta$ is a strain rate. Figure 1.6, showing a spring and a dashpot, is a pictorial representation of Eqs. (1.29)

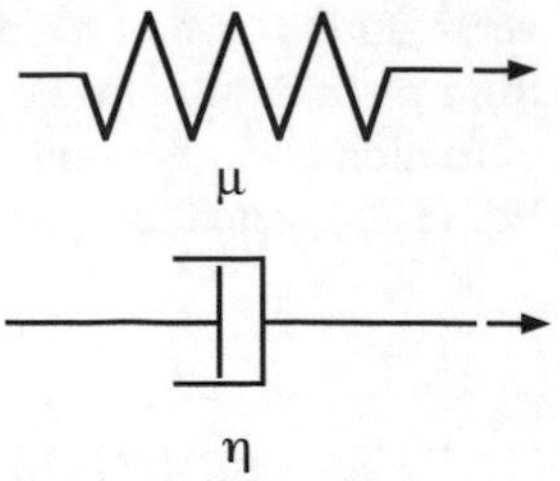

Fig. 1.6 Pictorial
representation of Eqs. (1.29)
and (1.30)

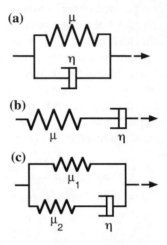

Fig. 1.7 Three simple
combinations (**a**, **b**, **c**) of the
two elements

and (1.30). Figure 1.7 shows three simple combinations (a, b, c) of the two
elements depicted in Fig. 1.6. These combinations may be translated into
differential equations and serve as simple models for so called viscoelastic
behavior (Wrana 2009). Important viscoelastic materials are the tread compounds in automobile tires. In the following we merely focus on sketch (a).
Its translation is

$$\sigma \equiv \sigma_\mu + \sigma_\eta = \mu u + \eta \dot{u} \qquad (u \equiv u_\mu = u_\eta). \tag{1.31}$$

We assume that the applied stress is $\sigma = \sigma_o \sin(\omega t + \delta)$, where ω is a
frequency, t is time, and δ is a phase. The attendant strain is $u = u_o \sin(\omega t)$
(This is a simple mathematical description of an experimental procedure in
what is called dynamic mechanical analysis.). Inserting this into Eq. (1.31) we
find the relations

$$\sigma_o \cos \delta = \mu u_o \equiv \mu' u_o \tag{1.32}$$

$$\sigma_o \sin \delta = \eta \omega u_o \equiv \mu'' u_o. \tag{1.33}$$

The two newly defined quantities μ' and μ'' are called storage and loss modulus, respectively. Their meaning becomes clear if we compute the work done during one full shear cycle, i.e.

$$\oint \sigma du = \int_0^{2\pi/\omega} \sigma \dot{u} dt = \pi \mu'' u_o^2. \tag{1.34}$$

Actually this is work per volume (cf. Eq. (1.4)). However, if we do the same calculation just for the first quarter cycle (form zero to maximum shear strain) the result is

$$\int_0^{\pi/(2\omega)} \sigma \dot{u} dt = \frac{1}{2} \mu' u_o^2 + \frac{1}{4} \pi \mu'' u_o^2. \tag{1.35}$$

The first term is the reversible part of the work, which does not contribute to the integral in the case of a full cycle. This term is analogous to the elastic energy stored in a stretched/compressed spring. The second term as well as the result in (1.34) cannot be recovered and is lost, i.e. producing heat. Models like ours only convey a crude understanding of loss or dissipative processes in viscoelastic materials. Considerable effort is spend by the R&D departments of major tire makers to understand and control loss on a molecular basis. In tire materials the moduli themselves strongly depend on the shear amplitude. Understanding and controlling this effect, the Payne effect, is one important ingredient for the improvement of tire materials, e.g. optimizing rolling resistance (Vilgis et al. 2009).

1.2 The Postulates of Kelvin and Clausius

The first law does not address the limitations of heat conversion into work or heat transfer between systems. The following two postulates based on experimental experience do just this. They are the foundation of what is called the second law of thermodynamics.[8]

---

[8] Here we follow Fermi (1956). Dover (Enrico Fermi, Nobel prize in physics for his contributions to nuclear physics, 1938).

1.2.1   Postulate of Lord Kelvin (K)

A complete transformation of heat (extracted from a uniform source) into work is impossible.[9]

1.2.2   Postulate of Clausius (C)

It is impossible to transfer heat from a body at a given temperature to a body at higher temperature as the only result of a transformation.[10]

Remark   At this point we use the "temperature" θ to characterize a reservoir as hotter or colder than another. The precise meaning of temperature is discussed in the following section.

These two postulates are equivalent. A way to prove this is by assuming that the first postulate is wrong. This is then shown to contradict the second postulate. Subsequently the same reasoning is applied starting with the second postulate, i.e. the assumption that the second postulate is wrong is shown to contradict the first.

First we assume (K) to be false. Figure 1.8 illustrates what happens. At the top is a reservoir at a temperature θ_1 surrendering heat q to a device (circle) which converts this exact amount of heat into work w. A process possible if (K) is false. At the bottom this setup is extended by a friction device (f) converting the work w into heat q, which is transferred to a second reservoir at $\theta_2 (> \theta_1)$. Thus the only overall result of the process is the transfer of heat from the colder to the hotter reservoir. We therefore contradict (C).

Now we assume that (C) is false. The upper part of Fig. 1.9 shows heat q flowing from the colder reservoir to the hotter reservoir—with no other effect. At the bottom this setup is extended. The heat q is used to do work leaving the upper reservoir unaltered. Clearly, this is in violation of (K). Therefore both postulates are equivalent. They have important consequences, which we explore below.

---

[9] Thomson, Sir (since 1866) William, Lord Kelvin of Largs, (since 1892), British physicist, *Belfast 26.6.1824, †Netherhall (near Largs, North Ayrshire) 17.12.1907; one of the founders of classical thermodynamics; among his achievements are the Kelvin temperature scale, the discovery of the Joule-Thomson effect in 1853 with J. P. Joule and the thermoelectric Thomson effect in 1856, as well as the development of an atomic model with J. J. Thomson in 1898.

[10] Rudolf Julius Emanuel Clausius, German physicist, *Köslin (now Koszalin) 2.1.1822, †Bonn 24.8.1888; one of the developers of the mechanical theory of heat; his achievements encompass the formulation of the second law and the introduction of the "entropy" concept.

Fig. 1.8 Assumption of
postulates: Kelvin

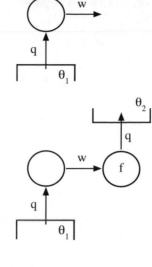

Fig. 1.9 Assumption of
postulates: Clausius

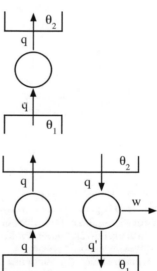

1.3   Carnot's Engine and Temperature

Consider a fluid undergoing a cyclic transformation shown in Fig. 1.10. The upper
graph shows the cycle in the P-V-plane, whereas the lower is a sketch illustrating
the working principle of a corresponding device. Here the amount of heat q_2 is
transferred from a heat reservoir at temperature θ_2 ($\theta_2 > \theta_1$) to the device. During

Fig. 1.10 Fluid undergoing a cyclic transformation

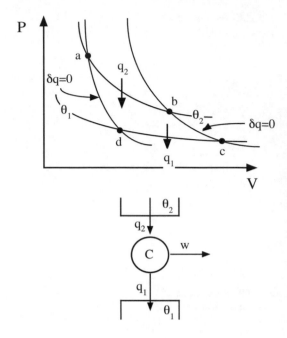

the transfer (path from a to b in the P-V-diagram) the temperature in the device is θ_2. This part of the process is an isothermal expansion. Then the device crosses via adiabatic[11] expansion to a second isotherm at temperature θ_1, the temperature of a

---

[11] A transformation of a thermodynamic system is adiabatic if it is reversible and if the system is thermally insulated. Definitions of an adiabatic process taken from the literature:

Pathria (1972): "Hence, for the constancy of S (Entropy) and N (number of particles), which defines an adiabatic process, ..."

Fermi (1956): "A transformation of a thermodynamic system is said to be adiabatic if it is reversible and if the system is thermally insulated so that no heat can be exchanged between it and its environment during the transformation."

Pauli (1973): "Adiabatic: During the change of state, no addition or removal of heat takes place; ..."

Chandler (1987): "... the change ΔS is zero for a reversible adiabatic process, and otherwise ΔS is positive for any natural irreversible adiabatic process."

Guggenheim (1986): "When a system is surrounded by an insulating boundary the system is said to be thermally insulated and any process taking place in the system is called adiabatic. The name adiabatic appears to be due to Rankine (Maxwell, Theory of Heat, Longmans 1871)."

Kondepudi and Prigogine (1998): "In an adiabatic process the entropy remains constant."

We note that for some authors "adiabatic" includes "reversibility" and for others, here Pauli, Chandler, and Guggenheim, "reversibility" is a separate requirement, i.e. during an "adiabatic" process no heat change takes place but the process is not necessarily reversible. (See also the discussion of the "adiabatic principle" in Hill (1956).)

second reservoir (path from b to c in the P-V-diagram).[12] Now follows an isothermal compression during which the device releases the amount of heat q_1 into the second reservoir (path from c to d in the P-V-diagram). The final part of the cycle consists of the crossing back via adiabatic compression to the first isotherm (path from d to a in the P-V-diagram). In addition to the heat transfer between reservoirs the device has done the work w. Any device able to perform such a cyclic transformation in both directions is called a Carnot engine.[13,14]

According to the first law, $\delta E = \delta q - \delta w$, applied to the Carnot engine we have $\Delta E = 0$ and thus $w = q_2 - q_1$. Our Carnot engine has a thermal efficiency, generally defined by

$$\eta = \frac{\text{work done}}{\text{heat absorbed}} = \frac{w}{q_2}, \tag{1.36}$$

which is

$$\eta = 1 - \frac{q_1}{q_2}. \tag{1.37}$$

Remark If the arrows in Fig. 1.10 are reversed the result is a heat pump, i.e. a device which uses work to transfer heat from a colder reservoir to a hotter reservoir. The efficiency of such a device is $1/\eta$. Here the aim is to use as little work as possible to transfer as much heat as possible.

Now we prove an interesting fact—the Carnot engine is the most efficient device, operating between two temperatures, which can be constructed! This is called Carnot's theorem. To prove Carnot's theorem we put the Carnot engine (C) in series with an arbitrary competing device (X) as shown in Fig. 1.11.

First we note that if we operate both devices many cycles we can make their total heat inputs added up over all cycles, q_2 and q'_2, equal (i.e., $q_2 = q'_2$ with arbitrary precision). After we have realized this we now reverse the Carnot engine (all arrows on C are reversed). Again we operate the two engines for as many cycles as it takes to fulfill $q_2 = q'_2$. This means that reservoir 2 is completely unaltered. But what are the consequences of all this?

According to the first law we have

$$w_{total} \overset{1.law}{=} q_{2,total} - q_{1,total} \tag{1.38}$$

---

[12] Do you understand why the slopes of the isotherms are less negative than the slopes of the adiabatic curves? You find the answer on p. 48.

[13] Nicolas Léonard Sadi Carnot, French physicist, *Paris 1.6.1796, †ibidem 24.8.1832; his calculations of the thermal efficiency for steam engines prepared the grounds for the second law.

[14] If you are interested in actual realizations of the Carnot engine and what they are used for visit http://www.stirlingengine.com.

Fig. 1.11 Proof of Carnot's theorem

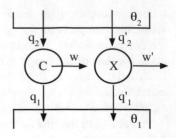

where

$$q_{2,total} = -q_2 + q'_2 = 0$$
$$q_{1,total} = -q_1 + q'_1.$$

Because the second reservoir is unaltered we must have

$$w_{total} \leq 0. \tag{1.39}$$

$w_{total} > 0$ violates Kelvin's postulate! However, this implies

$$q_{1,total} \geq 0$$
$$\Rightarrow q'_1 \geq q_1$$
$$\Rightarrow q'_1 q_2 \geq q_1 q'_2$$
$$\Rightarrow \frac{q'_1}{q'_2} \geq \frac{q_1}{q_2}.$$

And therefore

$$\eta_X = 1 - \frac{q'_1}{q'_2} \leq 1 - \frac{q_1}{q_2} = \eta_{Carnot}. \tag{1.40}$$

There is no device more efficient than Carnot's engine. Question: Do you understand what distinguishes the Carnot engine in this proof from its competitor? It is the reversibility. If the competing device also is fully reversible we can redo the proof with the two engines interchanged. We then find $\eta_{Carnot} \leq \eta_X$, and thus $\eta_{Carnot} = \eta_X$. We may immediately conclude the following corollary: All Carnot engines operating between two given temperatures have the same efficiency.

This in turn allows to define a temperature scale using Carnot engines. The idea is illustrated in Fig. 1.12. We imagine a sequence of Carnot engines all producing the same amount of work w. Each machine uses the heat given off by the previous engine as input. According to the first law

Fig. 1.12 Defining
temperature scale using
Carnot engines

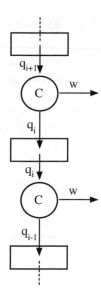

$$w = q_{i+1} - q_i. \tag{1.41}$$

We define the reservoir temperature θ_i via

$$\theta_i = xq_i, \tag{1.42}$$

where x is a proportionality constant independent of i. Thus the previous equation
becomes

$$xw = \theta_{i+1} - \theta_i. \tag{1.43}$$

We may for instance choose $xw = 1$ K, i.e. the temperature difference between
reservoirs is 1 K. We remark that this definition of a temperature scale is independent of the substance used. Furthermore the thermal efficiency of the Carnot
engine becomes

$$\eta_{Carnot} = 1 - \frac{\theta_1}{\theta_2} \tag{1.44}$$

$(\theta_2 > \theta_1)$. Notice that the efficiency can be increased by making θ_1 as low and θ_2 as
high as possible. Notice also that $\theta_1 = 0$ is not possible, because this violates the
second law. θ_1 can be arbitrarily close but not equal to zero. On p. 50 we compute
the thermal efficiency for the Carnot cycle in Fig. 1.10 using an ideal gas as
working medium. We shall see that for the ideal gas temperature $T \propto \theta$. Thus from
here on we use $\theta = T$.

1.4   Entropy

Some of you may have heard about the thermodynamic time arrow. Gases escape from open containers and heat flows from a hot body to its colder environment. Never has spontaneous reversal of such processes been observed. We call these irreversible processes. The world is always heading forward in time. Mathematically this is expressed by Clausius' theorem.

1.4.1   Theorem of Clausius

In any cyclic transformation throughout which the temperature is defined, the following inequality holds

$$\oint \frac{dq}{T} \leq 0.$$
(1.45)

The integral extends over one cycle of the transformation. The equality holds if the cyclic transformation is reversible.

Proof We make use of the assembly of Carnot engines and reservoirs shown in Fig. 1.13. The device called system successively visits all reservoirs indicated by the temperatures T_1 to T_n. After it has visited reservoir T_n it is in the same state as in the beginning.[15] According to Eq. (1.42) we may write

$$q_{i,0} = \frac{T_0}{T_i} q_i.$$

Thus the total heat surrendered by the reservoir at T_0 ($T_0 > T_i$ and $i = 1, 2, \ldots, n$) in one complete turn around of the system is

$$q_0 = \sum_{i=1}^{n} q_{i,0} = T_0 \sum_{i=1}^{n} \frac{q_i}{T_i}.$$

As before, when we compared the thermal efficiency of the Carnot engine to the X-machine, we use the first law, i.e.

$$0 = \Delta E = \underbrace{\sum_{i=1}^{n} q_{i,0}}_{=q_0} - \underbrace{\sum_{i=1}^{n} q_i}_{=0} - w_{total}.$$

---

[15] To achieve this not all Carnot engines operate in the same direction.

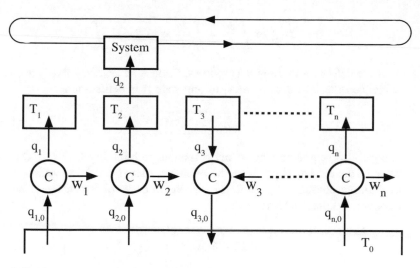

Fig. 1.13 Use of the assembly of Carnot engines and reservoirs

Because $w_{total} \leq 0$ if Kelvin's postulate is correct, we must have $q_0 \leq 0$ and consequently

$$\sum_{i=1}^{n} \frac{q_i}{T_i} \leq 0.$$

Taking the limit $n \to \infty$ and $q_i \to dq$ we have

$$\oint \frac{dq}{T} \leq 0.$$

If the cycle is reversed, then the signs of all q_i change and we have

$$\sum_{i=1}^{n} \left(-\frac{q_i}{T_i} \right) \leq 0.$$

Thus, for a reversible cycle the equal sign holds. This completes our proof.

1.4.2 Consequences of Clausius' Theorem

(i) Note first that (1.45) implies that $\int_A^B \frac{dq}{T}$ is independent of the path joining A and B if the corresponding transformations are reversible. If I and II are two distinct paths joining A and B we have

$$0 = \oint \frac{dq}{T} = \int_I \frac{dq}{T} - \int_{II} \frac{dq}{T} \Rightarrow \int_I \cdots = \int_{II} \cdots .$$

(ii) Next we define the entropy S as follows. Choose an arbitrary fixed state O as reference state. The entropy $S(A)$ of any state A is defined via

$$\boxed{S(A) \equiv \int_O^A \frac{dq}{T} .} \qquad (1.46)$$

The path of integration may be any reversible path joining O and A. Thus the value of the entropy depends on the reference state, i.e. it is determined up to an additive constant. The difference in the entropy of two states A and B, however, is completely defined:

$$S(B) - S(A) = \int_A^B \frac{dq}{T} .$$

Therefore

$$dS = \frac{dq}{T} \qquad (1.47)$$

for any infinitesimal reversible transformation.

1.4.3 Important Properties of the Entropy

(i) For an irreversible transformation from A to B:

$$\int_A^B \frac{dq}{T} \leq S(B) - S(A). \qquad (1.48)$$

Proof We construct a closed path consisting of the irreversible piece joining A and B and a reversible piece returning to A. Thus

$$0 \geq \oint \frac{dq}{T} = \int_{\text{irreversible path from A to B}} \frac{dq}{T} - \int_{\text{reversible path from A to B}} \frac{dq}{T} ,$$

and therefore

$$\int_{\text{irrev.}} \frac{dq}{T} \leq S(B) - S(A). \qquad (1.49)$$

(ii) The entropy of a thermally isolated system never decreases.

Proof Referring to the previous equation thermal isolation means $dq = 0$. It follows that

$$0 \le S(B) - S(A) \qquad \text{or} \qquad S(A) \le S(B). \tag{1.50}$$

This is the manifestation of the thermodynamic arrow of time.

All of the above follows from the two equivalent postulates by Kelvin and Clausius. They constitute the second law of thermodynamics. However, mathematical formulations of the second law are the Clausius theorem or the last two inequalities above.

(iii) Another important property of the entropy, as we have shown above, is its sole dependence on the state in which the system is in. Like the internal energy the entropy is a state function whereas q is not (cf. Remark 1 below!). Combining the first law[16] with Eq. (1.47) yields

$$\boxed{dS = \tfrac{1}{T} dE + \tfrac{P}{T} dV - \tfrac{1}{T} \vec{H} \cdot d\vec{m} - \sum_i \tfrac{\mu_i}{T} dn_i + \dots,} \tag{1.51}$$

where

$$\frac{\partial S}{\partial E}\bigg|_{V,\vec{m},n,\dots} = \frac{1}{T}, \tag{1.52}$$

$$\frac{\partial S}{\partial V}\bigg|_{E,\vec{m},n,\dots} = \frac{1}{T}P, \tag{1.53}$$

$$\frac{\partial S}{\partial \vec{m}}\bigg|_{E,V,n,\dots} = -\frac{1}{T}\vec{H}, \tag{1.54}$$

and

$$\frac{\partial S}{\partial n_i}\bigg|_{E,V,\vec{m},n_{j\neq i}\dots} = -\frac{1}{T}\mu_i. \tag{1.55}$$

Equation (1.52) may be viewed as a thermodynamic definition of temperature. Note also that here the $\vec{H}$-field is assumed to be constant and $\delta\vec{m} = \int_V dV \delta\vec{M}$. In the analogous electric case $\vec{H} \cdot d\vec{m}$ is replaced by $\vec{E} \cdot d\vec{p}$, where $\delta\vec{p} = \int_V dV \delta\vec{P}$. If the

---

[16] The type of work to be included of course depends on the problem at hand. The terms in the following equation represent an example.

field strengths and the moments are parallel, then we have $\vec{H} \cdot d\vec{m} = Hdm$ and $\vec{E} \cdot d\vec{p} = Edp$.

Notice the correspondence between the pairs (H, m), (E, p) and $(P, -V)$. In other words, we may convert thermodynamic relations derived for the variables $P, -V$ via replacement into relations for the variables H, m and E, p. Even more general is the mapping $(P, -V) \leftrightarrow (\vec{E}, V\vec{D}/(4\pi))$ or $(P, -V) \leftrightarrow (\vec{H}, V\vec{B}/(4\pi))$, where we assume homogeneous fields throughout the (constant) volume V.

Equation (1.51), including modifications thereof according to the types of work involved during the process of interest, is a very important result! For thermodynamics it is what Newton's equation of motion is in mechanics or the Schrödinger equation in quantum mechanics—except that here there is no time dependence.[17]

Remark 1 Thus far we have avoided the special mathematics of thermodynamics, e.g. what is a state function, what is its significance, and why is q not a state function etc. At this point it is advisable to study Appendix A, which introduces the mathematical concepts necessary to develop thermodynamics.

Remark 2 The discussion of state functions in Appendix A leads to the conclusion that Eq. (1.51) holds irrespective of whether the differential changes are due to a reversible or irreversible process![18]

---

[17] We return to this point in the chapter on non-equilibrium thermodynamics.

[18] We shall clarify the meaning of this in the context of two related equations starting on p. 64.

Chapter 2
Thermodynamic Functions

2.1 Internal Energy and Enthalpy

We consider the internal energy to be a function of temperature and volume, i.e. $E = E(T, V)$. This is sensible, because if we imagine a certain amount of material at a given temperature, T, occupying a volume, V, then this should be sufficient to fix its internal energy. Thus we may write

$$dE = \frac{\partial E}{\partial T}\Big|_V dT + \frac{\partial E}{\partial V}\Big|_T dV. \tag{2.1}$$

The coefficient of dT is called isochoric heat capacity or heat capacity at constant volume:

$$C_V \equiv \frac{\partial E}{\partial T}\Big|_V. \tag{2.2}$$

It is useful to define another state function, the enthalpy H, via

$$H = E + PV.$$

We find out on which variables H depends by computing its total differential:

$$dH = dE + d(PV)$$
$$= \frac{\partial E}{\partial T}\Big|_V dT + \frac{\partial E}{\partial V}\Big|_T dV + PdV + VdP.$$

Replacing dV via

$$dV = \frac{\partial V}{\partial T}\Big|_P dT + \frac{\partial V}{\partial P}\Big|_T dP,$$

© Springer Nature Switzerland AG 2022
R. Hentschke, *Thermodynamics*, Undergraduate Lecture Notes in Physics,
https://doi.org/10.1007/978-3-030-93879-6_2

$V = V(T, P)$, leads to

$$dH = \left(\frac{\partial E}{\partial T}\bigg|_V + \frac{\partial E}{\partial V}\bigg|_T \frac{\partial V}{\partial T}\bigg|_P + P\frac{\partial V}{\partial T}\bigg|_P \right) dT$$
$$+ \left(\frac{\partial E}{\partial V}\bigg|_T \frac{\partial V}{\partial P}\bigg|_T + P\frac{\partial V}{\partial P}\bigg|_T + V \right) dP.$$

Application of Eq. (A.1) with $A = E$ yields

$$(\ldots)dT = \left(\frac{\partial E}{\partial T}\bigg|_V + P\frac{\partial V}{\partial T}\bigg|_P + \frac{\partial E}{\partial T}\bigg|_P - \frac{\partial E}{\partial T}\bigg|_V \right) dT$$
$$= \frac{\partial (E + PV)}{\partial T}\bigg|_P dT.$$

Applying Eq. (A.2) again setting $A = E$ yields

$$(\ldots)dP = \left(\frac{\partial E}{\partial P}\bigg|_T + P\frac{\partial V}{\partial P}\bigg|_T + V \right) dP$$
$$= \frac{\partial (E + PV)}{\partial P}\bigg|_T dP.$$

Thus we find

$$dH = \frac{\partial H}{\partial T}\bigg|_P dT + \frac{\partial H}{\partial P}\bigg|_T dP \qquad (2.3)$$

and therefore

$$H = H(T, P).$$

Replacing the dependence on volume by a dependence on pressure is of great practical importance. From a theoretical point of view working at fixed volume usually is convenient. But experimenting with a closed apparatus, inside which a process leads to the buildup of uncontrolled pressure, is likely to produce uncomfortable feelings.

The coefficient of dT in Eq. (2.3),

$$C_P \equiv \frac{\partial H}{\partial T}\bigg|_P, \qquad (2.4)$$

is the isobaric heat capacity, i.e. the heat capacity at constant pressure. Two other useful quantities are

Table 2.1 Selected compounds and values for C_P, α_P, and κ_T

Compound	C_P [J/(g K)]	α_P [10^{-4} K^{-1}]	κ_T [10^{-5} MPa^{-1}]
Air (20 °C, 1 bar)	1.007	36.7	10^6
n-pentane (20 °C, 1 bar)	2.3	16	247
Ethanol (20 °C, 1 bar)	2.43	11	117
Water (20 °C, 1 bar)	4.18	2.06	45.9
Water (0 °C, 1 bar)	4.22	−0.68	50.9
Ice I_h (0 °C, 1 bar)	2.11	1.59	13.0
Iron (20 °C, 1 bar)	0.45	0.35	0.6

$$\alpha_P \equiv \frac{1}{V} \frac{\partial V}{\partial T}\bigg|_P, \tag{2.5}$$

the isobaric thermal expansion coefficient, and

$$\kappa_T \equiv -\frac{1}{V} \frac{\partial V}{\partial P}\bigg|_T, \tag{2.6}$$

the isothermal compressibility. Selected compounds and values for C_P, α_P, and κ_T are listed in Table 2.1. There is no need to discuss these quantities at this point. We shall encounter many examples illustrating their meaning.

2.2 Simple Applications

2.2.1 Ideal Gas Law

Here we consider a number of simple examples involving gases. Most of the following applications are based on assuming that the gases are ideal. This means that pressure, P, volume, V, and temperature, T, are related via

$$\boxed{PV = nRT.} \tag{2.7}$$

The quantity R is the gas constant

$$R = 8.31447 \, \text{m}^3 \text{Pa} \text{K}^{-1} \text{mol}^{-1}. \tag{2.8}$$

Figure 2.1 shows PV_{mol}/R, where $P = 10^5$ Pa is the pressure and V_{mol} is the molar volume of air, plotted versus temperature. The data are taken from HCP (Appendix C). The mass density c in the reference is converted to V_{mol} via $V_{mol} = m_{mol}/c$ using the molar mass $m_{mol} = 0.029$ kg. We note that air at these conditions is indeed quite ideal. Notice also that the line, which is a linear least squares fit to

Fig. 2.1 The ideal gas law

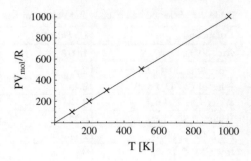

the data (crosses), intersects the axes at the origin. The temperature $T = 0$ K corresponds to $T = -273.15\,°C$.

Remark 1   For an ideal gas we easily work out

$$\alpha_P = \frac{1}{T} \quad \text{and} \quad \kappa_T = \frac{1}{P}. \tag{2.9}$$

Remark 2   Equation (2.7) is simple but nonetheless important, because it is used frequently throughout this text as a first and often satisfactory approximation.

$\partial E/\partial V|_T = 0$ for an Ideal Gas

First we prove that

$$\left.\frac{\partial E}{\partial V}\right|_T = 0 \tag{2.10}$$

for an ideal gas, because we shall rely on this equation a number of ties. Starting from Eq. (1.51) we have

$$TdS = dE + PdV, \tag{2.11}$$

because all other variables like n etc. are constant. Immediately it follows that

$$T\left.\frac{\partial S}{\partial V}\right|_T = \left.\frac{\partial E}{\partial V}\right|_T + P. \tag{2.12}$$

But this does not look like much progress. With some foresight we compute the following differential

$$\begin{aligned} d(E - TS) &= dE - TdS - SdT \\ &= -SdT - PdV. \end{aligned} \tag{2.13}$$

Thus we find

$$\frac{\partial(E - TS)}{\partial T}\bigg|_V = -S \tag{2.14}$$

and

$$\frac{\partial(E - TS)}{\partial V}\bigg|_T = -P. \tag{2.15}$$

Consequently

$$\frac{\partial}{\partial V}\frac{\partial(E - TS)}{\partial T}\bigg|_V\bigg|_T = -\frac{\partial S}{\partial V}\bigg|_T \tag{2.16}$$

as well as

$$\frac{\partial}{\partial T}\frac{\partial(E - TS)}{\partial V}\bigg|_T\bigg|_V = -\frac{\partial P}{\partial T}\bigg|_V, \tag{2.17}$$

and therefore

$$\frac{\partial S}{\partial V}\bigg|_T = \frac{\partial P}{\partial T}\bigg|_V. \tag{2.18}$$

For an ideal gas this becomes

$$T\frac{\partial S}{\partial V}\bigg|_T = P. \tag{2.19}$$

Inserting (2.19) into Eq. (2.12) completes the proof of the above statement.

Remark Integrating Eq. (2.19) using (2.7) we immediately obtain

$$S(T, V) - S(T, V_o) = nR\ln\frac{V}{V_o}. \tag{2.20}$$

This means that if an ideal gas is compressed (expanded) isothermally, i.e. $V < V_o$ ($V > V_o$), its entropy is decreased (increased).

Kinetic Pressure

Concrete thermodynamical calculations require concrete models. Here we consider a gas of point particles with masses m, i.e. atoms or molecules without internal structure or specific spatial extend and shape, confined to a volume V. The particles posses a momentum distribution $dN_{\vec{p}} = Nf(|\vec{p}|)d^3p$. N is the total particle number and $dN_{\vec{p}}$ is the fraction of particles whose momenta occupy a momentum space element d^3p. The quantity $f(|\vec{p}|)$ is the attendant momentum probability density.

Fig. 2.2 A particle reflected
from a wall

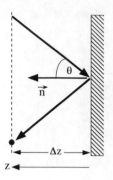

Figure 2.2 shows a particle reflected by one of its containment's walls. The sum of very many (simultaneous) momentum transfers, Δp_z, each contributing a force $f_z = \Delta p_z / \Delta t$, yields the pressure $P = F/A$, where F is the total force and A is the wall area.

- **Kinetic pressure:** First we want to show that

$$P = \frac{1}{V} \int{}' dN_{\vec{p}}(\vec{p} \cdot \vec{n})(\vec{v}_{\vec{p}} \cdot \vec{n}),   \tag{2.21}$$

where $\vec{p} = m\vec{v}_{\vec{p}}$ is the momentum of a particle with velocity $\vec{v}_{\vec{p}}$. The prime $'$ is a reminder that the angle θ is between 0 and $\pi/2$ for particles impinging on the wall.

We consider a single particle for which

$$f_z = \frac{\Delta p_z}{\Delta t} = \frac{2m|\vec{v}_{\vec{p}} \cdot \vec{n}|}{\Delta t}.$$

Here Δz is the thickness of a narrow layer adjacent to the wall in which the momentum transfer occurs. This layer is ill defined, because we consider point particles interacting with a completely smooth wall. Fortunately the final result does not require to specify Δz assumed to be the same for all particles. We define the collision time, i.e. the time a particle spends inside the layer, Δt, via $v_{\vec{p},z} = 2\Delta z/\Delta t$, i.e.

$$\frac{1}{\Delta t} = \frac{|\vec{v}_{\vec{p}} \cdot \vec{n}|}{2\Delta z}.$$

Combination of the two formulas yields

$$f_z = \frac{(\vec{p} \cdot \vec{n})(\vec{v}_{\vec{p}} \cdot \vec{n})}{\Delta z}.   \tag{2.22}$$

In order to obtain the total force on the wall exerted by the gas we must sum or integrate over all collisions, i.e.

$$F = \int' dN_{\vec{p}} \, \underbrace{\frac{\Delta zA}{V} f_z}_{*}$$

(2.23)

*: number of particles inside the surface layer volume ΔzA possessing momenta $\vec{p}$ in d^3p. Limiting θ to values between zero and $\pi/2$ includes particles colliding with the wall only. Thus we obtain Eq. (2.21).

- **Ordinary particles:** For an ideal gas[1] we can deduce the important result

$$E = \frac{3}{2} nRT.$$

(2.24)

Consequently the isochoric heat capacity of an ideal gas of point particles is given by

$$C_V = \frac{3}{2} nR.$$

(2.25)

In order to show Eq. (2.24) we express the internal energy of the gas via

$$E = \int' dN_{\vec{p}} \frac{\vec{p}^2}{2m}.$$

(2.26)

Notice that we continue to use the prime (consistent with the normalization of our above probability density $f(|\vec{p}|)$). It is now easy to deduce

$$P = \frac{2}{3} \frac{E}{V}.$$

(2.27)

This is because the θ-integration in the case of E is

$$\int_0^{\pi/2} d\theta \sin\theta = -\int_0^{\pi/2} d\cos\theta = \int_0^1 dx = 1$$

---

[1] Whether or not our assumption of point particles already implies ideality is a matter of definition of the inter particle interactions.

and thus

$$E = N \int_0^{2\pi} d\phi \int_{-\infty}^{\infty} dp p^2 \frac{p^2}{2m} f(p). \tag{2.28}$$

In the case of P we have instead

$$\int_0^{\pi/2} \cos^2 \theta \sin \theta d\theta = \int_0^1 x^2 dx = \frac{1}{3},$$

and therefore

$$P = \frac{N}{3V} \int_0^{2\pi} d\phi \int_{-\infty}^{\infty} dp p^2 \frac{p^2}{m} f(p). \tag{2.29}$$

Comparison of Eqs. (2.28) and (2.29) immediately yields (2.27), which in combination with the ideal gas law yields Eq. (2.24).

Finally we take another look at the entropy. We have

$$dS = \frac{1}{T} dE + \frac{P}{T} dV \tag{2.30}$$

(cf. Eq. (1.51)). Using (2.24), i.e. $dE = (3/2)nRdT$, and once again (2.7) we find

$$dS = nR \left(d \ln T^{3/2} + d \ln V \right). \tag{2.31}$$

Integration yields the generalization of Eq. (2.20)

$$S(T, V) - S(T_o, V_o) = nR \ln \left[\left(\frac{T}{T_o} \right)^{3/2} \frac{V}{V_o} \right]. \tag{2.32}$$

This is the ideal (point particle) gas entropy change as function of temperature and volume.

- **Photons:** Now let us assume that the gas particles are photons obeying the energy-momentum relation $\epsilon = cp$. Inserting this relation into (2.21) and (2.26) yields

$$P = \frac{1}{V} \int' dN_{\vec{p}} pc \cos^2 \theta = \frac{1}{3V} \int' dN_{\vec{p}} pc \tag{2.33}$$

and

$$E = \int' dN_{\vec{p}} p c. \tag{2.34}$$

Notice that $v_{\vec{p}}$ is replaced by the velocity of light, c. Comparison of the two equations produces

$$P = \frac{1}{3}\frac{E}{V}. \tag{2.35}$$

We can even relate the photon pressure to the gas temperature via

$$\left.\frac{\partial E}{\partial V}\right|_T = T\left.\frac{\partial P}{\partial T}\right|_V - P \tag{2.36}$$

derived in the previous example. Because the classical energy, E, of the photon gas is given by

$$\frac{E}{V} = \frac{1}{V}\int_V dV \frac{\vec{E}^2 + \vec{H}^2}{8\pi}, \tag{2.37}$$

where the argument of the integral is the electromagnetic energy density, we conclude that P does not depend on V and therefore

$$\left.\frac{\partial E}{\partial V}\right|_T = 3P.$$

The resulting differential equation is

$$4\frac{dT}{T} = \frac{dP}{P} \quad \text{or} \quad d\ln T^4 = d\ln P$$

and thus

$$\frac{P}{P_o} = \left(\frac{T}{T_o}\right)^4 \tag{2.38}$$

or

$$\frac{E}{V} = c_r T^4, \tag{2.39}$$

where c_r is a constant. This is Stefan's law of the energy density dependence on temperature in the case of black body radiation.

This also is a good place to mention the entropy of the black body, i.e. a volume containing (and possibly emitting) radiation in thermal equilibrium with some reservoir.[2] We start by inserting (2.39) into the thermodynamic definition of temperature, Eq. (1.52), i.e.

$$\frac{\partial S}{4c_r V T^3 \partial T}\bigg|_{V,n,...} = \frac{1}{T}. \tag{2.40}$$

Integration of this equation yields the entropy of the black body

$$S = \frac{4}{3}\frac{E}{T}, \tag{2.41}$$

which is proportional to T^3. Notice that $S = 0$ at $T = 0$. However, there is more to learn here. Subtracting the two differentials

$$TdS = 4c_r V T^3 dT + \frac{4}{3}c_r T^4 dV \text{ and } dE = 4c_r V T^3 dT + c_r T^4 dV \tag{2.42}$$

yields

$$TdS - dE = \frac{1}{3}c_r T^4 dV \overset{(2.35)}{=} PdV. \tag{2.43}$$

Comparing this result to (1.51) we conclude that the chemical potential of the photons vanishes, $\mu = 0$. We shall return to this conclusion later in this book (on page 253).

Figure 2.3 shows an experimental setup in the author's office allowing to verify Stefan's law. The red cube in the center is an oven (black body) emitting radiation through the aperture shown in the inset. The radiation energy is measured and converted into volts shown on the instrument panel on the right. The attendant oven temperature is shown by the instrument on the left. Figure 2.4 contains data taken by the author upon heating (up-triangles) and subsequent cooling (down-triangles) of the oven ($T_o = 25\,°C$). The solid line is a linear fit through the data points. Even though there is some room for improvement the result is clearly in accord with Eq. (2.39).

It is interesting to calculate the energy a black body looses per unit time due to radiation emanating from its surface. If δA is an area element on the black body's surface, a distant observer may look at δA from an angle θ. Here θ is the angle between the surface normal of δA and the direction of the observer. Thus the

---

[2] The term "black body" may be somewhat misleading, because a black body is not necessarily black. In fact the radiation spectrum of our sun measured above the atmosphere is very closely a black body spectrum. Here we merely deal with the temperature dependence of the total energy density of a black body. The spectrum is calculated in Sect. 5.3.

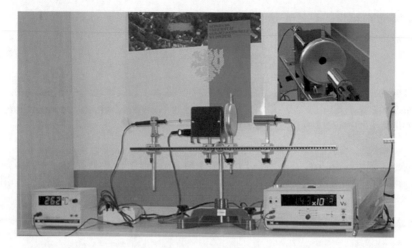

Fig. 2.3 Experimental setup in the author's office allowing to verify Stefan's law

Fig. 2.4 Data obtained with
the experimental setup
(including cooling of the
aperture plate) shown in the
previous figure

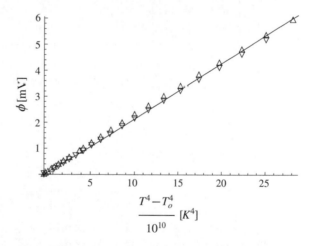

observer does not see the full δA but the projection $\delta A \cos \theta$ instead. Every volume
element on the surface contains the energy density E/V, and therefore emanates a
total flux density cE/V. Along a particular direction this is $cE/(4\pi V)$. This means
that the energy per time passing through the area element projection in the direction
of the observer is $cE/(4\pi V)\delta A \cos \theta$. Collecting together the energy per time
passing through δA towards all possible observer directions therefore yields

$$\frac{d\delta E}{dt} = -\frac{cE}{4\pi V}\delta A \int_0^{2\pi} d\varphi \int_0^{\pi/2} d\theta \sin \theta \cos \theta = -\frac{1}{4}\frac{cE}{V}\delta A \qquad (2.44)$$

or after integration over the full surface

$$\frac{dE}{dt} = -\frac{1}{4}\frac{cE}{V}A = -\sigma T^4 A. \tag{2.45}$$

The minus sign indicates that the energy of the black body diminishes. The quantity $\sigma = 5.67 \cdot 10^{-8}\,\mathrm{Wm^{-2}K^{-4}}$ is Stefan's constant, which we learn how to calculate in Sect. 5.1 (cf. Eq. (5.109)).[3]

> **Example—Earth's Equilibrium Temperature** Let us consider the last footnote from a different angle. The radiation energy emanating from the sun per unit time intercepted by the Earth is
>
> $$\frac{dE_{in}}{dt} = \sigma T_S^4 A_S f. \tag{2.46}$$
>
> Here the solar surface temperature is T_S and $A_S = 4\pi r_S^2$ is the Sun's surface area. The factor f is the solid angle covered by the earth divided by the total solid angle, i.e. $f = \pi r_E^2/(4\pi R_{SE}^2) = \frac{1}{4}(r_E/R_{SE})^2$. Since the earth has a certain average surface temperature, which we call the equilibrium temperature T_E, we expect that the incoming flux dE_{in}/dt is balanced by a corresponding energy flux emanating from the Earth's surface dE_{out}/dt, i.e.
>
> $$\frac{dE_{out}}{dt} = \sigma T_E^4 A_E, \tag{2.47}$$
>
> where $A_E = 4\pi r_E^2$. Hence
>
> $$\frac{dE_{in}}{dt}(1 - A) = \frac{dE_{out}}{dt}. \tag{2.48}$$
>
> The quantity A is called albedo (not to be confused with an area). It is a measures for how much of the sun's radiation is reflected from the irradiated body into space. Earth's albedo is ≈ 0.3, of which roughly three quarters are contributed by cloud. Solving Eq. (2.48) for T_E yields

---

[3] It is interesting to apply this formula to the sun. We use a solar surface temperature of 5780 K, a sun radius of $r_S \approx 7.0 \cdot 10^8$ m, an earth radius of $r_E \approx 6.4 \cdot 10^6$ m, and the mean sun-to-earth distance $R_{SE} \approx 1.5 \cdot 10^{11}$ m. With these numbers we calculate a radiation energy annually received by the earth's surface of about $1.6 \cdot 10^{15}$ MWh/y. At the time of this writing the entire world's electricity consumption is roughly $1.9 \cdot 10^{10}$ MWh/y (based on data collected between 2002 and 2010). In other words—a quadratic surface of about 40 by 40 km positioned in space near the earth would receive just this energy!

$$T_E = T_S \left(\frac{r_S^2 (1 - A)}{4 R_{SE}^2} \right)^{1/4}. \tag{2.49}$$

Substituting the previous numbers for T_S, r_S and R_{SE} we find $T_E \approx 254$ K. This is roughly 35 K less than the average surface temperature of about 288 K. We can apply (2.49) to other planets or moons as well if we use their albedo and their distance from the sun. In the case of Earth's moon, possessing an albedo of around 0.12, the equilibrium temperature is about 269 K whereas its average surface temperature is roughly 255 K. This is a smaller difference than in the case of Earth. We may suspect that Earth's atmosphere is responsible for most of the difference (Greenhouse effect), even though the example of the moon suggests that there are other factors as well—like the somewhat ill-defined concept of an average temperature. An extreme example is Venus. It has an equilibrium temperature of approximately 260 K but a surface temperature of 740 K.

Remark 1   In the late 1930s the famous physicist Paul A. Dirac[4] speculated that the large size of certain dimensionless numbers, constructed from combination of fundamental constants, may indicate their monotonous variation tied to the age of the universe. The concept of a changing, or rather expanding, universe had been developed a decade before mainly by Alexander Friedmann, Georges Lemaître, and Edwin Hubble—the former two studied solutions to Einstein's field equations of General Relativity and the latter measured the redshift of galaxies depending on their distance. The suspected changes should be slow, which would make them difficult to observe, since the "measurements" have to cover several hundred million years.

In 1948 Edward Teller published a paper in which he derives the dependence of T_E, on the gravitation constant G. G affects not only R_{SE} but also T_S and we do not want to present his derivation here (the interested reader is referred to Teller (1948)). He showed that $T_E \sim G^{2.25}$. Thus, if G had been 10% smaller 300 million years ago this would have resulted in a 20% higher T_E compared to the present—sufficiently high to affect the development of live on Earth as we know it. This, so his conclusion, can be viewed as evidence against Dirac's hypothesis.

---

[4] Paul Adrien Maurice Dirac, British physicist, *Bristol 8.8.1902, †Tallahassee (Florida) 20.10.1984; numerous seminal contributions to the development of quantum theory (e.g. Dirac equation); he shared the 1933 Nobel prize in physics with E. Schrödinger.

Remark 2   We can combine Eqs. (2.41) and (2.48) to obtain

$$\frac{dS_{out}/dt}{dS_{in}/dt} = (1-A)\frac{T_S}{T_E}. \tag{2.50}$$

This means that the long wavelength radiation emitted from the earth carries much more entropy than the incoming short wavelength radiation received from the sun, which, as we shall discuss in the last chapter, is of great significance for the development of life on earth.

Example—The Expanding Universe and Its Temperature   We want to apply what we have learned to the relation between the size of the universe and its temperature. This, however, requires one paragraph of cosmology before we can start.

In the 1920s the aforementioned American astronomer Edwin Hubble[5] carried out measurements which led him to conclude that far away galaxies recede from us with a velocity proportional to their distance D, i.e.

$$v = HD. \tag{2.51}$$

The proportionality constant H became the Hubble constant. Since Earth's location is not a special place in the universe, Hubble's law (2.51) is valid for any reference point. For simplicity's sake let us assume that our universe is an elastic band. On the band there are regularly spaced marks indicating galaxies. The spacing between neighboring marks or galaxies is a. In the following we shall call a the scale factor. If we sit on any one of the galaxies, then the distance from us to the fourth galaxy on either side is $4a$. More generally, the distance from us to the xth galaxy is $D(t) = xa(t)$. t is time and $a(t)$ means that the elastic band stretches or contracts with time. Note that x is a mere number and therefore does not depend on time. In the real universe this means that space itself expands or contracts. With $v = x\dot{a}$, where the dot indicates a time derivative, Hubble's law becomes $\dot{a}(t) = Ha(t)$. However, Hubble's constant is not really a constant but instead

$$H = \sqrt{\frac{8\pi G}{3}\rho(t)}, \tag{2.52}$$

---

[5] Edwin Powell Hubble, American astronomer, *Marshfield, Missouri, United States 20.11.1889, †San Marino, California, United States 28.9.1953.

where G is the gravitation constant and $\rho(t)$ is the energy density of the universe. Hence

$$\left(\frac{\dot{a}(t)}{a(t)}\right)^2 = \frac{8\pi G}{3}\rho(t). \tag{2.53}$$

This, apart from a term accounting for the "curvature of space" which is not important here, is the first Friedmann equation. The first Friedmann equation follows from General Relativity but there is also a simpler Newtonian approach which yields (2.53) (see for instance Chap. 8 in Hentschke and Hölbling 2020). In the following we consider three contributions to $\rho(t)$ called $\rho_r(t)$, the radiation energy density, $\rho_m(t)$, the mass energy density (note that according to the Special Theory of Relativity every mass can be converted into energy) and $\rho_v(t)$, a mysterious vacuum energy density (also called dark energy), i.e. $\rho(t) = \rho_r(t) + \rho_m(t) + \rho_v(t)$.

Now we return to thermodynamics, pursuing the dependence of the scale factor of the universe a on time as well as its relation to the temperature T of the universe. Note that a is not the size of the universe but nevertheless a measure of its size. Our starting point is the first law in the form

$$dE = -PdV. \tag{2.54}$$

The only contribution to the internal energy change of the universe is volume work. Equation (2.54) expressed in terms of ρ and a become

$$d(a^3\rho) = -Pda^3. \tag{2.55}$$

At this point we need an equation of state, i.e. an equation relating the pressure to the (energy) density. Inspired by Eqs. (2.27) and (2.35) we assume

$$P = \omega\rho, \tag{2.56}$$

where ω is unknown. Inserting this equation of state into Eq. (2.55) yields

$$d\ln\rho = -3(1+\omega)d\ln a, \tag{2.57}$$

which has the solution

$$\rho = \frac{\text{const}}{a^{3(\omega+1)}}. \tag{2.58}$$

In the following we discuss (2.58) for ρ_m, ρ_r and ρ_v separately. This means that the development of the universe can be described as succession of epochs. In each epoch ρ is dominated by either one of the three densities. If ρ

is for matter only, i.e. $\rho = \rho_m$, we expect $\rho_m \propto a^{-3}$ and thus $\omega_m = 0$. For radiation, as represented by photons, we conclude from our previous Eq. (2.35) that $\omega_r = 1/3$ and thus $\rho_r \propto a^{-4}$.

These conclusions are not trivial and we must discuss them. $\omega_m = 0$ in the case $\rho = \rho_m$ seemingly contradicts Eq. (2.27), according to which we would have expected $\omega_m = 2/3$. However, there are great differences between Eqs. (2.27) and (2.56) when $\rho = \rho_m$. First, the energy density E/V on the right hand side of Eq. (2.27) only includes the classical kinetic energy of the matter particles and not the much greater energy equivalent of their (rest) mass included in ρ_r. Second and even more importantly, Eq. (2.27) describes a gas of particles possessing a certain temperature T and therefore a non-zero pressure P. In the universe most mass is found in lumps, i.e. galaxies (including black holes, ...), which are cold in terms of their overall kinetic energy (even if they contain many hot stars), since they merely move with the expansion of the universe, and do not exert a pressure. Thus, $\omega_m = 0$ in this case is not a contradiction. Nevertheless, we shall return to $P_m = 0$ in the universe versus $P > 0$ in an ordinary gas in a separate example. Next is $\rho_r \propto a^{-4}$ and the question how the extra factor a^{-1} arises in comparison to $\rho_m \propto a^{-3}$? The radiation particles which make up ρ_r move with the speed of light c and have no rest mass. Each of them possesses an energy hc/λ, where h is Planck's constant. Their energy density is the product of their number density, which is proportional to a^{-3}, and hc/λ. Since hc is a constant, $\rho_r \propto a^{-4}$ requires that the radiation's wavelength $\lambda \propto a$. And this is indeed the case. While a photon travels towards us from a distant galaxy the scale factor, as we shall see, grows, i.e. the longer the photon must travel to reach us the longer its wavelength becomes. Or in other words, the more redshifted the photon is the greater is the distance to the galaxy in which it originated. This is what Hubble had exploited.

The last of the energy density contributions, and the most mysterious, is the vacuum energy density $\rho_v(t)$. Observation suggests that it is constant, i.e. $\rho_v(t) = \rho_o$. This means the vacuum energy does not thin out when space expands. It leads to $\omega_v = -1$ and thus $P_v = -\rho_o$. Since ρ_o is positive P_v is negative. This and the other relations between pressure, energy density and scale factor are compiled in the two upper rows in Table 2.2. The upper panel of Fig. 2.5 depicts the a-dependence of the three energy density components. In the early universe ρ_r dominates. It is followed by ρ_m since a^{-4} decreases

Table 2.2 Components of the energy density

	Matter	Radiation	Vacuum energy
ρ	$\rho_m \propto a^{-3}$	$\rho_r \propto a^{-4}$	$\rho_v = \rho_o$
p	$P_m = 0$	$P_r = \rho_r/3$	$P_v = -\rho_o$
a	$\sim t^{2/3}$	$\sim t^{1/2}$	$\sim \exp[\sqrt{...}t]$

Fig. 2.5 Summary of the
three energy density
components

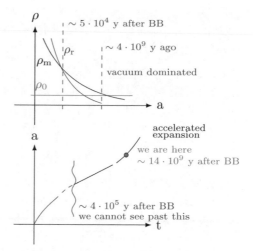

faster than a^{-3}. Finally both ρ_r and ρ_m decrease below $\rho_v = \rho_o$. The calculation of the various crossover times, e.g. $5 \cdot 10^4$ y after BB (Big Bang), requires additional tools which we cannot develop here (see for instance Hentschke and Hölbling 2020).

The third row in Table 2.2 lists the time dependence of the scale factor in the three epochs, i.e. matter dominates, radiation dominates and vacuum energy dominates. The respective $a(t)$ are obtained by inserting the first row energy densities into Eq. (2.53). For instance, inserting ρ_m yields $(\dot{a}/a)^2 \propto a^{-3}$. Using the ansatz $a(t) \propto t^q$ and equating the exponents of t on both sides of the resulting equation yields $q = 2/3$. For ρ_r we find $q = 1/2$ instead. In the case of ρ_v the right hand side of Eq. (2.53) is a constant and we find an exponential grows of a. Note that ... in Table 2.2, which is proportional to ρ_o, is very small. Nevertheless, measurements have shown that we are currently in this epoch. The lower panel in Fig. 2.5 shows a sketch of the scale factor's expansion during the three epochs.

Let us briefly address one obvious question. Just how much ρ_m is there in comparison to ρ_r or ρ_v in our current universe? Experimental evidence indicates that $\Omega_m \approx 0.3$, $\Omega_r \approx 9 \cdot 10^{-5}$ and $\Omega_v \approx 0.7$. Here Ω_i stands for the respective fraction of i compared to the current total energy density $\rho_{c,0} \approx 8 \cdot 10^{-10}$ J m^{-3}. In addition, $\Omega_m = \Omega_{m,b} + \Omega_{m,DM}$, where $\Omega_{m,b} \approx 0.05$ is the baryonic matter, i.e. the matter of which we know what it consists of, and the much larger rest consisting of dark matter.

One epoch, which we completely ignore here, was extremely short and occurred right after the Big Bang. It is called inflation. It is thought that the scale factor during this epoch grew very rapidly. We can understand this type of rapid growth if we assume that the scale factor during inflation was also growing as $a \sim \exp[\sqrt{\ldots} t]$, but that during inflation ..., i.e. ρ_o, was very much larger than today.

But what about the temperature of the universe? Matter, as we have discussed, is cold. But the radiation energy density is tied to T via Stefan's law (2.39), i.e. $\rho_r \propto T^4$. Since $\rho_r \propto a^{-4}$ as well, we conclude

$$T \propto \frac{1}{a}. \tag{2.59}$$

We can use this relation to actually compute one of the numbers in the lower panel in Fig. 2.5 i.e. $\sim 4 \cdot 10^5$ y after the Big Bang. Note that the universe was matter dominated during most of its existence, i.e. $a \propto t^{2/3}$. Hence

$$\frac{T_1}{T_2} = \left(\frac{t_2}{t_1}\right)^{2/3}. \tag{2.60}$$

Today the universe is $t_1 \approx 14 \cdot 10^9$ y old and its temperature is $T_1 \approx 3$ K. When its age was $t_2 \approx 4 \cdot 10^5$ y Eq. (2.60) implies the universe therefore had a temperature $t_2 \approx 3000$ K. This is a special temperature, because it was low enough for the protons and the electrons, which until then had been separate components of an optically opaque plasma, to combine into hydrogen atoms. From then on the universe was transparent. In an example in Chap. 3 we shall estimate this temperature, which in turn means that we can estimate the time at which this so called "recombination" took place (rather than assuming that we already know T_2 in order to be able to estimate t_2).

Example—$P_m = 0$ Versus the Ideal Gas Law $PV = NT > 0$ In the previous example we have motivated $P_m = 0$ with the lumpy masses suspended in space. But there was a time when the content of the universe could be described as a fairly hot "soup" of non-relativistic massive particles in equilibrium with photons (prior to or right around recombination). Shouldn't the particles under these conditions possess a pressure greater than zero—like a gas at a certain density and temperature?

Let us approximate the internal energy of the universe by a sum of three terms, i.e.

$$E = V\rho_{m,o} + \frac{3}{2}Nk_BT + c_rVT^4. \tag{2.61}$$

The first term is the volume of the universe multiplying the energy density contributed by the rest mass of the particles contained in it. The second term is the ideal gas contribution (2.24), where N is the number of the afore-mentioned particles. The last term accounts for the radiation according to Eq. (2.39). Note that vacuum energy is not included here.

According to the first law

$$dE = d\left(mc^2 N + \frac{3}{2}Nk_BT + c_r VT^4 \right) = -(P_m + P_r)dV. \qquad (2.62)$$

Here m is the particle mass (for simplicity we consider one type of particle only), mc^2 is its energy equivalent, and P_m and P_r are the partial pressures of the particles and the photons, respectively. Using $P_m V = Nk_BT$ and $P_r = \frac{1}{3}c_rT^4$ as well as $V \propto a^3$, we obtain after some algebra

$$d\ln T = -Y(x)d\ln a \quad \text{where} \quad Y(x) = \frac{1+3x/4}{1+3x/8}. \qquad (2.63)$$

Here $x = \frac{Nk_B}{Vc_rT^3} = \frac{1}{3}P_m/P_r$. Currently we use $P_m = 0$. Thus $x = 0$ and $Y(x) = 1$, which yields $T \propto a^{-1}$—our previous result! But if $P_m \gg P_r$ then $Y(x \to \infty) = 2$ and consequently $T \propto a^{-2}$.

Can $P_m \gg P_r$ occur? Using $N \approx N_\gamma\eta$, where N_γ is the number of photons, it is not difficult to see that

$$\frac{P_m}{P_r} \approx \eta. \qquad (2.64)$$

This result is independent of temperature. The quantity η is the baryon (composite particles made from quarks) to photon ratio obtained from the theory of nucleosynthesis in the early universe. According to this theory $\eta \sim 10^{-9}$ and we find that $P_m = 0$ is indeed a very good approximation.

However, in order to convince ourselves that the universe is quite different from our usual surroundings, we compute P_m/P_r inside an average office (office pressure is 1 bar; office temperature is 293 K). Assuming the office is a black body cavity, we can describe the cavity's radiation pressure via $P_r = \frac{1}{3}c_rT^4 \approx 1.9 \cdot 10^{-6}$ Pa, where we have used that Stefan's constant $\sigma = (c/4)c_r$ (cf. Eq. (2.45)). Therefore $P \approx P_m$ or $P_m/P_r \approx 5 \cdot 10^{10}$, i.e. $\eta_{\text{office}} \sim 10^{10}$.

Remark Our description of massive particles or radiation is somewhat loose. A massive particle may be non-relativistic or relativistic depending on temperature. In the ultra-relativistic case, i.e. mc^2 is much less than the total energy, their equation of state is the same as for radiation. ρ_m, according to the current cosmological standard model, consists mostly of "dark matter". It is not known what dark matter is. What is known is that it must have certain properties to not contradict certain observations. In particular the mass of dark matter particles (if dark matter is due to particles) is not known—which, in principle, could affect our above conclusions based on η. The universe contains neutrinos, which in current models contribute significantly to ρ_r. But neutrinos are believed to possess mass and measurements are underway to determine this mass. Thus, there are these and other open issues.

Remark—Black Hole Entropy In the early 1970s Hawking[6] (1976; and references therein) showed that a black hole of mass M should emit radiation possessing the spectral distribution of black body radiation[7] corresponding to the temperature

$$T_{bh} = \frac{1}{8\pi} \frac{\hbar c^3}{k_B G M}. \tag{2.65}$$

This formula, joining ideas and concepts from quantum theory and the theory of general relativity, is engraved in a plaque in Westminster Abbey marking Stephen Hawking's burial site (cf. Fig. 2.6). Here k_B is the so-called Boltzmann constant, G is the gravitational constant, $\hbar$ is Planck's constant divided by 2π, and c is the speed of light. The origin of this radiation is the separation of virtual particle pairs, spontaneously created just outside the event horizon of the black hole due to vacuum fluctuations, such that the partner with positive energy is traveling away from the horizon and the other in turn vanishes behind the horizon. This negative energy partner diminishes the energy of the black hole by an amount that is carried away by the other. In this sense a black hole may "evaporate" over time. As already mentioned, remarkably this radiation has the same distribution as the radiation of a black body—despite its different nature!

According to Eq. (1.52) the black hole should posses entropy, which we can calculate by integrating this equation ($S = \int_0^E dE' T(E')^{-1} + const$):

$$S_{bh} = 4\pi \frac{k_B G M^2}{\hbar c}. \tag{2.66}$$

We have used $E = Mc^2$ ($dE = c^2 dM$) and assumed that $const = 0$. Notice that Eq. (1.52) holds if the other variables ($V, \ldots$) are held fixed. In the case of the black hole the corresponding quantities are its charge and angular momentum. The entropy in Eq. (2.66) may be cast in a different form, i.e.

---

[6] Stephen William Hawking, British theoretical physicist, *Oxford 8.1.1942, †Cambridge 14.3.2018.

[7] This distribution is shown for one particular temperature in Fig. 5.13.

Fig. 2.6 Hawking's burial
site in Westminster Abbey,
London, UK

$$S_{bh} = \frac{k_B}{4} \frac{A}{\lambda_P^2}. \tag{2.67}$$

$A = 16\pi G^2 M^2/c^4$ is the horizon area[8] and λ_P^2 the square of the Planck length
$\lambda_P = (\hbar G/c^3)^{1/2} \sim 10^{-35}$ m. Equation (2.67) is the Bekenstein-Hawking entropy
formula. Bekenstein (1973) was the first to systematically explore that "a black hole
exhibits a remarkable tendency to increase its horizon surface area when undergoing any transformation" in analogy to the second law of thermodynamics (as
expressed via (1.50)). He concluded that a black hole should posses entropy proportional to A. His reasoning, predating the discovery of black hole radiation, was
based on the connection between entropy and information or rather the lack of
information.[9] The formula for S_{bh}, which he obtained, differs from (2.67) by a
numerical factor (he introduced λ_P^2 on dimensional grounds!).

---

[8] This formula may be obtained classically! The velocity necessary to escape from a mass M
starting at a distance R is $v_{esc} = (2GM/R_{bh})^{1/2}$. Substituting $v_{esc} = c$ and solving for R_{bh} yields
A via $A = 4\pi R_{bh}^2$.

[9] This connection is discussed in Sect. 5.1.1.

We may estimate the time t_{vap} (very roughly!) it should take for the black hole to evaporate via

$$-\frac{dM}{dt} \sim -\frac{dE}{dt} \sim T^4 A \sim \frac{1}{M^2}, \tag{2.68}$$

i.e.

$$\int_0^{t_{vap}} dt' \sim \int_0^M dM' M'^2. \tag{2.69}$$

Our final result is

$$t_{vap} \sim \frac{G^2 M^3}{\hbar c^4}, \tag{2.70}$$

where the factor $G^2/(\hbar c^4)$ follows from dimensional analysis. Table 2.3 compiles some numbers for M equal to the mass of the sun, the earth, and two arbitrary masses (one corresponding to the cosmic background temperature discussed in the context of Fig. (5.13), and illustrates the extreme numbers (The age of the universe is estimated at $5 \cdot 10^{17}$ s!). This shows that the radiation is significant only for black holes much smaller than those that are expected to form by the collapse of stars. Black holes possessing smaller masses could have formed in the early universe, but thus far no conclusive experimental evidence for black body radiation, especially for the brief intense flash indicating a vanishing black hole, has been found. A detailed discussion of the underlying theory is given in Susskind and Lindsay (2005).

Isotherms and Adiabatic Curves

Discussing the Carnot engine we had studied the thermodynamic cycle in the P-V-plane depicted in Fig. 2.7. The two curves labeled T_1 and T_2 are called isotherms ($T = constant$). The two other curves are adiabatic curves ($\delta q = 0$).

Starting from the ideal gas law, $PV = nRT$, we may show that the sketch is correct in so far as the isotherms are less steep than the adiabatic curves, i.e.

Table 2.3 Characteristic quantities for black holes with selected masses

	M (kg)	R_{bh} (m)	T_{bh} (K)	t_{vap} (s)
Sun	$2 \cdot 10^{30}$	$3 \cdot 10^3$	$6 \cdot 10^{-7}$	10^{70}
Earth	$6 \cdot 10^{24}$	$9 \cdot 10^{-3}$	$6 \cdot 10^{-2}$	10^{54}
	$4 \cdot 10^{22}$	$6 \cdot 10^{-5}$	2.75	10^{47}
	$7 \cdot 10^{11}$	10^{-15}	$2 \cdot 10^{11}$	10^{15}
	1	10^{-27}	10^{23}	10^{-20}

Fig. 2.7 Carnot cycle

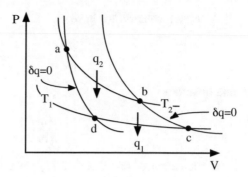

$$\left.\frac{\partial P}{\partial V}\right|_T > \left.\frac{\partial P}{\partial V}\right|_{\delta q=0}.$$

Notice that the combination of the first law, i.e. $dE = -PdV$ (if $\delta q = 0$), with Eq. (2.2) yields

$$-PdV = C_V dT. \tag{2.71}$$

From

$$dT = \left.\frac{\partial T}{\partial V}\right|_P dV + \left.\frac{\partial T}{\partial P}\right|_V dP$$

follows

$$dT = \frac{P}{nR}dV + \frac{V}{nR}dP.$$

This means

$$-PdV = C_V\left(\frac{P}{nR}dV + \frac{V}{nR}dP\right)$$

or

$$-\left(1 + \frac{C_V}{nR}\right)\frac{dV}{V} = \frac{C_V}{nR}\frac{dP}{P}$$

and therefore

$$\left.\frac{d\ln P}{d\ln V}\right|_{\delta q=0} = -1 - \frac{nR}{C_V}. \tag{2.72}$$

Along an isotherm we have

$$\frac{dP}{dV}\Big|_T = -\frac{nRT}{V^2} = -\frac{P}{V}$$

and therefore

$$\frac{d\ln P}{d\ln V}\Big|_T = -1. \tag{2.73}$$

Combination of (2.72) and (2.73) yields

$$\frac{d\ln P}{d\ln V}\Big|_T > \frac{d\ln P}{d\ln V}\Big|_{\delta q=0}$$

or

$$\frac{dP}{dV}\Big|_T > \frac{dP}{dV}\Big|_{\delta q=0}.$$

Efficiency of Engines with Ideal Gas as Working Substance

We study three examples: (a) the Carnot engine or cycles, (b) the Otto cycle, and
(c) the Diesel cycle.

(a) Figures 1.10 and 2.7 both show the Carnot cycle. Assuming that the working
medium is an ideal gas we want to compute the thermal efficiency for this cycle.
We consider the work done by the gas along the different parts of the cycle:

$$a \rightarrow b: \quad w_{a\rightarrow b} = \int_{V_a}^{V_b} P dV \overset{\delta T=0}{=} nRT_2 \ln\frac{V_b}{V_a}$$

$$b \rightarrow c: \quad w_{b\rightarrow c} = \int_{V_b}^{V_c} P dV \overset{\delta q=0}{=} -C_V \int_{T_2}^{T_1} dT = -C_V(T_1 - T_2)$$

$$c \rightarrow d: \quad w_{c\rightarrow d} = \int_{V_c}^{V_d} P dV \overset{\delta T=0}{=} nRT_1 \ln\frac{V_d}{V_c}$$

$$d \rightarrow a: \quad w_{d\rightarrow a} = \int_{V_d}^{V_a} P dV \overset{\delta q=0}{=} -C_V \int_{T_1}^{T_2} dT = -C_V(T_2 - T_1).$$

The total work done by the gas is

$$w = w_{a\rightarrow b} + w_{b\rightarrow c} + w_{c\rightarrow d} + w_{d\rightarrow a} = w_{a\rightarrow b} + w_{c\rightarrow d}. \tag{2.74}$$

Now we compute the heat input q_2. Notice that for an ideal gas, as we just have
seen, $\Delta E = \Delta E(T)$. The path from a to b is along an isotherm however and

therefore $0 = \Delta E_{a \to b} = q_2 - w_{a \to b}$, i.e. $q_2 = w_{a \to b}$. We thus obtain for the thermal efficiency

$$\eta = \frac{w}{q_2} = 1 + \frac{w_{c \to d}}{w_{a \to b}} = 1 - \frac{T_1 \ln[V_c/V_d]}{T_2 \ln[V_b/V_a]}. \tag{2.75}$$

Integrating Eq. (2.71) for the ideal gas yields

$$\frac{V}{V'} = \left(\frac{T'}{T}\right)^{C_V/(nR)} \tag{2.76}$$

along an adiabatic curve. We can use this to express V_d via V_a and V_c via V_b. The result is

$$\eta = 1 - \frac{T_1}{T_2} \tag{2.77}$$

in accord with Eq. (1.44).

(b) Figure 2.8 shows the Otto cycle. The contributions from the different parts of the cycle are

$$
\begin{aligned}
a \to b &: q_{a \to b} = 0 & -w_{a \to b} &= C_V(T_b - T_a) \\
b \to c &: w_{b \to c} = 0 & q_{b \to c} &= C_V(T_c - T_b) \\
c \to d &: q_{c \to d} = 0 & -w_{c \to d} &= C_V(T_d - T_c) \\
d \to a &: w_{d \to a} = 0 & q_{d \to a} &= C_V(T_a - T_d).
\end{aligned}
$$

The thermal efficiency is

$$\frac{w}{q_{b \to c}} = \frac{-C_V(T_b - T_a) - C_V(T_d - T_c)}{C_V(T_c - T_b)} = 1 - \frac{T_d - T_a}{T_c - T_b}.$$

Fig. 2.8 Otto cycle

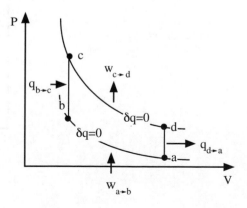

Along the adiabatic curves we have

$$\frac{T_a}{T_b} = \left(\frac{V_b}{V_a}\right)^{(nR)/C_V} \quad \text{and} \quad \frac{T_d}{T_c} = \left(\frac{V_c}{V_d}\right)^{(nR)/C_V}.$$

With $V_b = V_c$ and $V_a = V_d$ follows

$$\frac{T_a}{T_b} = \frac{T_d}{T_c} \quad \text{or} \quad T_d = \frac{T_a T_c}{T_b}$$

and therefore

$$\eta = 1 - \frac{T_a}{T_b}. \tag{2.78}$$

(c) Figure 2.9 shows the Diesel cycle. This cycle is similar to the Otto cycle. The difference is that the isochor, i.e. the line of constant volume from b to c, is changed to an isobar, a line of constant pressure. The contributions from the different parts of the cycle are in this case

$$
\begin{aligned}
a \rightarrow b &: q_{a \rightarrow b} = 0 \qquad\qquad &-w_{a \rightarrow b} &= C_V(T_b - T_a) \\
b \rightarrow c &: w_{b \rightarrow c} = P_b(V_c - V_b) \quad &q_{b \rightarrow c} &= C_V(T_c - T_b) + w_{b \rightarrow c} \\
c \rightarrow d &: q_{c \rightarrow d} = 0 \qquad\qquad &-w_{c \rightarrow d} &= C_V(T_d - T_c) \\
d \rightarrow a &: w_{d \rightarrow a} = 0 \qquad\qquad &q_{d \rightarrow a} &= C_V(T_a - T_d).
\end{aligned}
$$

Here the thermal efficiency is

$$\frac{w}{q_{b \rightarrow c}} = \frac{-C_V(T_b - T_a) + P_b(V_c - V_b) - C_V(T_d - T_c)}{C_V(T_c - T_b) + P_b(V_c - V_b)}.$$

Fig. 2.9 Diesel cycle

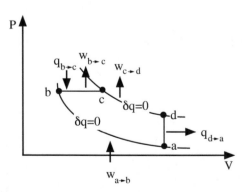

Fig. 2.10 Two cycles in the T-S-plane

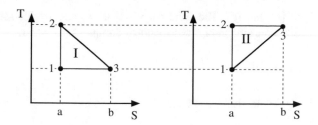

Using the ideal gas law together with Eq. (2.76) this may be rewritten as

$$\eta = 1 - \left(\frac{V_b}{V_a}\right)^{nR/C_V} \frac{1}{1+nR/C_V} \frac{(V_c/V_b)^{1+nR/C_V} - 1}{V_c/V_b - 1}. \tag{2.79}$$

Notice that there is the following relation between the efficiencies of Otto and Diesel cycles:

$$1 - \eta = \left(\frac{V_b}{V_a}\right)^{\gamma-1} g \tag{2.80}$$

where

$$g = \begin{cases} \frac{1}{\gamma}\frac{x^\gamma-1}{x-1} & \text{Diesel} \\ 1 & \text{Otto} \end{cases} \tag{2.81}$$

with $x = V_c/V_b$ and $\gamma = 1 + nR/C_V$. One can show (exercise[10]) that $g_{Diesel} > 1$ for $x > 1$ and $\gamma > 1$. This means that a reversible Otto cycle with an ideal gas as working substance is more efficient than a reversible Diesel cycle at the same compression ratio, V_b/V_a. However, real Diesel engines can operate at greater compression ratios and therefore greater efficiencies than Otto engines.

Remark In Fig. 1.11 we compare the Carnot engine to a competing X-engine. A conclusion in the attendant discussion is that $\eta_{Carnot} = \eta_X$ if X is reversible. This may inspire the idea to replace X with reversible Diesel and Otto engines. The result is that both engines should have the same efficiency in contradiction to the above calculation. Where is the mistake? If we apply Eq. (1.37), i.e. $\eta = 1 - q_1/q_2$, to our two engines using $q_1 = -q_{d\to a}$ and $q_2 = q_{b\to c}$, then we can indeed use the same $\eta = 1 - (-q_{d\to a})/q_{b\to c}$ in both cases. The respective results agree with the results

---

[10] Idea: (a) expansion in terms of $x - 1$ near $x = 1$ shows that $g > 1$ in this limit; (b) comparison of x-derivatives of denominator and numerator of g.

of (b) and (c). The difference is that the two cycles are not the same, leading to different efficiencies for identical compression ratios.

Cycles in the *T*-*S*-Plane

Figure 2.10 shows two reversible cycles in the *T*-*S*-plane (scales are identical). Which of the two has the greater thermal efficiency? The basic equation is

$$dE = TdS - dw. \tag{2.82}$$

In both cases $E = E(T, S)$ and $\oint_{cycle} dE = 0$, because E is a state function. The work done during one complete cycle is

$$w = \oint_{cycle} dw = \oint_{cycle} TdS \tag{2.83}$$

(clockwise). The heat absorbed is obtained by integration along the parts of the cycle for which $dS > 0$, i.e.

$$q_{in} = \oint_{cycle, >} dw = \oint_{cycle, >} TdS. \tag{2.84}$$

Thus the two thermal efficiencies are

$$\eta = \frac{w}{q_{in}} = \frac{\text{area } 1-2-3-1}{\text{area a-2-3-b-a}}. \tag{2.85}$$

While the areas in the numerators are equal, the area a-2-3-b-a is larger for cycle II and therefore

$$\eta_I > \eta_{II}. \tag{2.86}$$

Temperature Profile of the Troposphere

At high altitude the air temperature may be much lower than the ground temperature as some of us know from traveling on airplanes. How can we explain this?

We consider an air parcel rising in the atmosphere. According to the first law the differential change of its internal energy is

$$dE = \delta q - PdV = C_V dT + \frac{\partial E}{\partial V}\bigg|_T dV.$$

Assuming that the air is an ideal gas the last term on the right is zero as we have just shown. We also assume that the bubble does not exchange heat and thus rises adiabatically, i.e. $\delta q = 0$. Again we simply have

$$-PdV = C_V dT. \tag{2.87}$$

We want the temperature, T, expressed in terms of the height, h, the air bubble has risen. The quantity which we may connect easily to h is the pressure—as we shall see. Therefore we use the ideal gas law to replace dV via

$$-PdV = -nRdT + nRTd\ln P. \tag{2.88}$$

Combination of Eqs. (2.87) and (2.88) yields

$$zd\ln T = d\ln P, \tag{2.89}$$

where $z = C_V/(nR) + 1$.

In order to express P in terms of h we consider a column of air parallel to the gravitational field of the earth as shown in Fig. 2.11. The pressure at the bottom of the column element (solid cube), $P(h)$, is related to the pressure at its top, $P(h + \delta h)$, via

$$P(h) = P(h + \delta h) + \frac{\delta m_{air}g}{A}. \tag{2.90}$$

Here δm_{air} is the mass of the air contained in the column element and g is the gravitational acceleration. Via the expansion $P(h + \delta h) \approx P(h) + (dP(h)/dh)\delta h$ we find

Fig. 2.11 A column of air parallel to the gravitational field

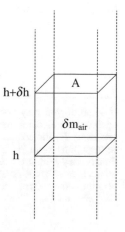

$h+\delta h$

A

δm_{air}

h

$$dP(h) = -cgdh, \qquad (2.91)$$

where $c = \delta m_{air}/(A\delta h)$ is the mass density in the column element. c of course depends on h, the position of the element in the column. Assuming that the column element contains n moles of air we write $c = nm_{mol}/(nRT/P)$ using the ideal gas law. Here $m_{mol} \approx 0.21 m_{O_2} + 0.78 m_{N_2} \approx 0.029$ kg is the molar mass of air, where $m_{O_2} \approx 0.032$ kg and $m_{N_2} \approx 0.028$ kg are the molar masses of oxygen and nitrogen. Thus Eq. (2.91) turns into

$$d\ln P = -\frac{T_o dh}{TH_o}, \qquad (2.92)$$

with

$$H_o = \frac{RT_o}{m_{mol}g}, \qquad (2.93)$$

i.e. $H_o \approx 29.2 T_o \, \mathrm{mK^{-1}}$, where T_o is the air temperature at $h = 0$. Usually H_o is close to 8500 m.

Combination of Eq. (2.89) with Eq. (2.92) yields after integration

$$T = T_0 \left(1 - \frac{h}{zH_o} \right). \qquad (2.94)$$

According to this equation the temperature drops linearly with increasing altitude. However, we need z before we can compute concrete numbers. We may look up z in a table. In Ref. HCP we find $C_P = 1.007 \, \mathrm{JK^{-1}g^{-1}}$ at $T = 300$ K and $P = 10^5$ Pa. For an ideal gas $C_P = C_V + nR$ (cf. p. 74) and therefore $z \approx 3.5$.[11] Thus, according to Eq. (2.94) the temperature reaches absolute zero at around $3 \cdot 10^4$ m.

Before we compare this to the experimental data we want a corresponding pressure profile $P(h)$, which is readily obtained by inserting Eq. (2.94) into Eq. (2.92). We find

$$d\ln P = -z \frac{dh}{zH_o - h}. \qquad (2.95)$$

---

[11] Some of you may already know that $C_V/(nR) = 3.5 = 7/2$ on the basis of the so called equipartition theorem, because every degree of freedom contributes $1/2$ to $C_V/(nR)$. Every O_2- and every N_2-molecule, the majority of what air consists of, has three center of mass kinetic degrees of freedom ($3 \cdot 1/2$; cf. Eq. (2.25)). In addition both have two axes of rotation ($2 \cdot 1/2$). Finally they both are one-dimensional oscillators ($2 \cdot 1/2$). A more detailed, i.e. quantum theoretical, calculation reveals that these two degrees of freedom do not contribute at the temperatures considered here. Therefore $C_V/(nR) = 5/2$ to good approximation and thus $C_P/(nR) \approx 7/2$.

If $zH_o \gg h$ we can neglect h in the denominator and the pressure profile becomes

$$P = P_0 \exp[-h/H_o], \qquad (2.96)$$

where P_0 is the pressure at $h = 0$. Equation (2.96) is called barometric formula. Integration of the full Eq. (2.95) yields

$$P = P_0(T/T_o)^z \qquad (2.97)$$

instead.

Fig. 2.12 Summary of results

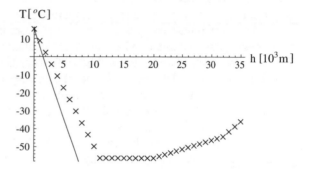

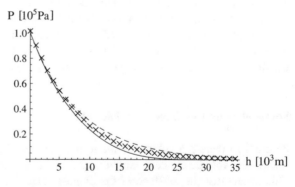

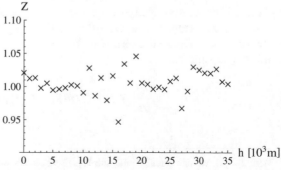

Fig. 2.13 A volume element
experiencing a pressure
difference

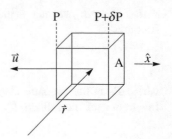

Figure 2.12 summarizes our results (solid lines—Eqs. (2.94) and (2.97); dashed line—Eq. (2.96)) and compares them to data (crosses) taken from the literature (here: "http://www.usatoday.com/weather/wstdatmo.htm"; "Source: Aerodynamics for Naval Aviators"). Notice that the temperature data are not direct measurements but rather data points computed from simple formulas describing the average temperature profile at different heights. Our calculation applies to the troposphere, i.e. to a maximum altitude of roughly 10,000 m. Beyond the troposphere other processes determine the temperature of the atmosphere. We see that our result somewhat underestimates the actual temperature. The middle graph shows the pressure profile. We notice that the two theoretical models (2.96) (isothermal case[12]) and (2.97) (adiabatic case) bracket the true pressure profile. The bottom graph shows the compressibility factor, $Z = PV/(nRT)$, versus h. The data points scatter because of scatter in the density values. Nevertheless the graph shows that our assumption of ideal gas behavior is very reasonable.

Before moving on we want to estimate one interesting number—the total mass of the earth's atmosphere, M_{atm}. Notice that the ground pressure is $P_o = M_{atm}g/(4\pi R_E^2)$, where $R_E \approx 6.37 \cdot 10^6$ m is the earth's radius. With $P_o = 1$ bar the total mass of the atmosphere is $M_{atm} \approx 5.2 \cdot 10^{18}$ kg.

Speed of Sound in Gases and Liquids

Figure 2.13 depicts a volume element in a medium. The medium can be a gas or a liquid. The volume element experiences a pressure difference along the x-direction. This means that the left face of the element experiences the pressure P while the right face is under slightly higher pressure $P + \delta P$. Here $P = \bar{P}$ is a constant average pressure in the medium, whereas $\delta P = \delta P(\vec{r}, t)$ depends on position and time.

In order to derive an expression for the speed of sound we work from the continuity equation

$$\frac{\partial}{\partial t} c(\vec{r}, t) + \vec{\nabla}(\vec{u}(\vec{r}, t) c(\vec{r}, t)) = 0. \tag{2.98}$$

---

[12] With $T = T_o$ we can directly integrate Eq. (2.92) to obtain the barometric equation.

Here $c(\vec{r}, t)$ is the mass density inside the volume element. Again we assume $c(\vec{r}, t) = \bar{c} + \delta c(\vec{r}, t)$. The pressure modulation causes a corresponding slight spatial and temporal variation of the density relative to its average, $\bar{c}$. The quantity $u(\vec{r}, t)$ is the instantaneous velocity of the volume element due to the pressure gradient. Using the approximation $\vec{u}(\vec{r}, t)c(\vec{r}, t) \approx \vec{u}(\vec{r}, t)\bar{c}$ and taking another partial derivative with respect to time yields

$$\frac{\partial^2}{\partial t^2} \delta c(\vec{r}, t) + \vec{\nabla}\left(\bar{c}\frac{\partial}{\partial t}\vec{u}(\vec{r}, t) \right) \approx 0. \tag{2.99}$$

The term in brackets is the average mass density times the acceleration of the volume element, which can be expressed via the pressure gradient according to the equation of motion $\bar{c}\partial\vec{u}(\vec{r}, t)/\partial t = -\vec{\nabla}P(\vec{r}, t)$. Therefore Eq. (2.99) becomes

$$\frac{\partial^2}{\partial t^2} \delta c(\vec{r}, t) - \vec{\nabla}^2 \delta P(\vec{r}, t) \approx 0. \tag{2.100}$$

In a final step we express $\delta P(\vec{r}, t)$ in terms of $\delta c(\vec{r}, t)$ using the thermodynamic definition of the adiabatic compressibility,

$$\kappa_S = -\frac{1}{V}\frac{\partial V}{\partial P}\bigg|_S, \tag{2.101}$$

defined analogously to the isothermal compressibility in Eq. (2.6). Adiabatic in the present context means that the density changes in the volume element are fast and no heat is transferred during the fluctuation. On p. 66 we work out in detail the relation between κ_S and κ_T. But for the moment we make use of $-\delta V/V = \delta c/c$ and obtain $\delta P \approx (\bar{c}\kappa_S)^{-1}\delta c$. This immediately yields the (density) wave equation

$$\frac{\partial^2}{\partial t^2} \delta c(\vec{r}, t) - \frac{1}{\bar{c}\kappa_S}\vec{\nabla}^2 \delta c(\vec{r}, t) \approx 0. \tag{2.102}$$

The velocity of the waves, the sound waves, is

$$v_s = \frac{1}{\sqrt{\bar{c}\kappa_S}}. \tag{2.103}$$

First we want to apply this formula to air and we ask: What is the speed of sound in this medium? Air is an ideal gas for our purpose. When we work out κ_S in detail (beginning on p. 66) we also show that in an ideal gas $\kappa_S = (z - 1)/(zP)$. The quantity $z = C_V/(nR) + 1$ was introduced in the context of Eq. (2.89). As in the previous example, the temperature profile of the troposphere, we use $C_V/(nR) = 5/2$. Thus in air

$$v_s = \sqrt{\frac{7RT}{5m_{mol}}}, \tag{2.104}$$

where $m_{mol} \approx 0.029$ kg is the air's molar mass. At $T = -100\,°C$ we calculate $v_s = 264$ m/s, while for $T = 25\,°C$ we find $v_s = 346$ m/s. Both numbers are in excellent agreement with v_s-values tabulated in HCP.

And what is the speed of sound in water? Here we need to know κ_S. Below we shall show that $\kappa_S = \kappa_T - TV\alpha_P^2/C_P$ (cf. Eq. (2.165)). In Table 2.1 we find the necessary values for water at for instance $20\,°C$. We also find that $\kappa_S \approx \kappa_T$ in this case, and we obtain $v_s = 1481$ m/s at this temperature—again in very good agreement with the corresponding value in HCP.

Joule-Thomson Coefficient

The release of propane gas from its metal can causes significant cooling of the latter. However, there may be situations when a gas leak causes heating and the possible danger of explosion. The so called Joule-Thomson coefficient,

$$\mu_{JT} = \left.\frac{\partial T}{\partial P}\right|_H, \tag{2.105}$$

is the quantity which tells us whether the temperature will increase or decrease in such a process. Here $\mu_{JT} > 0$ means cooling (refrigerator) whereas $\mu_{JT} < 0$ means heating.

The general process is depicted in Fig. 2.14. A gas initially is under pressure P_1 and confined to a volume V_1. In an adiabatic process ($\delta q = 0$) the gas is pushed through a throttle and expands into the volume V_2, where the pressure is P_2. According to the first law the internal energy change is $\Delta E = E_2 - E_1 = \Delta w$. The net amount of work is $\Delta w = P_1 V_1 - P_2 V_2$, because $P_1 V_1$ is the work done to the system and $-P_2 V_2$ is the work done by the system. Overall we find $E_1 + P_1 V_1 = E_1 + P_1 V_1$ and therefore $\Delta H = 0$. The process is said to be isenthalpic. This is why the derivative in Eq. (2.105) is at constant enthalpy.

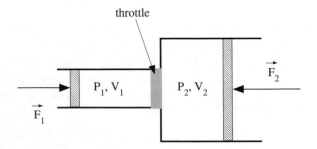

Fig. 2.14 A gas being pushed through a throttle

For concrete computations Eq. (2.105) may be transformed into

$$\mu_{JT} = \frac{1}{C_P} \left(T \frac{\partial V}{\partial T} \Big|_P - V \right) \tag{2.106}$$

or

$$\mu_{JT} = \frac{TV}{C_P} \left(\alpha_P - \frac{1}{T} \right). \tag{2.107}$$

Let us find out how to get from Eqs. (2.105) to (2.106).
We start with

$$\mu_{JT} = \frac{\partial T}{\partial P} \Big|_H \overset{(A.3)}{=} -\frac{\partial T}{\partial H} \Big|_P \frac{\partial H}{\partial P} \Big|_T$$
$$= -\frac{1}{C_P} \frac{\partial H}{\partial P} \Big|_T. \tag{2.108}$$

The quantity $\partial H / \partial P|_T$ is similar to the quantity $\partial E / \partial V|_T$ calculated on p. 30, i.e. the general approach is analogous. Via Eq. (2.11) it follows immediately that

$$TdS = dH - VdP$$

and thus

$$T \frac{\partial S}{\partial P} \Big|_T = \frac{\partial H}{\partial P} \Big|_T - V. \tag{2.109}$$

Similar to the derivation of $\partial E / \partial V|_T$ we now compute

$$d(H - TS) = dH - TdS - SdT$$
$$= -SdT + VdP. \tag{2.110}$$

Here we find

$$\frac{\partial (H - TS)}{\partial T} \Big|_P = -S \tag{2.111}$$

and

$$\frac{\partial (H - TS)}{\partial P} \Big|_T = V. \tag{2.112}$$

Using the interchangeability of partial derivatives, i.e.

$$\frac{\partial}{\partial P} \frac{\partial (H - TS)}{\partial T}\bigg|_P\bigg|_T = -\frac{\partial S}{\partial P}\bigg|_T \tag{2.113}$$

and

$$\frac{\partial}{\partial T} \frac{\partial (H - TS)}{\partial P}\bigg|_T\bigg|_P = \frac{\partial V}{\partial T}\bigg|_P, \tag{2.114}$$

we finally obtain

$$\frac{\partial S}{\partial P}\bigg|_T = -\frac{\partial V}{\partial T}\bigg|_P. \tag{2.115}$$

Combination of this equation with Eqs. (2.108) and (2.109) completes the proof. If we assume that $V \propto T^k$ then the Joule-Thomson coefficient becomes

$$\mu_{JT} \propto \frac{1}{C_P}\left(kTT^{k-1} - T^k\right) = \frac{1}{C_P}(k-1)T^k.$$

Because $C_P > 0$ (we shall show this), we find that for the ideal gas $\mu_{JT} = 0$. If the gas is not ideal we have $\mu_{JT} > 0$ for $k > 1$, which means cooling, and $\mu_{JT} < 0$ for $k < 1$, which means heating. We can now go ahead and measure k in order to find out what will happen. Below we return to the Joule-Thomson coefficient in the context of the van der Waals theory. The latter provides insight as to why a gas will do one or the other.

2.3   Free Energy and Free Enthalpy

In the preceding section we found the two quantities $E - TS$ (cf. Eq. (2.13)) and $H - TS$ (cf. Eq. (2.110)) to be rather useful. We therefore define the two new functions called the free energy[13]

$$F = E - TS \tag{2.116}$$

and the free enthalpy[14]

$$G = H - TS. \tag{2.117}$$

---

[13] Or also Helmholtz free energy; Hermann Ludwig Ferdinand von Helmholtz, German physiologist and physicist, *31.8.1821 Potsdam, Germany; †8.9.1894 Charlottenburg, Germany.

[14] Or also Gibbs free energy; Josiah Willard Gibbs, American scientist, *11. 2.1839 New Haven, Connecticut, †28.4.1903 New Haven, Connecticut; the founder of modern thermodynamics.

Computing their total differentials we find

$$dF = dE - d(TS) = dE - SdT - TdS$$
$$\stackrel{(1.51)}{=} -SdT - PdV + \mu dn + \ldots \tag{2.118}$$

and

$$dG = dH - d(TS) = dH - SdT - TdS$$
$$\stackrel{(1.51)}{=} -SdT + VdP + \mu dn + \ldots, \tag{2.119}$$

where

$$\frac{\partial F}{\partial T}\bigg|_{V,n,\ldots} = -S, \tag{2.120}$$

$$\frac{\partial F}{\partial V}\bigg|_{T,n,\ldots} = -P, \tag{2.121}$$

$$\frac{\partial F}{\partial n}\bigg|_{T,V,\ldots} = \mu, \tag{2.122}$$

and analogously

$$\frac{\partial G}{\partial T}\bigg|_{P,n,\ldots} = -S, \tag{2.123}$$

$$\frac{\partial G}{\partial P}\bigg|_{T,n,\ldots} = V, \tag{2.124}$$

$$\frac{\partial G}{\partial n}\bigg|_{T,P,\ldots} = \mu. \tag{2.125}$$

Obviously $F = F(T, V, n, \ldots)$ whereas $G = G(T, P, n, \ldots)$. The two are related via

$$G = F + PV, \tag{2.126}$$

i.e. they are Legendre transforms of each other ($G = F - V \partial F / \partial V|_T$ and $F = G - P \partial G / \partial P|_T$).

Remark 1 F is called a thermodynamic potential with respect to the variables T, V, n, ... The same is true for G with respect to T, P, n, ... In general we call a thermodynamic quantity a thermodynamic potential if all other thermodynamic quantities can be derived from partial derivatives with respect to its variables.

Remark 2 By straightforward differentiation and knowing that S and E are state functions it is easy to show that F and G are state functions also.

2.3.1 Relation to the Second Law

According to the first law we have

$$dE = \delta q - \delta w. \tag{2.127}$$

Combination of this with Clausius' statement of the second law (cf. Eq. (1.48)) yields

$$dE - TdS \leq -\delta w \tag{2.128}$$

or

$$dF \mid_T + \delta w \leq 0. \tag{2.129}$$

If δw stands for volume work only, then we may deduce

$$dF \mid_{T,V} \leq 0. \tag{2.130}$$

From this follows (cf. Sect. A.3) the attendant relation for the free enthalpy, i.e.

$$dG \mid_{T,P} \leq 0. \tag{2.131}$$

An illustration in the case of the free enthalpy is shown in Fig. 2.15. A system initially may be prepared in a state corresponding to the solid circle. It lowers its free enthalpy as much as possible, which brings it down to a point on the surface

Fig. 2.15 An illustration of the relation of G to the second law

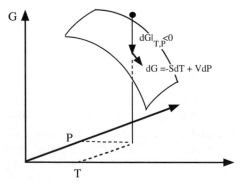

shown in the sketch, where dG is given by Eq. (2.119). This surface represents the states of lowest free enthalpy above the T-P-plane.

Of course there may be other types of work, not just volume work, involving variables, X, other than V, which are controlled from outside the system (the examples below explain what is meant here). In this case Eq. (2.130) and Fig. (2.13) still apply if ... $|_{T,V}$ and ... $|_{T,P}$ are replaced by ... $|_{T,V,X}$ and ... $|_{T,P,X}$.

This book contains a number of applications of both (2.130), e.g. phase separation in the context of van der Waals theory, and (2.131), e.g. chemical reactions. Nevertheless, already at this point we want to look at three instructive examples.

Example—Capillary Rise Figure 2.16 shows a tube of radius R with one end submerged in a liquid. In the figure the liquid has risen to a height h against the pull of gravity. We want to calculate the equilibrium height of the liquid in the tube.

This system comprises the liquid, the tube, and the earth. Ignoring all other effects we use $\delta F |_T = (\delta E - T\delta S) |_T$, and find that $\delta F |_T$ may be expressed by the following types of work involved when h changes by the small amount δh:

$$\delta F |_T = \gamma_{TL}\delta A - \gamma_{TA}\delta A + cgh\delta V. \tag{2.132}$$

Here $\delta A = 2\pi R\delta h$ and $\delta V = \pi R^2 \delta h$, i.e.

$$\delta F |_T = \left[2\pi R(\gamma_{TL} - \gamma_{TA}) + \pi R^2 cgh\right]\delta h. \tag{2.133}$$

The quantities γ_{TL} and γ_{TA} are the surface tensions of the tube-liquid and tube-air interface, respectively (cf. Eq. (1.13)). The last term in (2.132) is the negative of the work done by the system when raising the liquid volume δV to the height h. Notice that c is the mass density of the liquid, and g is the magnitude of the gravitational acceleration.

In the above equations h is a parameter not controlled from outside the system. We are not doing work to the overall system nor do we extract work. The equilibrium height may be affected from the outside only by altering the

Fig. 2.16 Capillary rise

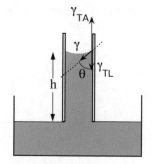

temperature, changing the surface tension or the liquid density. But here T is held constant together with the overall volume of the system. Thus we may apply Eq. (2.130) to the above free energy, i.e. we find the equilibrium value of h by minimization of the above free energy with respect to h. This means we simply set the term in square brackets equal to zero:

$$0 = (\gamma_{TL} - \gamma_{TA})2\pi R + cgh\pi R^2. \tag{2.134}$$

Solving for h yields the equilibrium height

$$h = \frac{2(\gamma_{TA} - \gamma_{TL})}{cgR}. \tag{2.135}$$

Because h depends on the sign of $\gamma_{TA} - \gamma_{TL}$, it may be positive or negative. The difference $\gamma_{TA} - \gamma_{TL}$ can be expressed in terms of the liquid-air surface tension γ and the contact angle θ (cf. Fig. 2.16) via the "force balance" $\gamma_{TA} - \gamma_{TL} = \gamma \cos \theta$. This is Young's equation, which we shall discuss in more detail in the next chapter. Hence

$$h = \frac{2\gamma \cos \theta}{cgR} \tag{2.136}$$

(e.g. $\gamma = 0.0728$ N/m for water-air at 20 °C). A nice discussion of capillarity and wetting phenomena can be found in de Gennes et al. (2004) (Pierre-Gilles de Gennes, Nobel Prize in physics for his contributions to the theory of polymers and liquid crystals, 1991).

Example—Dielectric Liquid in a Plate Capacitor The next, partially related problem is illustrated in Fig. 2.17. A plate capacitor is in contact with a liquid possessing again the mass density c and the dielectric constant ε_r. If a constant voltage, ϕ, is applied to the capacitor, the liquid rises to a certain

Fig. 2.17 Dielectric liquid in a plate capacitor

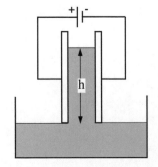

equilibrium height between the capacitor plates. This is what we want to calculate.

However, first we must discuss the relation of the free energy, F, to the electric work expressed in Eq. (1.21) **...to be continued...**

We start from

$$
\delta F = \delta(E - TS) = \delta E - S\delta T - T\delta S
$$

$$
= \delta E - S\delta T - \left(\delta E - \int dV \frac{\vec{E} \cdot \delta\vec{D}}{4\pi} + \ldots \right)
$$

$$
= -S\delta T + \int dV \frac{\vec{E} \cdot \delta\vec{D}}{4\pi},
$$

i.e.

$$
\delta F = -S\delta T + \int dV \frac{\vec{E} \cdot \delta\vec{D}}{4\pi}. \tag{2.137}
$$

Momentarily we use only the electric field contribution from Eq. (1.21). The dots in the second line indicate that there are other types of work in general, which here either do not occur (e.g., chemical work) or can be neglected (e.g., volume change) or will be added later (work against the gravitational field). Also notice that

$$
\frac{\vec{E}}{4\pi} = \frac{\partial f}{\partial \vec{D}} \Big|_{T,V,\ldots}, \tag{2.138}
$$

where f is a free energy density.

Thus we find that in the present case we have $F = F(T, \vec{D})$. However, this is not appropriate here—why? Notice that the scalar potential (voltage drop), ϕ is related to the (mean) electric field in the capacitor via

$$
\vec{E} = -\vec{\nabla}\phi. \tag{2.139}
$$

The equation applies to both the filled and the empty part of the capacitor. This means that if we hold the voltage on the capacitor plates constant, we also hold the $\vec{E}$-field between the plates constant. If we want to minimize $F = F(T, \vec{D}; h)$ with respect to h, in order to compute the equilibrium height, we must do this keeping T and $\vec{D}$ fixed. But $\vec{D}$ is not the same as $\vec{E}$, which is constant in the present setup. Thus we need a new function $\tilde{F} = \tilde{F}(T, \vec{E}; h)$ instead.

We try the Legendre transformation

$$\tilde{F} = F - \int dV \frac{\vec{E} \cdot \vec{D}}{4\pi} \tag{2.140}$$

and obtain

$$\delta \tilde{F} = \delta F - \int dV \frac{\delta \vec{E} \cdot \vec{D}}{4\pi} - \int dV \frac{\vec{E} \cdot \delta \vec{D}}{4\pi}$$
$$= -S\delta T - \int dV \frac{\delta \vec{E} \cdot \vec{D}}{4\pi},$$

i.e.

$$\delta \tilde{F} = -S\delta T - \int dV \frac{\delta \vec{E} \cdot \vec{D}}{4\pi}. \tag{2.141}$$

Indeed we have $\tilde{F} = \tilde{F}(T, \vec{E}; h)$. We now continue our example.

> ... The explicit $\vec{E}$-dependent contribution to $\tilde{F}\,|_T$ is obtained by integration of
> the differential relation $\partial \tilde{F}/\partial E|_T = -\varepsilon_r EV/(4\pi)$ with respect to E from zero
> to the actual field strength. Notice that we assume a constant field strength
> between the capacitor plates as well as $\vec{D} = \varepsilon_r \vec{E}$ (This equation is used in
> several places throughout this book, which means that the attendant calcu-
> lations do depend on its validity!). Notice also that the field strength is the
> same with and without the dielectric. Analogous to Eq. (2.132) we collect all
> relevant work terms expressing $\delta \tilde{F}\,|_T$ via
>
> $$\delta \tilde{F}\,|_T = -\epsilon_r \int_0^E \frac{d\vec{E}' \cdot \vec{E}'}{4\pi} \delta V + \int_0^E \frac{d\vec{E}' \cdot \vec{E}'}{4\pi} \delta V + cgh\delta V \tag{2.142}$$
>
> The first term is due to a volume increase, δV, of liquid between the
> capacitor plates. The second term is due to the corresponding reduction of
> vacuum (or air). Following the same reasoning as in the previous example,
> here the system includes the capacitor, the liquid, and the earth, we find the
> equilibrium height by setting the right side of the above equation equal to
> zero:
>
> $$0 = -\frac{1}{8\pi}(\varepsilon_r - 1)E^2 + cgh. \tag{2.143}$$

Solving for h yields

$$h = \frac{(\varepsilon_r - 1)\vec{E}^2}{8\pi cg}.$$
(2.144)

Remark Somebody may comment that the introduction if $\tilde{F}$ is not necessary, because $\vec{D}$ and $\vec{E}$ here are related via a constant ε_r. This is not true. Using F instead of $\tilde{F}$ changes the sign of the first term in brackets in Eq. (2.143) and there would be no sensible solution.

Before leaving this subject we study the same problem from another angle. We detach the capacitor from its voltage supply. This means that we switch from $\phi = const$ to a situation where $Q = const$. Here $\pm Q$ is the charge on the capacitor plates. What happens to Eq. (2.144)?

Figure 2.18 illustrates the situation. At (a) we have the boundary conditions

$$(\vec{E}_I - \vec{E}_{II}) \times \vec{n} = 0 \qquad \text{and} \qquad (\vec{D}_I - \vec{D}_{II}) \cdot \vec{n} = 0,$$
(2.145)

where $\vec{n}$ is a unit vector perpendicular to the interface between liquid and vacuum (or air) at (a). At (b) we have instead

$$\vec{E}_I \times \vec{n} = 0 \qquad \text{and} \qquad \vec{D}_I \cdot \vec{n} = 4\pi\sigma_I.$$
(2.146)

Now $\vec{n}$ is perpendicular to the capacitor plates. If we move (b) up into region II the boundary conditions become

$$\vec{E}_{II} \times \vec{n} = 0 \qquad \text{and} \qquad \vec{D}_{II} \cdot \vec{n} = 4\pi\sigma_{II}.$$
(2.147)

σ_I and σ_{II} are the surface charge densities in these respective regions. From (2.145) we can see that $E_{x,I} = E_{x,II}$. Because $\vec{D}_I = \varepsilon_r \vec{E}_I$ and $\vec{D}_{II} = \vec{E}_{II}$ we can conclude from (2.146) and (2.147) that

Fig. 2.18 Different boundaries in the case of the liquid inside the plate capacitor

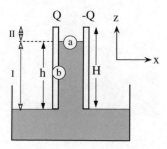

$$\sigma_I = \varepsilon_r \sigma_{II}, \tag{2.148}$$

i.e. the surface charge densities in the two regions are not equal. We can use this result to obtain

$$Q = \sigma_I L_y h + \frac{\sigma_I}{\varepsilon_r} L_y (H - h) \qquad \text{or} \qquad \sigma_I = \frac{\varepsilon_r Q / L_y}{(\varepsilon_r - 1)h + H}, \tag{2.149}$$

where L_y is the width of the capacitor in y-direction.

We now work out

$$\int_{V_{capacitor}} dV \frac{\vec{E} \cdot \vec{D}}{8\pi} \bigg|_{filled} - \int_{V_{capacitor}} dV \frac{\vec{E} \cdot \vec{D}}{8\pi} \bigg|_{empty}$$

$$= \frac{A}{8\pi} \left(h \varepsilon_r E_{x,I}^2 + (H - h) E_{x,II}^2 \right) - \frac{A}{8\pi} H E_{x,II}^2 \tag{2.150}$$

$$= \frac{A}{8\pi} (\varepsilon_r - 1) h E^2.$$

In the last step we have use $E_{x,I} = E_{x,II} = E$. This shows that we can treat this case analogous to the fixed potential case above. The result of course is the same as before. However, this time we must express E^2 via the total charge Q.

Example—Euler Instability The third example in this series is illustrated in Fig. 2.19. The figure shows a thin plate in a vice subject to a compressive force, f_c. This systems encompasses the plate only. The pressure exerted by the vice is controlled from the outside. In this case there is an elastic $F_{el}|_T$, i.e.

$$F_{el} \big|_T = \frac{\varepsilon h^3}{24} \int_{plate} dx dy \left(\frac{\partial^2 \zeta}{\partial x^2} \right)^2.$$

The quantity h is the plate's thickness, ε is the elastic modulus of the plate's material. The function $\zeta = \zeta(x)$ describes the shape of the plate when looked at along the y-direction. The expression on the right side of the equation is the work done by the plate's internal forces, when it is bent into the shape described by $\zeta(x)$. This result of continuum theory of elasticity may be found in Landau et al. (1986).

What is the equilibrium shape of the plate depending on the applied compressive force? To find the answer we now employ relation (2.129), i.e.

$$\delta F_{el} \big|_T + \delta w \leq 0. \tag{2.151}$$

Fig. 2.19 Buckling of a thin plate

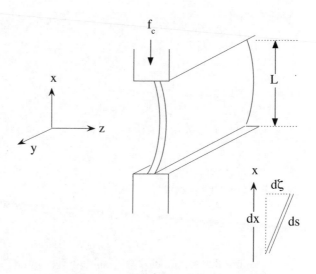

Here w is the negative of the work done by the vice, i.e. $w_c = -w$. We compute w_c by adding up the total displacement parallel to the direction of the force from infinitesimal increments (cf. the inset in Fig. 2.19):

$$ds - dx = \sqrt{dx^2 + (d\zeta(x))^2} - dx \approx dx\left(1 + \frac{1}{2}\left(\frac{\partial \zeta(x)}{\partial x}\right)^2 - 1\right).$$

.Thus

$$w_c = \frac{1}{2}f_c \int dx \left(\frac{\partial \zeta}{\partial x}\right)^2 = \frac{1}{2}\sigma \int_V dV \left(\frac{\partial \zeta}{\partial x}\right)^2.$$

V is the volume of the plate, and σ is the stress equal to $f_c/(hL_y)$, where L_y is the extend of the plate in y-direction.

We search for the minimum of the left side in (2.151) by "offering" suitable shape alternatives to the plate, i.e. we carry out a variation (or minimization) in terms of ζ:

$$0 = \delta_\zeta(F_{el} - w_c) = \delta_\zeta \int dV \left[\frac{\varepsilon}{2}z^2\left(\frac{\partial^2 \zeta}{\partial x^2}\right)^2 - \frac{\sigma}{2}\left(\frac{\partial \zeta}{\partial x}\right)^2\right].$$

We write $\frac{1}{h}\int_{-h/2}^{h/2} dz z^2 = \frac{1}{12}h^2 \equiv \mathcal{I}$ and thus

$$\delta_\zeta \frac{1}{2} \int dxdy \left[\varepsilon \mathcal{I} \left(\frac{\partial^2 \zeta}{\partial x^2} \right)^2 - \sigma \left(\frac{\partial \zeta}{\partial x} \right)^2 \right] = 0. \tag{2.152}$$

Carrying out the variation we obtain

$$\int dxdy \left[\varepsilon \mathcal{I} \left(\frac{\partial^2 \zeta}{\partial x^2} \right) \left(\frac{\partial^2}{\partial x^2} \delta\zeta \right) - \sigma \left(\frac{\partial \zeta}{\partial x} \right) \left(\frac{\partial}{\partial x} \delta\zeta \right) \right] = 0.$$

Partial integration using $\delta\zeta(x) = 0$ at $\pm L/2$ finally yields the differential equation for the "shape function":

$$\varepsilon \mathcal{I} \frac{\partial^4 \zeta}{\partial x^4} + \sigma \frac{\partial^2 \zeta}{\partial x^2} = 0.$$

Making the Ansatz $\zeta(x) = \zeta_0 \sin(qx)$ we find

$$\varepsilon \mathcal{I} q^4 - \sigma q^2 = 0 \quad \text{or} \quad \sigma_{crit} = \varepsilon \mathcal{I} q_{min}^2.$$

The planar solution is $q = 0$, whereas the bent plate corresponds to $q > 0$. The smallest possible non-vanishing q-value, q_{min}, defines the critical stress limit, σ_{crit}, at which the transition from planar to bent occurs spontaneously. This value of q_{min} depends on the boundary conditions. Here we have $q_n = (\pi/L)n$, where $n = 1, 2, \ldots$ and therefore

$$\sigma_{crit} = \frac{\pi^2 \varepsilon \mathcal{I}}{L^2}. \tag{2.153}$$

In the literature this phenomenon is called Euler buckling. The problem may be modified by embedding the plate into an elastic medium. This results in larger values for q_{min} depending on the medium's stiffness (Young's modulus).

Remark Notice that the above system has the freedom to decide to which side it buckles. This phenomenon is a spontaneous symmetry breaking.

2.3.2   Maxwell Relations

Equating the right sides of (2.16) and (2.17) as well as the right sides of (2.113) and (2.114) we have used that both F and G are state functions. The resulting formulas, (2.18) and (2.115), are examples of so called Maxwell relations.

It is easy to construct more Maxwell relations via the following recipe. Take any state function g and any pair of variables, x and y, it depends upon. If $g = pdx + qdy$ then

$$\left.\frac{\partial p}{\partial y}\right|_x = \left.\frac{\partial q}{\partial x}\right|_y \tag{2.154}$$

yields a Maxwell relation. In general we may use the differential relations in Appendix A.2 to generate even more differential relations between thermodynamic quantities, i.e. more Maxwell relations.

Example—Relating C_V and C_P   A nice exercise useful for practicing the "juggling" of partial derivatives, which is so typical for thermodynamics, is the derivation of the general relation between C_V and C_P. We start from

$$C_P - C_V = \left.\frac{\partial H}{\partial T}\right|_P - \left.\frac{\partial E}{\partial T}\right|_V. \tag{2.155}$$

Using $G = H - TS$ and

$$-S = \left.\frac{\partial G}{\partial T}\right|_P = \left.\frac{\partial H}{\partial T}\right|_P - S - T\left.\frac{\partial S}{\partial T}\right|_P \tag{2.156}$$

as well as $F = E - TS$ and

$$-S = \left.\frac{\partial F}{\partial T}\right|_V = \left.\frac{\partial E}{\partial T}\right|_V - S - T\left.\frac{\partial S}{\partial T}\right|_V \tag{2.157}$$

yields

$$C_V = T\left.\frac{\partial S}{\partial T}\right|_V \tag{2.158}$$

and

$$C_P = T\left.\frac{\partial S}{\partial T}\right|_P. \tag{2.159}$$

With

$$\left.\frac{\partial S}{\partial T}\right|_P \overset{(A.1)}{=} \left.\frac{\partial S}{\partial T}\right|_V + \left.\frac{\partial S}{\partial V}\right|_T \left.\frac{\partial V}{\partial T}\right|_P \tag{2.160}$$

we have

$$
\begin{aligned}
C_P - C_V &= T \left.\frac{\partial S}{\partial V}\right|_T \left.\frac{\partial V}{\partial T}\right|_P \\
&\overset{(2.18)}{=} T \left.\frac{\partial P}{\partial T}\right|_V \left.\frac{\partial V}{\partial T}\right|_P \\
&\overset{(A.3)}{=} -T \left.\frac{\partial P}{\partial V}\right|_T \left(\left.\frac{\partial V}{\partial T}\right|_P\right)^2 \\
&\overset{(2.5),(2.6)}{=} TV \frac{\alpha_P^2}{\kappa_T}.
\end{aligned}
\tag{2.161}
$$

Using (2.9) we find for an ideal gas

$$
C_P - C_V = nR,
\tag{2.162}
$$

which we had used in the calculation of the air temperature profile on p. 56.

Example—Adiabatic Compressibility Another exercise similar to the previous one is the derivation of κ_S, the adiabatic compressibility. This quantity is defined in Eq. (2.101), i.e.

$$
\kappa_S = -\frac{1}{V} \left.\frac{\partial V}{\partial P}\right|_S.
$$

We transform the right side via

$$
\left.\frac{\partial V}{\partial P}\right|_S \overset{(A.1)}{=} \underbrace{\left.\frac{\partial V}{\partial T}\right|_P}_{\overset{(2.5)}{=} V\alpha_P} \left.\frac{\partial T}{\partial P}\right|_S + \underbrace{\left.\frac{\partial V}{\partial P}\right|_T}_{\overset{(2.6)}{=} -V\kappa_T}.
\tag{2.163}
$$

The remaining unknown derivative is

$$
\left.\frac{\partial T}{\partial P}\right|_S \overset{(A.3)}{=} -\underbrace{\left.\frac{\partial T}{\partial S}\right|_P}_{\overset{(2.159)}{=} \frac{T}{C_P}} \left.\frac{\partial S}{\partial P}\right|_T = \left.\frac{\partial T}{\partial S}\right|_P \underbrace{\left.\frac{\partial}{\partial P}\left.\frac{\partial G}{\partial T}\right|_P\right|_T}_{= \left.\frac{\partial V}{\partial T}\right|_P = V\alpha_P}.
\tag{2.164}
$$

Putting everything together yields

$$\kappa_S = \kappa_T - TV\frac{\alpha_P^2}{C_P} \tag{2.165}$$

or if we combine this equation with (2.161)

$$\frac{\kappa_S}{\kappa_T} = \frac{C_V}{C_P}. \tag{2.166}$$

Again we can calculate κ_S for an ideal gas. The result is

$$\kappa_S = \frac{z-1}{z}\frac{1}{P}. \tag{2.167}$$

The quantity $z = C_V/(nR) + 1$ was introduced in the context of Eq. (2.89) (cf. its discussion in the footnote on p. 56).

Example—Electric Field Effect on C_V What is the effect of an electric field on C_V? We find the answer by working from

$$\frac{\partial}{\partial\vec{E}}\underbrace{\left.\frac{\partial^2\tilde{F}}{\partial T^2}\right|_{\rho,\vec{E}}}_{\overset{(2.146)}{=}-T^{-1}C_{V,\vec{E}}}\bigg|_{T,\rho} = \frac{\partial^2}{\partial T^2}\underbrace{\left.\frac{\partial\tilde{F}}{\partial\vec{E}}\right|_{T,\rho}}_{\overset{(2.129)}{=}-V\vec{D}/(4\pi)}\bigg|_{\rho,\vec{E}}, \tag{2.168}$$

i.e.

$$\left.\frac{\partial C_{V,\vec{E}}}{\partial\vec{E}}\right|_{T,\rho} = \frac{TV}{4\pi}\left.\frac{\partial^2\vec{D}}{\partial T^2}\right|_{\rho,\vec{E}}, \tag{2.169}$$

where $\rho = N/V$ and assuming constant fields throughout V. With $\vec{D} = \varepsilon_r\vec{E}$ we may easily integrate this equation from zero to the final field strength, which yields

$$C_V(E) = C_V(0) + \frac{TVE^2}{8\pi}\left.\frac{\partial^2\varepsilon_r}{\partial T^2}\right|_{\rho}. \tag{2.170}$$

Remark A similar strategy helps to find corresponding expressions for κ_T or α_P.

Example—Electrostriction Here we want to show the validity of

$$\frac{1}{\rho}\frac{\partial \rho}{\partial E}\bigg|_{\mu,T} = \frac{E}{4\pi}\frac{\partial \varepsilon_r}{\partial P}\bigg|_{E,T} \tag{2.171}$$

(cf. Frank 1955). Imagine a plate capacitor completely submerged in a dielectric liquid. The dielectric constant of the liquid is ε_r. The size of the capacitor is small compared to the extend of the liquid reservoir. This means that far from the capacitor the latter has no effect on the chemical potential, μ, of the liquid, i.e. μ is constant. In addition the temperature, T, is held constant as well. The mean electric field, $\vec{E}$, between the capacitor plates will affect the density, $\rho = N/V$, inside the capacitor. This is the expression on the left. The change does depend on electric field strength, E, and the derivative of ε_r with respect to pressure, P. Our starting point is relation (A.3), i.e.

$$\frac{1}{\rho}\frac{\partial \rho}{\partial E}\bigg|_{\mu,T} = -\underbrace{\frac{1}{\rho}\frac{\partial \rho}{\partial \mu}\bigg|_{E,T}}_{\stackrel{(*)}{=}\rho\kappa_T}\frac{\partial \mu}{\partial E}\bigg|_{\rho,T}. \tag{2.172}$$

Note that (*) follows via $\rho\partial/\partial\rho = -V\partial/\partial V$ and using $\partial G/\partial V|_{T,N,E} = -1/\kappa_T$ derived below (cf. (2.195)). We continue with

$$\frac{\partial \mu}{\partial E}\bigg|_{\rho,T} = \frac{\partial}{\partial E}\frac{\partial \tilde{F}}{\partial N}\bigg|_{T,V,E}\bigg|_{\rho,T} = \frac{\partial}{\partial N}\frac{\partial \tilde{F}}{\partial E}\bigg|_{\rho,T}\bigg|_{T,V,E} \stackrel{(2.141)}{=} -\frac{E}{4\pi}\frac{\partial \varepsilon_r}{\partial \rho}\bigg|_{E,T}. \tag{2.173}$$

The final ingredient is

$$\rho\frac{\partial \varepsilon_r}{\partial \rho}\bigg|_{E,T} \stackrel{(A.2)}{=} \rho\frac{\partial P}{\partial \rho}\bigg|_{E,T}\frac{\partial \varepsilon_r}{\partial P}\bigg|_{E,T} = \frac{1}{\kappa_T}\frac{\partial \varepsilon_r}{\partial P}\bigg|_{E,T}. \tag{2.174}$$

Combination of the last three equations yields the desired result.

In order to estimate the magnitude of this effect we integrate Eq. (2.171) assuming that the derivative on the right side can be replaced by its value at zero field strength, i.e.

$$\frac{\Delta \rho}{\rho} \approx \frac{E^2}{8\pi}\frac{\partial \varepsilon_r}{\partial P}\bigg|_{o,T}. \tag{2.175}$$

The transition to SI-units is accomplished by replacing E with $\sqrt{4\pi\varepsilon_o}E$, i.e. in SI-units Eq. (2.175) becomes $\Delta\rho/\rho \approx (\varepsilon_o/2)E^2\partial\varepsilon_r/\partial P|_{o,T}$. Because ε_o is a small constant ($\sim 10^{-11}$ in these units), an appreciable effect requires rather high fields and/or conditions for which the derivative becomes large.[15]

2.4 Extensive and Intensive Quantities

An important concept is the characterization of thermodynamic quantities as either intensive or extensive.

Imagine two identical containers filled with the same kind and amount of of gas at the same temperature. What happens if we bring the two containers in contact and allow the gas to fill the combined containers as if it was one. Obviously temperature and pressure do not change. The quantities T and P therefore are called intensive. Volume on the other hand is doubled. Mathematically this means $V \propto n$. Such quantities are said to be extensive. n itself is therefore extensive. Thus far:

- intensive: $T, P, \ldots$
- extensive: $V, n, \ldots$

The ratio of two extensive quantities, e.g. n/V, again is intensive of course. Another intensive quantity is the chemical potential, μ. Whether one mole of material is added to a large system or to twice as large a system should not matter.[16] This however has implications for the free enthalpy, G. According to Eq. (2.119) we have for a one-component system

$$dG|_{T,P,\ldots} = \mu dn, \qquad (2.176)$$

where ... stands for other intensive variables in addition to T and P. Because μ is intensive and dn is extensive, we conclude that dG is extensive also. By adding (or integrating over) sufficiently many differential amounts of material, $n = \int dn$, we find the important relation

$$G(T, P, n, \ldots) = \mu n. \qquad (2.177)$$

For a K-component system this becomes

$$\boxed{G(T, P, n_1, \ldots, n_K, \ldots) = \sum_{i=1}^{K} \mu_i n_i.} \qquad (2.178)$$

---

[15] In liquid water $\partial \ln \varepsilon_r/\partial P|_{o,T} \approx 5 \cdot 10^{-5}/\text{bar}$.

[16] Momentarily we talk about one-component systems and not about mixtures.

Equating dG on the left with $d(\sum_{i=1}^{K} \mu_i n_i)$ on the right, i.e.

$$-SdT + VdP + \ldots + \sum_{i=1}^{K} \mu_i dn_i = \sum_{i=1}^{K} (\mu_i dn_i + n_i d\mu_i), \qquad (2.179)$$

yields

$$\boxed{-SdT + VdP + \ldots - \sum_{i=1}^{K} n_i d\mu_i = 0,} \qquad (2.180)$$

the Gibbs-Duhem equation. The significance of this equation will become clear in many examples to come.

Remark 1 Suppose we consider the potential energy $\mathcal{U}$ of a system consisting of N pairwise interacting molecules. Disregarding their spatial arrangement we may write $\mathcal{U} \sim (N^2/2)V^{-1} \int_a^\infty dr r^{2-n}$. The factor $N^2 \approx N(N-1)/2$ is the number of distinct pairs, V is the volume, a is a certain minimum molecular separation, and r^{-n} is the leading distance dependence of the molecular interaction. If $n > 3$ the integral is finite and consequently $\mathcal{U}/V \propto \rho^2$, where ρ is the number density molecules. This means that $\mathcal{U}$ (and also E) is extensive by our above definition. However, if $n \leq 3$ the situation is more complex. Now it is necessary to include the spatial and orientational correlations between the molecules. In general these conspire to yield an extensive $\mathcal{U}$. An exception is gravitation, where $\mathcal{U}/V \sim \rho^2 V^{2/3}$ including an additional shape dependence.

Remark 2 Looking at the two Eqs. (2.176) and (2.177) we may wonder whether one can apply the same argument to

$$dF\big|_{T,V,\ldots} = \mu dn,$$

valid according to Eq. (2.118). This immediately leads to $F = G$, which clearly is incorrect! The point is that we cannot keep the volume constant and simultaneously add up increments dn to the full n. Therefore this procedure does not work for $dF\big|_{T,V,\ldots}$.

Example—Partial Molar Volume Consider a binary liquid mixture (A and B) at constant temperature and pressure. The volume change due to a differential change of the composition is

$$dV = \frac{\partial V}{\partial n_A}\bigg|_{T,P,n_B} dn_A + \frac{\partial V}{\partial n_B}\bigg|_{T,P,n_A} dn_B = v_A dn_A + v_B dn_B. \qquad (2.181)$$

The quantities v_A and v_B are called partial molar volumes. In this sense μ_i is the partial molar free enthalpy of component i.

Now we argue as in the case of Eq. (2.176). V is extensive and so is n. Therefore we may ad up dV's to the full V at constant T and P and thus

$$V = v_A n_A + v_B n_B. \tag{2.182}$$

However, the quantity of interest here is not V but the volume difference upon mixing, ΔV. The volumes of the pure substances are $v_i^* n_i$, where $v_i^* = \partial V / \partial n_i |_{T,P,n_j=0(i\neq j)}$ $(i,j = 1,2)$ and thus

$$\Delta V = (v_A - v_A^*)n_A + (v_B - v_B^*)n_B. \tag{2.183}$$

Notice that v_A and v_B are not independent. To see this we simply must realize that we can carry out the steps from Eq. (2.178) to the Gibbs-Duhem Eq. (2.180) with G replaced by V and μ_i replaced by v_i. At constant T and P this means

$$dv_A n_A + dv_B n_B = 0. \tag{2.184}$$

In addition we notice that this may immediately be extended to more than two components. And this is not the end, because the same reasoning applies to every extensive quantity $\Phi = \Phi(T, P, n_1, n_2, \ldots)$, i.e.

$$\Delta \Phi = \sum_i^K (\phi_i - \phi_i^*)n_i, \tag{2.185}$$

and

$$\sum_{i=1}^K n_i d\phi_i = 0 \qquad (T, P = \text{constant}), \tag{2.186}$$

where $\phi_i = \partial \Phi / \partial n_i |_{T,P,n_j(i\neq j)}$ are the respective partial molar quantities. Here Φ stands for extensive thermodynamic quantities like V, H, C_P, ...

2.4.1 Homogeneity

Before leaving this subject we look at it briefly from another angle. In mathematics a function $f(x_1, x_2, \ldots, x_n)$ is said to be homogeneous of order m if the following condition is fulfilled:

$$f(\lambda x_1, \lambda x_2, \ldots, \lambda x_n) = \lambda^m f(x_1, x_2, \ldots, x_n). \tag{2.187}$$

Thus we may consider the extensive quantities free energy and free enthalpy as first-order homogeneous functions in V, n_i and n_i respectively, i.e.

$$F(T, \lambda V, \lambda n_1, \lambda n_2, \ldots) = \lambda F(T, V, n_1, n_2, \ldots) \tag{2.188}$$

$$G(T, P, \lambda n_1, \lambda n_2, \ldots) = \lambda G(T, P, n_1, n_2, \ldots). \tag{2.189}$$

Differentiating (2.188) on both sides with respect to λ yields

$$\frac{dF}{d\lambda} = \frac{\partial F}{\partial(\lambda V)}\bigg|_{T,n_1,n_2,\ldots} V + \sum_{i=1}^{K} \frac{\partial F}{\partial(\lambda n_i)}\bigg|_{T,V,n_{k(\neq i)}} n_i = F. \tag{2.190}$$

For $\lambda = 1$ this becomes

$$F = \frac{\partial F}{\partial V}\bigg|_{T,n_1,n_2,\ldots} V + \sum_{i=1}^{K} \frac{\partial F}{\partial n_i}\bigg|_{T,V,n_{k(\neq i)}} n_i \overset{(2.121),(2.178)}{=} -PV + G \tag{2.191}$$

in agreement with Eq. (2.126). This also implies

$$\mu_i = \frac{\partial F}{\partial n_i}\bigg|_{T,V,n_{k(\neq i)}}, \tag{2.192}$$

i.e. the generalization of Eq. (2.122) to more than one component. Differentiating (2.189) on both sides with respect to λ and setting $\lambda = 1$ reproduces Eq. (2.178). Clearly, we may apply the same idea to other extensive thermodynamic functions like $S(E, V, n_1, n_2, \ldots)$, i.e. $\lambda S(E, V, n_1, n_2, \ldots) = S(\lambda E, \lambda V, \lambda n_1, \lambda n_2, \ldots)$, $E(T, V, n_1, n_2, \ldots)$, i.e. $\lambda E(T, V, n_1, n_2, \ldots) = E(T, \lambda V, \lambda n_1, \lambda n_2, \ldots)$, or others.

Likewise we may consider the intensive quantities as zero-order homogeneous functions in their extensive variables, e.g.

$$P(T, V, n_1, n_2, \ldots) = P(T, \lambda V, \lambda n_1, \lambda n_2, \ldots). \tag{2.193}$$

Differentiating with respect to λ on both sides and subsequently setting $\lambda = 1$ yields

$$0 = \frac{\partial P}{\partial V}\bigg|_{T,n_1,n_2,\ldots} V + \sum_{i=1}^{K} \frac{\partial P}{\partial n_i}\bigg|_{T,V,n_{k(\neq i)}} n_i. \tag{2.194}$$

Using $P = -\partial F/\partial V|_{T,n_1,n_2,...}$ and changing the order of differentiation we find

$$\frac{\partial G}{\partial V}\bigg|_{T,n_1,n_2,...} = -\frac{1}{\kappa_T}. \tag{2.195}$$

This easily is verified by insertion of (2.126) and subsequent differentiation.

The concept of homogeneity does not produce otherwise unattainable relations, but it is an elegant means to compute them. We revisit homogeneity in a generalized form in the context of continuous phase transitions in Sect. 4.2, where again it proves useful.

Chapter 3
Equilibrium and Stability

3.1 Equilibrium and Stability via Maximum Entropy

3.1.1 Equilibrium

The first row of boxes shown in Fig. 3.1 depicts a number of identical systems differing only in their internal energies, E_v, volumes, V_v, and mass contents, n_v. The boundaries of the systems allow the exchange of these quantities between the systems upon contact. The second row of boxes in Fig. 3.1 illustrates this situation. All (sub-)systems combined form an isolated system. We ask the following question: What can be said about the quantities x_v, where x represents E, V or n, after we bring the boxes into contact and allow the exchanges to occur?

According to our experience the exchange is an irreversible spontaneous process and therefore relation (1.50) applies to the entropy of the overall system. We can expand the entropy of the combined systems, S, in a Taylor series in the variables E_v, V_v, and n_v with respect to its maximum, i.e.

$$
\begin{aligned}
S = S^o &+ \sum_v \left(\Delta E_v \frac{\partial S_v}{\partial E_v} \Big|_{V_v,n_v}^o + \Delta V_v \frac{\partial S_v}{\partial V_v} \Big|_{E_v,n_v}^o + \Delta n_v \frac{\partial S_v}{\partial n_v} \Big|_{E_v,V_v}^o \right) \\
&+ \frac{1}{2} \sum_{v',v} \left(\Delta E_{v'} \frac{\partial}{\partial E_{v'}} \Big|_{V_{v'},n_{v'}}^o + \Delta V_{v'} \frac{\partial}{\partial V_{v'}} \Big|_{E_{v'},n_{v'}}^o + \Delta n_{v'} \frac{\partial}{\partial n_{v'}} \Big|_{E_{v'},V_{v'}}^o \right) \\
&\times \left(\Delta E_v \frac{\partial S_v}{\partial E_v} \Big|_{V_v,n_v}^o + \Delta V_v \frac{\partial S_v}{\partial V_v} \Big|_{E_v,n_v}^o + \Delta n_v \frac{\partial S_v}{\partial n_v} \Big|_{E_v,V_v}^o \right),
\end{aligned} \tag{3.1}
$$

where $S = \sum_\mu S_\mu(E_\mu, V_\mu, n_\mu)$. The quantity S^o is the maximum value of the entropy. Notice that this quantity is somewhat hypothetical. The usefulness of this approach relies on the differences between time scales on which certain processes take place.

© Springer Nature Switzerland AG 2022
R. Hentschke, *Thermodynamics*, Undergraduate Lecture Notes in Physics,
https://doi.org/10.1007/978-3-030-93879-6_3

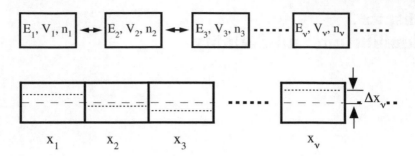

Fig. 3.1 Identical systems initially differing only in their internal energies, volumes, and mass content

Leaving a cup of hot coffee on the table, we expect to find coffee at room temperature upon our return several hours later. If we come back after some weeks of vacation the coffee has vanished, i.e. the water has evaporated. Only the dried remnants of the coffee remain inside the cup. After waiting for a much longer time, how long depends on numerous things including the material of the coffee cup, the cup itself has crumbled into dust. However, if we are interested merely in the initial cooling of the coffee to room temperature, we may neglect evaporation and we may certainly neglect the deterioration of the cup itself.

In this sense we shall use the expression of equilibrium. For all practical purposes equilibrium is understood in a "local" sense, i.e. the time scale underlying the process of interest is much shorter than the time scale underlying other processes influencing the former. In the case at hand equilibrium means that all variables, E_v, V_v, and n_v, have assumed the values E_v^o, V_v^o, and n_v^o corresponding to maximum entropy. However, we may impose deviations from these values in each subsystem, Δx_v, as illustrated in the bottom part of Fig. 3.1. The long dashed line indicates the equilibrium value(s), which is the same in all (identical) systems. The short dashed lines indicate the imposed deviations from equilibrium in each system, Δx_v. Because the whole system is isolated, we have the condition(s)

$$\sum_v \Delta x_v = 0. \tag{3.2}$$

Equation (3.1) is nothing but a Taylor expansion of S to second order in the Δx_v, which we can freely and independently adjust except for the condition(s) (3.2).

Here the value of S^o is of no interest to us. But already the linear terms, i.e the first sum, leads to important conclusions. If for the moment we consider two subsystems only, i.e. $v = 1, 2$, then the condition of maximum entropy yields

$$0 = \Delta E_1 \left(\frac{\partial S_1}{\partial E_1} \Big|^o_{V_1, n_1} - \frac{\partial S_2}{\partial E_2} \Big|^o_{V_2, n_2} \right)$$
$$+ \Delta V_1 \left(\frac{\partial S_1}{\partial V_1} \Big|^o_{E_1, n_1} - \frac{\partial S_2}{\partial V_2} \Big|^o_{E_2, n_2} \right) \tag{3.3}$$
$$+ \Delta n_1 \left(\frac{\partial S_1}{\partial n_1} \Big|^o_{E_1, V_1} - \frac{\partial S_2}{\partial n_2} \Big|^o_{E_2, V_2} \right),$$

where we have used $S = \sum_v S_v(E_v, V_v, n_v)$ and Eq. (2). Via the Eqs. (1.52) and (1.53), and (1.55) this becomes

$$\Delta E_1 \left(\frac{1}{T_1} - \frac{1}{T_2} \right) + \Delta V_1 \left(\frac{P_1}{T_1} - \frac{P_2}{T_2} \right) - \Delta n_1 \left(\frac{\mu_1}{T_1} - \frac{\mu_2}{T_2} \right) = 0.$$

Because ΔE_1, ΔV_1, and Δn_1 are arbitrary, we conclude that

$$T = T_1 = T_2 \tag{3.4}$$

$$P = P_1 = P_2 \tag{3.5}$$

$$\mu = \mu_1 = \mu_2 \tag{3.6}$$

at equilibrium.

These conditions of course may be generalized to an arbitrary number of subsystems. The latter in general are different regions in space within a large system. In some cases different regions in space may contain distinct phases. An example is ice in one region of space and liquid water in an adjacent region. One and the same material may occur in different phases depending on thermodynamic conditions. A phase is a homogeneous state of matter. Each phase usually differs from another phase by certain clearly distinguishable bulk properties. Ice, for instance, has a lower symmetry than liquid water. At coexistence, defined by the above conditions, ice has a lower density than liquid water etc. Changing from one phase to another often, but not always, is accompanied by a discontinuous change of certain thermodynamic quantities. We shall discuss phase transformations in detail below.

Equation (3.6) is derived for a one-component system. Of course we can extend our reasoning to a K-component system, which yields

$$\mu_i^{(1)} = \mu_i^{(2)} \tag{3.7}$$

($i = 1, \ldots, K$). Here $^{(1)}$ and $^{(2)}$ are the subsystem indices. Again $^{(1)}$ and $^{(2)}$ may refer to different phases, i.e. at equilibrium the chemical potential of each component is continuous across the phase boundary. In particular if there are Π phases, each considered to be a subsystem, we find

$$\mu_i^{(\nu)} = \mu_i^{(\mu)}, \tag{3.8}$$

where $\nu, \mu = 1, 2, \ldots, \Pi$.

3.1.2 Gibbs Phase Rule

In general there are K components in Π different phases (solid, liquid, gas, ...) at constant T and P. We may ask: What is the maximum number of coexisting phases at equilibrium? Or to be pictorial, is the situation in Fig. 3.2 possible, where a one-component system contains four coexisting phases—"Gas", "Liquid", "Solid", and "Flubber"?

We assume a system containing K components and Π coexisting phases. Each phase may be considered a subsystem in the above sense. The state of each phase ν is then determined by its temperature $T^{(\nu)}$, its pressure $P^{(\nu)}$, and its composition $\{n_1^{(\nu)}, n_2^{(\nu)}, \ldots, n_K^{(\nu)}\}$. All in all we must specify

$$\Pi(K + 2) \tag{3.9}$$

quantities.

On the other hand, equilibrium, as we have just discussed, imposes certain constraints. In the case of two subsystems (now phases) and just one component we had to fulfill the Eqs. (3.4) to (3.6). In the case of Π phases and K components we have

Fig. 3.2 Hypothetically coexisting phases

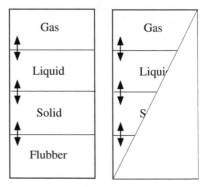

$$T^{(1)} = T^{(2)} = \ldots = T^{(\Pi)}$$
$$P^{(1)} = P^{(2)} = \ldots = P^{(\Pi)}$$
$$\mu_1^{(1)} = \mu_1^{(2)} = \ldots = \mu_1^{(\Pi)}$$
$$\mu_2^{(1)} = \mu_2^{(2)} = \ldots = \mu_2^{(\Pi)}$$
$$\vdots$$
$$\mu_K^{(1)} = \mu_K^{(2)} = \ldots = \mu_K^{(\Pi)}$$

and therefore

$$(\Pi - 1)(K + 2) \qquad (3.10)$$

constraints. In addition there are constraints having to do with the total amount of material in each phase. In the one-component system illustrated in Fig. 3.2 we may for instance insert a diagonal partition without physical effect—it is a key assumption that the shape of our container has no influence on the type of phases present. This means that the total amount of material in a phase v,

$$n^{(v)} = \sum_{i=1}^{K} n_i^{(v)},$$

does not affect the phase coexistence. This yields

$$\Pi \qquad (3.11)$$

constraints. The net number of adjustable quantities, i.e. the number of overall adjustable quantities, (3.9), minus the number of constraints, (3.10) and (3.11), is called the number of degrees of freedom, Z. Thus

$$Z = K - \Pi + 2 \geq 0 \qquad (3.12)$$

Applied to our above system we find $1 - 4 + 2 < 0$. This means that four phases cannot coexist simultaneously in a one-component system. The maximum number of coexisting phases in a one-component system is three—but thermodynamics does not specify which three phases. Relation (3.12) is Gibbs phase rule.

Remark 1 Our reasoning is based on subsystems whose state is determined by temperature, pressure, and composition. However, we may also include external electromagnetic fields requiring recalculation of the degrees of freedom.

Remark 2 Below we shall discuss in more detail what we mean by component. This in turn will affect the statement of the phase rule (cf. p. 110).

3.1.3 Stability

We now return to Eq. (3.1) and focus on the second order term. According to our discussion the linear term vanishes at equilibrium. In addition, in the second order term, the cross contributions $(v \neq v')$ also vanish and therefore $\Delta S = S - S^o$ is given by

$$
\begin{aligned}
\Delta S &= \frac{1}{2}\sum_v \left[\Delta E_v \frac{\partial}{\partial E_v}\Big|_{V_v,n_v}^o + \Delta V_v \frac{\partial}{\partial V_v}\Big|_{E_v,n_v}^o + \Delta n_v \frac{\partial}{\partial n_v}\Big|_{E_v,V_v}^o \right] \\
&\quad \times \left\{ \Delta E_v \underbrace{\frac{\partial S_v}{\partial E_v}\Big|_{V_v,n_v}^o}_{=1/T} + \Delta V_v \underbrace{\frac{\partial S_v}{\partial V_v}\Big|_{E_v,n_v}^o}_{=P/T} + \Delta n_v \underbrace{\frac{\partial S_v}{\partial n_v}\Big|_{E_v,V_v}^o}_{=-\mu/T} \right\} \\
&= \frac{1}{2}\sum_v [\ldots]\Delta S_v \\
&= \frac{1}{2}\sum_v \left(-\frac{1}{T}\Delta S_v[\ldots]T + \frac{1}{T}[\ldots](T\Delta S_v) \right) \\
&= \frac{1}{2T}\sum_v (-\Delta S_v \Delta T_v + \Delta P_v \Delta V_v - \Delta \mu_v \Delta n_v),
\end{aligned}
\tag{3.13}
$$

where we have used

$$
T\Delta S = \Delta E + P\Delta V - \mu \Delta n \tag{3.14}
$$

(cf. Eq. (1.51)). Equation (3.13) quite generally expresses the entropy fluctuations via the corresponding fluctuations in the subsystems.

Now we choose the variables T, V, and n, which yields

$$(\ldots) = - \left[\frac{\partial S_v}{\partial T_v}\Big|^o_{V_v,n_v} \Delta T_v + \frac{\partial S_v}{\partial V_v}\Big|^o_{T_v,n_v} \Delta V_v + \frac{\partial S_v}{\partial n_v}\Big|^o_{T_v,V_v} \Delta n_v \right] \Delta T_v$$

$$+ \left[\frac{\partial P_v}{\partial T_v}\Big|^o_{V_v,n_v} \Delta T_v + \frac{\partial P_v}{\partial V_v}\Big|^o_{T_v,n_v} \Delta V_v + \frac{\partial P_v}{\partial n_v}\Big|^o_{T_v,V_v} \Delta n_v \right] \Delta V_v$$

$$- \left[\frac{\partial \mu_v}{\partial T_v}\Big|^o_{V_v,n_v} \Delta T_v + \frac{\partial \mu_v}{\partial V_v}\Big|^o_{T_v,n_v} \Delta V_v + \frac{\partial \mu_v}{\partial n_v}\Big|^o_{T_v,V_v} \Delta n_v \right] \Delta n_v$$

$$= + \left[-\frac{C_V}{T}\Delta T_v - \frac{\partial P_v}{\partial T_v}\Big|^o_{V_v,n_v} \Delta V_v + \frac{\partial \mu_v}{\partial T_v}\Big|^o_{V_v,n_v} \Delta n_v \right] \Delta T_v$$

$$+ \left[\frac{\partial P_v}{\partial T_v}\Big|^o_{V_v,n_v} \Delta T_v - \frac{1}{V^o \kappa_T^o}\Delta V_v - \frac{\partial \mu_v}{\partial V_v}\Big|^o_{T_v,n_v} \Delta n_v \right] \Delta V_v$$

$$+ \left[-\frac{\partial \mu_v}{\partial T_v}\Big|^o_{V_v,n_v} \Delta T_v - \frac{\partial \mu_v}{\partial V_v}\Big|^o_{T_v,n_v} \Delta V_v - \frac{\partial \mu_v}{\partial n_v}\Big|^o_{T_v,V_v} \Delta n_v \right] \Delta n_v$$

$$= -\frac{C_V}{T}\Delta T_v^2 - \frac{1}{V\kappa_T}\Delta V_v^2 - \frac{\partial \mu_v}{\partial n_v}\Big|^o_{T_v,V_v} \Delta n_v^2 - 2\frac{\partial \mu_v}{\partial V_v}\Big|^o_{T_v,n_v} \Delta n_v \Delta V_v.$$

We need to transform this equation one last time using

$$\Delta V_v = \frac{\partial V_v}{\partial T_v}\Big|^o_{P_v,n_v} \Delta T_v + \frac{\partial V_v}{\partial P_v}\Big|^o_{T_v,n_v} \Delta P_v + \frac{\partial V_v}{\partial n_v}\Big|^o_{T_v,P_v} \Delta n_v$$

$$\equiv \Delta V_{n,v} + \frac{\partial V_v}{\partial n_v}\Big|^o_{T_v,P_v} \Delta n_v.$$

The quantity $\Delta V_{n,v}$ is the volume fluctuation at constant mass content. We obtain

$$(\ldots) = -\frac{C_V}{T}\Delta T_v^2 - \frac{1}{V\kappa_T}\Delta V_{n,v}^2 - \frac{\partial \mu_v}{\partial n_v}\Big|^o_{T_v,P_v} \Delta n_v^2. \tag{3.15}$$

According to the second law ΔS must be negative, because otherwise the fluctuations would grow spontaneously in order to increase the entropy. Therefore we find

$$\boxed{C_V \geq 0 \qquad \kappa_T \geq 0 \qquad \frac{\partial \mu}{\partial n}\Big|^o_{T,P} \geq 0} \tag{3.16}$$

for the isochoric heat capacity, C_V, the isothermal compressibility, κ_T, and the quantity $\frac{\partial \mu}{\partial n}\big|_{T,P}^o$. These relations are sometimes denoted as thermal stability, mechanical stability, and chemical stability.[1] Consequently we also have

$$\frac{\partial^2 G}{\partial T^2}\bigg|_{P,n} = -\frac{1}{T}C_P \leq 0 \quad \frac{\partial^2 G}{\partial P^2}\bigg|_{T,n} = -V\kappa_T \leq 0 \tag{3.17}$$

$$\frac{\partial^2 F}{\partial T^2}\bigg|_{V,n} = -\frac{1}{T}C_V \leq 0 \quad \frac{\partial^2 F}{\partial V^2}\bigg|_{T,n} = \frac{1}{V\kappa_T} \geq 0 \tag{3.18}$$

$(0 \leq C_V \leq C_P!)$.

Remark The above condition for chemical stability can be generalized to K components by replacing the one-component terms, Δn_v..., in the derivation with their multicomponent versions, $\sum_k \Delta n_{v,k}$. .... The result is

$$\sum_{j,k=1}^{K} \frac{\partial \mu_j}{\partial n_k}\bigg|_{T,P}^o \Delta n_j \Delta n_k \geq 0. \tag{3.19}$$

The simplest way to get rid of the subsystem indices v is to consider two subsystems only, i.e. $\Delta n_{1,k} = -\Delta n_{2,k} = \Delta n_k$.

3.2   Chemical Potential and Chemical Equilibrium

3.2.1   Chemical Potential of Ideal Gases and Ideal Gas Mixtures

Pure gas: Based on the Gibbs-Duhem equation (2.180) we may write for the chemical potential of a one-component gas A at fixed temperature T

$$\mu_A^{(g)}\left(T, P_A^*\right) - \mu_A^{(g)}\left(T, P_A^\circ\right) = \frac{1}{n}\int_{P_A^\circ}^{P_A^*} VdP. \tag{3.20}$$

The various indices do have the following meaning. The index $^{(g)}$ reminds us that we talk about gases. The index * indicates that the gas is a pure or one-component gas and not one component in a mixture of gases. The other index $^\circ$

---

[1] The conditions (3.16) are are mathematical statements of Le Châtelier's principle, i.e. driving a system away from its stable equilibrium causes internal processes tending to restore the equilibrium state.

indicates a reference pressure. In thermodynamics the chemical potential is not an absolute quantity. We rather compute differences between chemical potentials—often there is some standard state, defined by specifying temperature and pressure values, with respect to which the difference is calculated. If our gas is ideal the above equation becomes

$$\mu_A^{(g)}\left(T, P_A^*\right) - \mu_A^{(g)}\left(T, P_A^\circ\right) = RT \int_{P_A^\circ}^{P_A^*} \frac{dP}{P},$$

i.e.

$$\mu_A^{(g)}\left(T, P_A^*\right) = \mu_A^{(g)}\left(T, P_A^\circ\right) + RT \ln \frac{P_A^*}{P_A^\circ}. \tag{3.21}$$

Mixture: For a mixture of K components at fixed T the Gibbs-Duhem equation yields

$$\sum_{i=1}^{K} n_i d\mu_i = V dP. \tag{3.22}$$

Here we can make use of <u>Dalton's law</u> stating that for a mixture of ideal gases

$$P = \sum_{i=1}^{K} P_i \quad \text{and} \quad P_i = RT \frac{n_i^{(g)}}{V}. \tag{3.23}$$

The quantities P and V are the total pressure and the total volume. The quantities P_i are called partial pressures. P_i is the contribution to the total pressure due to the presence of $n_i^{(g)}$ moles of gas i. Thus Eq. (3.22) becomes

$$\sum_{i=1}^{K} n_i^{(g)}\left(d\mu_i^{(g)} - RTd \ln n_i^{(g)}\right) = 0. \tag{3.24}$$

In general this will be true only if $d\mu_i^{(g)} - RTd \ln n_i^{(g)} = 0$ for all i, which after integration yields

$$\mu_A^{(g)}(T, P_A) - \mu_A^{(g)}\left(T, P_A^*\right) = RT \ln \frac{n_A^{(g)}}{n^{(g)}}. \tag{3.25}$$

The chemical potential difference on the left is between a mixture of a certain composition $\{n_i^{(g)}\}_{i=1}^K$, where $n_A^{(g)} = P_A V/(RT)$, and a pure A-gas at given $P_A^* = P = RTn^{(g)}/V$. Because $n^{(g)} = \sum_{i=1}^K n_i^{(g)}$, we may write

$$\mu_A^{(g)}(T, P_A) = \mu_A^{(g)}\left(T, P_A^*\right) + RT \ln \frac{P_A}{P_A^*} \tag{3.26}$$

or

$$\mu_A^{(g)}(T, P_A) = \mu_A^{(g)}\left(T, P_A^*\right) + RT \ln x_A^{(g)}, \tag{3.27}$$

where we have used the definition

$$x_A^{(g)} = \frac{n_A^{(g)}}{n^{(g)}}. \tag{3.28}$$

The quantity $x_A^{(g)}$ is the mole fraction of component A and

$$\sum_{i=1}^K x_i^{(g)} = 1. \tag{3.29}$$

Even though we use P_A as the pressure argument in the chemical potentials on the left sides of Eqs. (3.26) and (3.27), the total pressure still is P, i.e. the total pressure is the same on both sides of these equations. Eqs. (3.26) and (3.27) describe the difference between the molar chemical potential of the A-component in an ideal mixture and the molar chemical potential of A in the pure and ideal A-gas at temperature T and identical overall pressure P.

Combining Eq. (3.26) with (2.178) we may write down the free enthalpy of mixing for a K-component ideal gas, i.e.

$$\Delta_m G(T, P, n_1^{(g)}, \dots, n_K^{(g)}, \dots) = n^{(g)} RT \sum_{i=1}^K x_i^{(g)} \ln x_i^{(g)}. \tag{3.30}$$

3.2.2   Chemical Potential in Liquids and Solutions

Pure liquid: Thus far we have dealt with gases. The new situation is illustrated in Fig. 3.3. A pure gas A coexisting with its liquid. This may be achieved by partly filling a container with the liquid of interest. After closing the container tightly an equilibrium between liquid and gaseous A develops according to the conditions

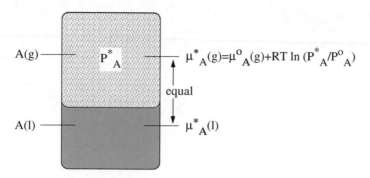

Fig. 3.3 A pure gas A coexisting with its liquid

(3.5) and (3.6).[2] In particular according to Eq. (3.6) the chemical potentials of A must be the same in the gas and in the liquid, i.e.

$$\mu_A^{(l)}(T, P_A^*) = \mu_A^{(g)}(T, P_A^*). \tag{3.31}$$

Here the index $^{(l)}$ indicates the liquid state and P_A^* now denotes the equilibrium gas pressure at coexistence. Assuming that the gas above the liquid is ideal we may obtain the right side of Eq. (3.31) by integrating from a reference pressure P_A° just as in the case of Eq. (3.26), i.e.

$$\mu_A^{(l)}(T, P_A^*) = \mu_A^{(g)}(T, P_A^*) = \mu_A^{(g)}(T, P_A^\circ) + RT \ln \frac{P_A^*}{P_A^\circ}. \tag{3.32}$$

This result applies along the coexistence curve separating gas and liquid in the T-P-plane. There should be such a curve according to the phase rule (3.12) applied to a one-component system. In this case $Z = 1 - 2 + 2 = 1$. We may vary one degree of freedom, T or P, which then fixes P or T, respectively. Figure 3.4 shows a partial sketch of the coexistence curve for a one-component system.[3] Equation (3.32) applies to point (a) but not to point (b) inside the liquid region. However, we can calculate the chemical potential at point (b) in the liquid via

$$\mu_A^{(l)}(T, P_A^{(b)}) = \mu_A^{(l)}(T, \underbrace{P_A^{(a)}}_{\equiv P_A^*}) + \frac{1}{n^{(l)}} \int_{P_A^{(a)}}^{P_A^{(b)}} V(P) dP. \tag{3.33}$$

---

[2] Thermodynamics does not predict the states of matter or describe their structure. Their existence, here gas and liquid, is an experimental fact, which we use at this point.

[3] We shall show how to calculate this curve on the basis of a microscopic interaction model—the van der Waals theory.

Fig. 3.4 Partial gas-liquid coexistence curve in a one-component system

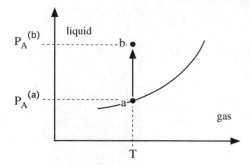

We do not know $V(P)$ in the liquid. But we do know from experience that the volume of a liquid changes little, compared to a gas, when the pressure is increased. Thus we simply Taylor-expand $V(P)$ around $V(P_A^{(a)})$, i.e.

$$V(P) \approx V(P_A^{(a)})\left(1 - \kappa_T(P_A^{(a)})(P - P_A^{(a)})\right), \tag{3.34}$$

where κ_T is the isothermal compressibility defined via Eq. (2.6). Typical liquid compressibilities are in the $(\text{GPa})^{-1}$-range. This means that the second term usually can be neglected, i.e.

$$\mu_A^{(l)}(T, P_A^{(b)}) \approx \mu_A^{(l)}(T, \underbrace{P_A^{(a)}}_{\equiv P_A^*}) + \frac{1}{n^{(l)}} V(P_A^{(a)})(P_A^{(b)} - P_A^{(a)}). \tag{3.35}$$

Example—Relative Humidity   An experiment is carried out at temperature T pressure P and 50 % relative humidity—what does this mean? The relative humidity, φ, is defined via

$$P_D(T) = \varphi P_{sat}(T). \tag{3.36}$$

$P_{sat}(T)$ is the saturation pressure of water at T, which is the pressure $P_A^{(a)}$ in Fig. 3.4. $P_D(T)$, on the other hand, is the partial water vapor pressure in air at T and relative humidity $\varphi \cdot 100\%$.

Before we come to the actual problem, we want to get a feeling for relative humidity, i.e. we ask: what is the mass of water contained in cubic meter of air if the relative humidity is 40 %? We look up the vapor pressure from a suitable table, e.g. HCP. At $T = 0\,°C$ and $T = 20\,°C$ we find $P_{sat} = 0.006$ bar and $P_{sat} = 0.023$ bar, respectively. Using the ideal gas law, $P_{sat}V = nRT$, we obtain the corresponding masses of water vapor, i.e. 4.8 and 17.3 g/m³, on the coexistence line. The water content at $\varphi = 0.4$ is therefore 1.9 and 6.9 g/m³. This is quite small compared to an approximate mass density of air of $1000 \, g/m^3$. Figure 3.5 shows the gas-liquid coexistence or saturation line for water (solid line). The data are from HCP. On this line the relative humidity is 100%. The dashed lines correspond to lines of constant humidity as indicated. The horizontal arrow indicates the cooling of air originally at 25% relative humidity at constant pressure until the saturation line is reached. The temperature at which this happens and the water vapor starts to condense is called dew point. The vertical arrow indicates a portion of a drying process. Dry air increases its moisture contents. Subsequently it may be cooled and upon reaching the saturation line the vapor in the air condenses. By moving along the saturation line towards lower partial water pressure more water is removed form the air. Eventually heating of the air restores it to its starting point at low relative humidity.

However, our real problem is the following. We are interested in the difference between the chemical potential of water in the gas phase, $\mu_{H_2O}^{(g)}(T, P)$, at a relative humidity φ and the chemical potential in the liquid phase of pure water, $\mu_{H_2O}^{(l)}(T, P)$, under the same conditions (for example $T = 40 \, °C$ and $P = 1$ bar). Such a question may arise when the water uptake in a material is measured by one experiment at fixed T, P, and φ in the gas phase or by another experiment via submerging the same material in liquid water at otherwise identical conditions. According to Eqs. (3.26) and (3.35) we have

$$\mu_{H_2O}^{(g)}(T, P_D) \approx \mu_{H_2O}^{(g)}(T, P_{sat}) + RT \ln \frac{P_D}{P_{sat}}$$

and

$$\mu_{H_2O}^{(l)}(T, P) \approx \mu_{H_2O}^{(l)}(T, P_{sat}) + \frac{1}{n^{(l)}} V(P)(P - P_{sat}).$$

With $\mu_{H_2O}^{(g)}(T, P_{sat}) = \mu_{H_2O}^{(l)}(T, P_{sat})$ and using Eq. (3.36) we obtain for the difference $\Delta \mu_{H_2O}(T) \equiv \mu_{H_2O}^{(l)}(T, P) - \mu_{H_2O}^{(g)}(T, P_D)$

$$\Delta \mu_{H_2O}(T) \approx -RT \ln \varphi. \tag{3.37}$$

Notice that the neglected term, i.e. $\frac{1}{n^{(l)}} V(P_{sat})(P - P_{sat})$, is small. With a liquid water molar volume of $18 \, cm^3$ and $P_{sat}(40 \, °C) = 0.0737$ bar we obtain $\approx 1.7 \times 10^{-3} \, kJ \, mol^{-1}$. Because $\varphi = 0.5$, i.e. 50 % relative humidity, we find finally

$$\Delta \mu_{H_2O} \approx 1.8 \, kJ \, mol^{-1}.$$

Fig. 3.5 Water saturation
line including lines of
constant humidity

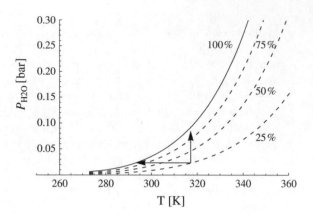

Remark 1 In the preceding discussion we have implicitly assumed a
one-component system, i.e. neat water. However, relative humidity belongs to our
everyday life. This means that we deal with air, a mixture which includes gaseous
water as one particular component. In addition ordinary liquid water also contains a
certain amount of each of the gaseous components which can be found in the air.
Therefore we must ask, how much is the saturation line of neat water shown in
Fig. 3.5 affected by the presence of other components. We shall answer this
question on page 100 after we have discussed solutions.

Remark 2   cloud base: We want to estimate the lowest altitude of the visible
portion of a cloud, i.e. the cloud base. The idea is as follows. Our study of the
temperature profile of the troposphere (see p. 54) has resulted in the Eqs. (2.94) and
(2.97) allowing to relate pressure, temperature, and corresponding height above see
level for air. If we apply these two formulas to the partial pressure of water vapor in
air at a given humidity, we can estimate at what temperature (if at all) the partial
pressure in air will become equal to the saturation pressure of water. The resulting
temperature may then be used to compute the height at which this happens. This is
the height when the water vapor condenses an thus defines the cloud base.
Neglecting the effect of the different molar weights of water and (dry air), we
compute the partial water pressure via $P = P_o(T/T_o)^{3.5}$, where $P_o = \varphi P_{sat}^{H2O}(T_o)$.
$T_o = 20\,°C$ is the ground temperature. The two dashed curves in Fig. 3.6 are for
$\varphi = 0.5$ and $\varphi = 0.7$, i.e. 50 and 70 % relative humidity, respectively. The solid
line in Fig. 3.6 is the saturation line for water. The two temperatures at which the
curves intersect are converted into heights, i.e. ≈ 2100 and ≈ 1100 m. We notice that

Fig. 3.6 Cloud base for different humidities

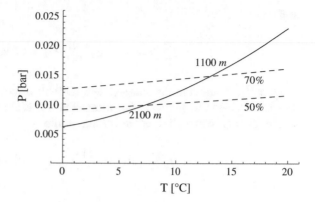

Fig. 3.7 A binary solution of the components A and B in equilibrium with its gas

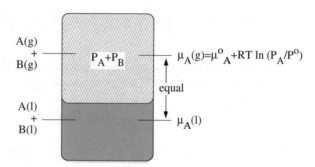

the cloud base is lower when the humidity is greater. Taking into account the molar weight difference mentioned above decreases these values by roughly 10 %. A sensitive quantity is T_o, i.e. decreasing T_o also decreases the cloud base.

Solution: A more complex system is shown in Fig. 3.7. A binary solution of the components A and B is in equilibrium with its gas. The coexistence conditions are

$$\mu_A^{(l)}(T, P_A, P_B) = \mu_A^{(g)}(T, P_A, P_B) \tag{3.38}$$

and

$$\mu_B^{(l)}(T, P_A, P_B) = \mu_B^{(g)}(T, P_A, P_B). \tag{3.39}$$

Assuming that the gas phase is an ideal mixture we make use of Eq. (3.26) to express this as

$$\mu_A^{(l)}(T, P_A) = \mu_A^{(g)}(T, P_A^*) + RT \ln \frac{P_A}{P_A^*} \tag{3.40}$$

and

$$\mu_B^{(l)}(T, P_B) = \mu_B^{(g)}(T, P_B^*) + RT \ln \frac{P_B}{P_B^*}. \tag{3.41}$$

Note that $*$ refers to the same system with all B-moles in (3.38) replaced by A-moles and vice versa in (3.9). This means that P_i^* is the vapor pressure in the pure i system at coexistence. Therefore we may write

$$\mu_A^{(l)}(T, P_A) = \mu_A^{(l)}(T, P_A^*) + RT \ln \frac{P_A}{P_A^*} \tag{3.42}$$

and

$$\mu_B^{(l)}(T, P_B) = \mu_B^{(l)}(T, P_B^*) + RT \ln \frac{P_B}{P_B^*}. \tag{3.43}$$

Using partial pressures is somewhat inconvenient. There are two limiting laws, which are very useful here. The first is <u>Raoult's law</u>,

$$P_A = P_A^* x_A^{(l)}, \tag{3.44}$$

valid if $x_A^{(l)} \gg x_B^{(l)}$, i.e. the liquid is a (very) dilute solution of solute B in solvent A.[4] Notice that $x_A^{(l)} + x_B^{(l)} = 1$. Inserting Raoult's law in Eq. (3.42) yields

$$\mu_A^{(l)}(T, P_A) = \mu_A^{(l)}(T, P_A^*) + RT \ln x_A^{(l)}. \tag{3.45}$$

Raoult's law is an example for a colligative property of the solution. Colligative properties of solutions depend on the number of molecules in a given amount of solvent and not on the particular identity of the solute.

The second useful law is <u>Henry's law</u>,

$$P_B = K_B x_B^{(l)}, \tag{3.46}$$

where K_B is called Henry's constant. Henry's constant is not a universal constant. It depends on the system of interest and on temperature. Henry's law is valid in the same limit as Raoult's law, i.e. $x_B \ll 1$. Inserting Eq. (3.46) into (3.43) yields

$$\mu_B^{(l)}(T, P_B) = \mu_B^{(l)}(T, P_B^*) + RT \ln \frac{K_B}{P_B^*} + RT \ln x_B^{(l)}. \tag{3.47}$$

---

[4] The meaning of A and B can of course be interchanged.

Table 3.1 Henry's law applied to three gases in water

Component	T [K]	K_H [10^9 Pa]	c_{H_2O} [g/m^3]	c_{air} [g/m^3]
O_2	288.15	3.7	10	284
	298.15	4.4	8.6	275
	308.15	5.1	7.4	266
N_2	288.15	7.3	17	924
	298.15	8.6	14	893
	308.15	9.7	13	864
CO_2	288.15	0.12	78	71
	298.15	0.16	59	68
	308.15	0.21	45	66

The first two terms may be absorbed into the definition of a new hypothetical reference state with the chemical potential $\bar{\mu}_B$

$$\mu_B^{(l)}(T, P_B) = \bar{\mu}_B(T) + RT \ln x_B^{(l)}. \tag{3.48}$$

We call this reference state hypothetical, because $\mu_B^{(l)}(T, P_B) = \bar{\mu}_B(T)$ requires $x_B = 1$, a concentration at which Henry's law does not apply.

Table 3.1 compiles Henry's constant for oxygen, nitrogen, and carbon dioxide in water (based on solubility data in HCP). In the third column c_{H_2O} is the mass density of the respective component in water in equilibrium with air at a pressure of 1 atm. The last column shows the corresponding mass density in air.

Figure 3.8 shows the partial pressures in the two-component vapor-liquid system acetone-chloroform at $T = 308.15$ K (Ozog and Morrison 1983). Solid lines are polynomial fits to the data points. The long dashed lines illustrate Raoult's law applied to the two components while the short dashed lines illustrate Henry's law.

Fig. 3.8 Partial pressures in the two-component vapor-liquid system acetone-chloroform at T = 308.15 K

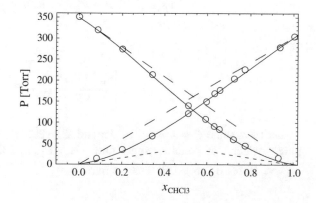

Remark We now return to the question on p. 96—how much is the saturation line of neat water shown in Fig. 3.5 affected by the presence of other components?

When water vapor coexists with liquid water we have

$$\mu_{H_2O}^{(l)}(T, P_{sat}, P) = \mu_{H_2O}^{(g)}(T, P_{sat}, P)$$
$$\overset{(3.40)}{=} \mu_{H_2O}^{(g)*}(T, P) + RT \ln \frac{P_{sat}}{P}. \tag{3.49}$$

Here P is the overall gas pressure and P_{sat} is the partial water pressure at coexistence. The star indicates pure water. Let us assume we change P by an amount dP. This will alter the two sides of the above equation but the changes still will be the same on both sides: still

$$\underbrace{\frac{\partial \mu_{H_2O}^{(l)}(T, P_{sat}, P)}{\partial P}\bigg|_T}_{=v_{H_2O}^{(l)}(T, P_{sat}, P)} dP = \underbrace{\frac{\partial \mu_{H_2O}^{(g)}(T, P = P_{sat}^*)}{\partial P}\bigg|_T}_{=v_{H_2O}^{(g)*}(T, P)=RT/P} dP + RTd \ln \frac{P_{sat}}{P}. \tag{3.50}$$

Here $v_{H_2O}^{(l)}(T, P_{sat}, P)$ is the partial molar volume of water in contact with air at pressure P, and $v_{H_2O}^{(g)*}(T, P)$ is the same quantity for pure water vapor at pressure P. Under standard conditions (1 bar) we do not make a big mistake if we replace $v_{H_2O}^{(l)}(T, P_{sat}, P)$ by the same quantity for pure water, $v_{H_2O}^{(l)*}(T, P)$. Thus we obtain

$$v_{H_2O}^{(l)*}(T, P)dP \approx RTd \ln P_{sat}. \tag{3.51}$$

Now we integrate the left side from P_{sat}^*, the saturation pressure of pure water at T to the ambient pressure of air (including water vapor), P. The corresponding integration limits on the right side are also P_{sat}^* and P_{sat}. This yields

$$v_{H_2O}^{(l)*}(T, P)(P - P_{sat}^*) \approx RTd \ln \frac{P_{sat}}{P_{sat}^*} \tag{3.52}$$

or

$$\frac{P_{sat}}{P_{sat}^*} \approx \exp \left[\frac{v_{H_2O}^{(l)*}(T, P)}{RT} (P - P_{sat}^*) \right]. \tag{3.53}$$

Let us assume $P = 1$ bar $= 10^5$ Pa and $T = 293$ K. The saturation pressure of pure water at this temperature is $P_{sat}^* = 2338.8$ Pa. In addition $v_{H_2O}^{(l)*} \approx 18 \times 10^{-6} m^3$. And thus we find

$$\frac{P_{sat}}{P^*_{sat}} \approx \exp\left[7 \times 10^{-4}\right]. \tag{3.54}$$

This means that the saturation pressure of water under ambient conditions is scarcely different from the saturation pressure of pure water at the same temperature.

3.3 Applications Involving Chemical and Mechanical Equilibrium

3.3.1 Osmotic Pressure

Figure 3.9 shows a beaker containing the pure liquid A. Immersed in the liquid is a tube with its lower end closed to the liquid by a membrane. The membrane allows A to permeate into the tube and vice versa. Inside the tube there is a binary mixture of two components A and B. The latter however is held back inside the tube by the membrane. What happens? As far as A is concerned the two subsystems, the pure solvent outside the tube and the binary mixture inside the tube, do exchange A-moles and therefore chemical equilibrium requires

$$\mu^*_A(T,P) = \mu_A(T,P+\Pi,x_A). \tag{3.55}$$

The left side is the chemical potential of pure A outside the tube. The right side is the chemical potential of A inside the tube. Because we do not consider the gas phase we may omit the index $^{(l)}$. The temperature is the same in both subsystems. The outside pressure is P, whereas the inside pressure is different i.e. $P+\Pi$. Why?

Initially the tube may contain B only. Chemical equilibrium therefore requires flow of A across the membrane into the tube. For simplicity we assume that the initial surface level inside and outside the tube is the same and the density of A and B is the same as well. The pressure difference across the membrane, Π, can then be determined by measuring h (at equilibrium) and computing the force of gravitation exerted by the mass of material above the surface level of the surrounding A solvent. The reason for the sustained pressure difference is the membrane, which does not allow the chemical equilibration of B on both sides.

Fig. 3.9 Sketch of a simple osmotic pressure experiment

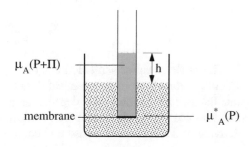

Making use of the Gibbs-Duhem Eq. (2.180) we have

$$\mu_A^*(T, P + \Pi) \approx \mu_A^*(T, P) + \frac{1}{n_A^*} V^*(P)(P + \Pi - P). \tag{3.56}$$

Again it is assumed that incompressibility of the liquid is a good approximation (cf. Eq. (3.34)). Combination of (3.55) and (3.56) then yields

$$\mu_A^*(T, P + \Pi) - \frac{1}{n_A^*} V^*(P)\Pi \approx \mu_A(T, P + \Pi, x_A), \tag{3.57}$$

and according to Eq. (3.45), if $x_A \gg x_B$, we obtain

$$\mu_A^*(T, P + \Pi) - \frac{1}{n_A^*} V^*(P)\Pi \approx \mu_A^*(T, P + \Pi) + RT \ln x_A. \tag{3.58}$$

At first glance this may seem strange, because in Eq. (3.45) the pressure arguments of the chemical potentials are P_A and P_A^*, which are presumably different. In general the chemical potential of a particular species does depend on temperature, total pressure, and composition. In Eq. (3.45) the total pressure is the same on both sides of the equation and does not show up explicitly in the lists of arguments. The composition dependence is expressed in terms of different partial pressures via (originally)Dalton's law. Here the total pressure is included in the argument of the chemical potential and the composition is expressed in mole fractions. Having explained this we can write down the final result for Π given by

$$\Pi \approx \frac{RT}{V_{A,mol}} x_B, \tag{3.59}$$

where we have used the molar volume of A in the liquid state at pressure P and temperature T, i.e. $V_{A,mol} = V^*(P)/n_A^*$, and $\ln x_A = \ln(1 - x_B) \approx -x_B$.

One final transformation of this equation is useful. Inside the tube we have

$$x_B = \frac{n_B}{n_A + n_B} \approx \frac{n_B}{n_A}. \tag{3.60}$$

In addition $V_{solution} \approx V_{A,mol} n_A$ and thus

$$\Pi \approx \frac{n_B RT}{V_{solution}}. \tag{3.61}$$

This is the so called van't Hoff equation (Jacobus van't Hoff, first Nobel prize in chemistry for his work on chemical dynamics and osmotic pressure, 1901). Note that the osmotic pressure only depends on the molar concentration of component B and temperature (under the approximations we have made in the course of the derivation). Here osmotic pressure is another example for a colligative property.

Remark—Reverse Osmosis   According to our derivation leading to Eq. (3.61), it should be possible to apply extra pressure to the tube in Fig. 3.9 and by doing so reduce its solute content. A technical example is desalination of sea water, which is forced through a membrane using a pressure exceeding the osmotic pressure. This process is called reverse osmosis.

Example—Osmotic Pressure in Hemoglobin Solutions   As an application we consider the following problem. Use the osmotic pressure data ($p_{obs.}$) from table X in Adair (1928) to estimate the molar mass of (sheep) hemoglobin. Gilbert S. Adair was a pioneer of macromolecular biochemistry and succeeded in determining the correct molecular weight of hemoglobin from osmotic pressure measurements. He also supplied the horse hemoglobin crystals which allowed Max Perutz (Nobel prize in chemistry for his work on the structure of globular proteins, 1962) to obtain the first hemoglobin X-ray structures.

   We may rewrite van't Hoff's equation as

$$\frac{\Pi}{c} \approx \frac{RT}{m_{Hb}}.$$

Here $c = (n_{Hb}/V)m_{Hb}$ and m_{Hb} is the molar mass of component B (Hb: hemoglobin). Figure 3.10 shows the data from the above reference plotted in the original units. The solid line is a fit on the basis of a theory explained later in this book. Notice first that van't Hoff's equation describes the data only at very low hemoglobin concentration. This is expected, because we have used the approximation $x_A \gg x_B$. The deviation from van't Hoff's equation arise due to non-ideality, which basically means that there are complex solute-solute interactions—something we have no information about at this point. However, we can still determine m_{Hb}, i.e.

Fig. 3.10 Concentration dependent osmotic pressure in hemoglobin solutions

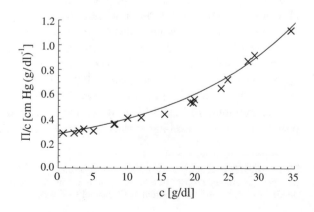

$$m_{Hb} \approx \frac{RT}{\Pi/c},$$

in the limit $c \to 0$. From the figure we extract the value $\Pi/c \approx 0.3$ cm Hg/(g/dl). In addition $T = 0\,°C = 273.15$ K. After converting the units,

$$1\mathrm{cm\,H}g = 1333.224\,\mathrm{Pa}$$
$$1\mathrm{g/dl} = 10\,\mathrm{kg/m^3},$$

we obtain $m_{Hb} \approx 57$ kg/mol. This is roughly 10% below the exact value—but not bad at all.

The example also shows that van't Hoff's equation is valid at small concentrations only. We continue our discussion of osmotic pressure on page 125 and a second time in Sect. 4.4.1 dealing with extensions of different origin. In particular we shall discuss the so called scaled particle theory behind the solid line through the data in Fig. 3.10 beginning on p. 213. This theory allows to estimate the size of Hb (≈ 5.5 nm in diameter) based on its osmotic pressure data.

3.3.2 Equilibrium Adsorption

Consider a gas in contact with a solid surface. Molecules from the gas may adsorb onto and subsequently desorb from the surface. Eventually an equilibrium develops characterized by a constant coverage depending on temperature and the pressure in the bulk gas. Coverage here refers to the net amount of gas adsorbed. We obtain the net amount adsorbed by counting the gas molecules in a column-shaped volume perpendicular to the surface. This column continues out into the bulk gas, where the surface is no longer felt by the gas molecules. Subsequently we subtract the (average) number of gas molecules present in an identical column when the surface is removed (The number is equal to the volume of the column multiplied by the bulk density of the gas). Just how long the column has to be, in order for it to extend into the bulk gas, depends on the interaction forces between the gas molecules and the surface as well as on thermodynamic conditions. In some cases the "interfacial thickness" to good approximation is just one molecular layer. One speaks of monolayer or even sub-monolayer coverage. In other cases the interface is "thicker" and more "diffuse".

The examples in Fig. 3.11 show computer simulation generated gas density profiles above an adsorbing surface at $z = 0$.[5] The units used here are so-called

---

[5] The system is methane gas adsorbing on the graphite basal plane located at $z = 0$. A computer program generating profiles like these is included in the appendix. The theoretical background needed to understand the program is discussed in Chap. 6.

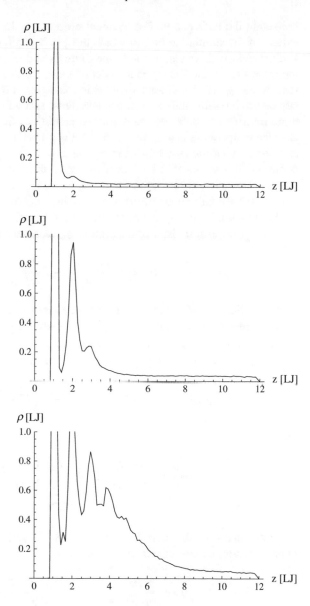

Fig. 3.11 Computer simulation generated gas density profiles above an adsorbing surface at different temperatures

Lennard-Jones units[6]—but this is of no particular interest to us at this point. What is shown is the gas number density, $\rho(z)$, as function of distance, z, from the surface. In the top panel we recognize a first peak at $z \approx 1$ (cut off at 1) and a second smaller one at $z \approx 2$. Beyond the second peak the density levels off (with fluctuations)

---

[6] In these units the gas pressure in Fig. 3.11 is $P = 0.04$. The temperatures from top to bottom are $T = 2.0, 1.2, 1.05$.

indicating the bulk phase. The eventual drop at $z = 12$ is merely due to the finite extend of the simulation box to which the gas is confined. The "gap" between the surface and the first peak is due to the finite extend of the atoms in the surface and the molecules in the gas—ρ is a center of mass number density. This figure shows that at the given conditions there exists a dense layer of adsorbed molecules adjacent to the solid surface. A much less dense second layer is followed by a rather rapid transition to bulk behavior. Altered conditions do change the picture. In this case the temperature is reduced. In fact we approach the saturation line of methane at constant pressure (the transition temperature at this pressure in LJ-units is just about 1). We notice that the adsorbed layer thickness increases as more peaks emerge. However, at this point these graphs merely serve as illustration to bear in mind when we talk about adsorption on solid surfaces.[7]

An important quantity characterizing the interaction of the molecules with the surface is the isosteric heat of adsorption q_{st}, defined via

$$q_{st} = T \frac{\partial \mu_s}{\partial T}\bigg|_{V_s,N_s} - T \frac{\partial \mu_b}{\partial T}\bigg|_{P_b}. \tag{3.62}$$

The indices s and b refer to the surface and the bulk, respectively. Note that $\ldots|_{V_s,N_s}$ means "at constant coverage", whereas $\ldots|_{P_b}$ means "at constant (bulk) pressure". The temperature is the same in both cases.

Using the equilibrium condition $\mu_s = \mu_b$ we may write

$$d\mu_s\big|_{V_s,N_s} = \frac{\partial \mu_s}{\partial T}\bigg|_{V_s,N_s} dT = \frac{\partial \mu_b}{\partial T}\bigg|_{P_b} dT + \frac{\partial \mu_b}{\partial P_b}\bigg|_T dP_b, \tag{3.63}$$

which yields

$$\frac{\partial \mu_s}{\partial T}\bigg|_{V_s,N_s} = \frac{\partial \mu_b}{\partial T}\bigg|_{P_b} + \underbrace{\frac{\partial \mu_b}{\partial P_b}\bigg|_T}_{V_b/N_b = \rho_b^{-1}} \frac{\partial P_b}{\partial T}\bigg|_{V_s,N_s}. \tag{3.64}$$

Combination of this equation with Eq. (3.62) yields another, and perhaps the most common, expression for q_{st}:

$$q_{st} = \frac{T}{\rho_b} \frac{\partial P_b}{\partial T}\bigg|_{V_s,N_s} = -\frac{P_b}{\rho_b T} \frac{\partial \ln P_b}{\partial(1/T)}\bigg|_{V_s,N_s}. \tag{3.65}$$

At very low gas pressure one may assume that N_s is proportional to P_b, i.e.

$$N_s = k_H P_b + \mathcal{O}(P_b^2). \tag{3.66}$$

---

[7] We return to Fig. 3.11 in an example in Sect. 5.3.

The leading term is a "surface version" of Henry's law (3.46). In this approximation Eq. (3.65) becomes

$$q_{st}^{(o)} = R \frac{\partial \ln k_H}{\partial (1/T)} \bigg|_{V_s, N_s}. \tag{3.67}$$

Here $q_{st}^{(o)}$ is the molar isosteric heat of adsorption in the limit of vanishing coverage. Experimentally this quantity may be determined by measuring the amount of adsorbed gas (e.g., by weighing the sample) at a given (low) pressure. The general relation $N_s(T, P)$ versus P is called adsorption isotherm, the low pressure slope of which again is k_H. In an example in Sect. 5.3 we return to the isosteric heat of adsorption and discuss one explicit method how to calculate it theoretically.

3.3.3 Law of Mass Action

In the following we discuss an important application of (2.131), i.e.

$$dG\big|_{T,P} \leq 0. \tag{3.68}$$

At equilibrium we can use the equal sign and based on Eq. (2.119) (with μdn replaced by $\sum_{i=1}^{K} \mu_i dn_i$) we have

$$\sum_{i=1}^{K} \mu_i(T, P) dn_i = 0. \tag{3.69}$$

This equation requires some thought. If $K = 1$ then Eq. (3.69) implies $dn = 0$. The inequality (3.68) applies to cases where, aside from keeping T and P at fixed values, we leave the system alone. In particular we do not change its mass content.[8] If $K > 1$ there exists however the possibility of a suitable relation between the dn_i, developed by the system itself, allowing Eq. (3.69) to hold without requiring $dn_i = 0 \forall i$. For instance we may replace dn_i in Eq. (3.69) via

---

[8] Potentially this may be disturbing. According to the steps leading from Eq. (2.176) to Eq. (2.177) one may be led to conclude that $G = 0$ all the time and everything falls apart. However this reasoning confuses two very different situations. Inequality (3.68) means that we prepare a system subject to certain thermodynamic conditions T and P and leave this system alone until no further change is observed. This fixes the equilibrium value of the free enthalpy, G, for a particular pair T, P. Repeating this procedure for many T, P-pairs we map out the equilibrium values of G above the T-P-plane (cf. Fig. 2.15). With this function $G = G(T, P)$ or $G = G(T, P, n)$ we now can do calculations, differentiating or integrating, involving T, P, n and possibly other variables. This is how we have obtained the Eqs. (2.176) and (2.177). Therefore there is no problem here!

$$dn_i = v_i d\xi, \tag{3.70}$$

where of course not all v_i do have the same sign. It turns out that chemical reaction equilibria may be described in this fashion. For a chemical reaction the dn_i obey according to experimental evidence

$$\frac{dn_i}{dn_j} = \frac{v_i}{v_j}, \tag{3.71}$$

where v_i and v_j are integers.

We proceed writing the chemical potentials of the components as

$$\mu_i(T,P) = \bar{\mu}_i(T,P) + RT \ln a_i. \tag{3.72}$$

This particular form is analogous to the special limiting forms (3.27) and (3.45). The quantity a_i, which is called activity of component i, contains interaction and mixing contributions to the chemical potential of component i, i.e. all effects due to the interactions of this component with all other components. Often the activity is expressed via

$$a_i = \gamma_i x_i, \tag{3.73}$$

where γ_i is the activity coefficient. This is the usual terminology in condensed phases. In the gas phase the fugacity,

$$f_i = \gamma_i' P, \tag{3.74}$$

where γ_i' is the fugacity coefficient and P is the pressure, replaces a_i. We see that the special limiting forms (3.27) and (3.45) correspond to $\gamma_i = 1$. The reference chemical potential, $\bar{\mu}_i(T,P)$, may be identified if we let γ_i and x_i approach unity.

Combining Eqs. (3.69), (3.70), and (3.72) we obtain

$$\sum_{i=1}^{K} (\bar{\mu}_i(T,P) + RT \ln a_i) v_i = 0 \tag{3.75}$$

or

$$\prod_{i=1}^{K} a_i^{v_i} = K(T,P), \tag{3.76}$$

where

$$K(T,P) = \exp\left[-\frac{\sum_{i=1}^{K} v_i \bar{\mu}_i(T,P)}{RT} \right]. \tag{3.77}$$

Equation (3.76) is called law of mass action and $K(T, P)$, not to be confused with the index K, the number of components, is the equilibrium constant. The equilibrium constant is not really a constant. It depends on T and P. By convention $v_i < 0$ for reactants and $v_i > 0$ for products.

The law of mass action in its present form provides little insight. Therefore we study the special case of a gas phase reaction assuming that the gas is ideal. Combining (3.26) and (3.21) we obtain

$$\mu_i^{(g)}(T, P_i) = \mu_i^{(g)}(T, P_i^o) + RT \ln \frac{P_i}{P_i^o}. \tag{3.78}$$

Here P_i is the partial pressure of component i, and P_i^o is a standard pressure, which remains constant during the reaction. In addition we have

$$x_i^{(g)} = \frac{P_i}{P^{(g)}} \tag{3.79}$$

(cf. Eq. (3.28)), where $P^{(g)}$ is the total pressure. Combination of (3.78) and (3.79) yields

$$\mu_i^{(g)}(T, P_i) = \mu_i^{(g)}(T, P_i^o) + RT \ln \frac{P^{(g)} x_i^{(g)}}{P_i^o}. \tag{3.80}$$

Inserting this into Eq. (3.69) we find the law of mass action

$$K(T, P^o) = P^{\Sigma_i v_i} \prod_i x_i^{v_i}, \tag{3.81}$$

where $-RT \ln P_i^o$ is absorbed into $K(T, P^o)$. Notice that we have omitted the index (g), and we assume that $P^o = P_i^o \forall i$. This equilibrium constant is independent of the gas pressure P and the mole fractions x_i.

Example—A Chemical Reaction In the following simple example of a chemical reaction,

$$2H_2 + O_2 \rightleftharpoons 2H_2O, \tag{3.82}$$

we have $v_{H_2} = -2$, $v_{O_2} = -1$, and $v_{H_2O} = 2$. Thus Eq. (3.81) becomes

$$K(T, P^o) = \frac{1}{P} \frac{x_{H_2O}^2}{x_{H_2}^2 x_{O_2}}. \tag{3.83}$$

Increasing the total pressure (at constant temperature) shifts the reaction equilibrium to the right. Analogously we can see what happens if the concentrations are changed.

Remark What is different if in addition to H_2, O_2, and H_2O another inert gas is present or there is an excess of one or more of the aforementioned components? The corresponding mole fractions do not appear explicitly on the right side of Eq. (3.83), but they do enter into the pressure, P.

At this point we may ask: What is a component? Thermodynamic knows nothing about atoms, molecules, and details of the interactions/reactions between them. But we know that there are even smaller building blocks than atoms—electrons, protons, and neutrons. And this is not the end. So what is a component? In principle we may apply thermodynamics on all levels. For the above it is important, however, that there exists a meaningful chemical potential for everything we want to call component. That is a component must exist long enough (on average) under well defined thermodynamic conditions like equilibrium T and P.

This requires us to rethink our derivation of the phase rule. Consider the following example for a chemical reaction:

$$3A \rightleftharpoons A_3. \tag{3.84}$$

If we consider A and A_3 as components, then the phase rule (3.12) allows up to four coexisting phases. However, we have an additional equilibrium constraint imposed by Eq. (3.69), reducing the degrees of freedom by one and the maximum number of coexisting phases to three. The modified phase rule therefore is

$$Z = K - Q - \Pi + 2 \quad (\geq 0), \tag{3.85}$$

where Q is the number of additional constraints imposed via Eq. (3.69). Notice that Q is not necessarily one all the time. There may be independent chemical reactions occurring simultaneously in which case the summation in Eq. (3.69) breaks up into independent parts, e.g.

$$3A \rightleftharpoons A + A_2 \rightleftharpoons A_3.$$

Here we have a system containing three components according to our definition. But for each reaction we have to fulfill Eq. (3.69). Therefore $K = 3$ and $Q = 2$.

Example—Critical Micelle Concentration Figure 3.12 shows a sketch of a system containing typo-amphiphilic molecules. Amphiphilic molecules consist of two covalently bonded moieties—one, depicted as zigzag-line, does

Fig. 3.12 Sketch illustrating
the reversible assembly of
amphiphilic molecules into
micelles

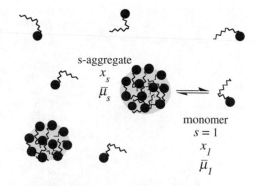

not like to be in contact with water, not shown explicitly, whereas the other, depicted as solid circle, does like to be in contact with water. An example of such a molecule is Hexaethylene-glycol-dodecylether $(C_{12}H_{25}(OCH_2CH_2)_6OH)$. In this case it is the $C_{12}H_{25}$-moiety that does not like to be in contact with water. The natural thing to happen therefore is a clustering of the zigzag "tails" into droplets shielded on the outside by their water loving "head" groups. In a sense this is a phase separation, which we study in the next chapter, on a molecular scale. Because of this the molecular "shape" strongly couples to the "shape" of the drop or aggregate and in fact determines it (The aggregates we have in mind can be spherical, cylindrical or transform into layered structures with complicated topology. It also is possible to extend this approach to vesicles. But this is not our topic here.). The type of droplet aggregate we just described is called a micelle. However, our current approach covers other types of aggregates as well.

As our starting point we choose the "chemical reaction equation"

$$sA_1 \rightleftharpoons A_s. \tag{3.86}$$

Here A_s denotes a s-aggregate containing s molecules or monomers A_1. We put "chemical reaction" in quotes, because the bonding forces between monomers considered here are different from chemical bonds within molecules. In principle s can be any integer number and therefore Eq. (3.86) represents many "reaction equations". Expressing this in terms of the chemical potential yields

$$s\mu_1 = \mu_s. \tag{3.87}$$

Assuming low monomer concentration we may use Eq. (3.48), i.e.

$$s\bar{\mu}_1 + sRT \ln x_1 = \bar{\mu}_s + RT \ln(x_s/s). \tag{3.88}$$

Note that the quantity x_s/s is the mole fraction s-aggregates. Therefore x_s is the mole fraction of monomers in s-aggregates. We may solve for x_s, i.e.

$$x_s = s(x_1 e^{\alpha})^s \qquad (3.89)$$

with

$$\alpha = \frac{1}{RT}\left(\bar{\mu}_1 - \frac{1}{s}\bar{\mu}_s\right). \qquad (3.90)$$

We assume that $\bar{\mu}_s$ is an extensive quantity in terms of s and that therefore α is independent of s (see also the next example). Equation (3.89) has an interesting consequence. To see this we note that the total monomer mole fraction is given by

$$x = x_1 + \sum_{s=m}^{\infty} x_s = x_1 + \sum_{s=m}^{\infty} s(x_1 e^{\alpha})^s. \qquad (3.91)$$

Here x_1 is the mole fraction of free monomers, whereas the sum is the mole fraction due to all other monomers bonded inside aggregates. We note that m is a minimum aggregate size. In the case of spherical micelles for instance, it accounts for the fact that a certain number of head groups are required to form a closed surface avoiding contact of the tail groups with water. This number may be large—say $m \approx 50$—depending of course on the type of monomer. But $m = 2$ also is possible. This is the case of linear aggregates (chains of monomers—These monomers may be disk-shaped with flexible tails on their perimeter. In water the disk-like cores tend to form stacks. It also is possible to apply this idea to dipolar molecules forming chains due to dipole-dipole interaction.). The right side of Eq. (3.91) is bounded, because $x \leq 1$. In particular this requires $x_1 e^{\alpha} < 1$, because the sum $\sum_{s=m}^{\infty} sq^s$ diverges at $q = 1$ (geometric series!). Putting in some numbers we find $\sum_{s=50}^{\infty} sq^s \approx 4 \times 10^{-3}$ if $q = 0.8$ and $\sum_{s=50}^{\infty} sq^s \approx 3$ if $q = 0.9$, i.e. for $x_1 < 0.8 e^{-\alpha}$ virtually all of x is due to free monomers. Addition of monomers at this point leads to their assembly into aggregates. Figure 3.13 illustrates this for different combinations of assumed values for m and α (Notice the change of scale in the third panel.). Because of the sharpness of the "transition" in the typical case of large m the threshold concentration

$$x_{CMC} \approx e^{-\alpha} \qquad (3.92)$$

is called critical aggregate concentration or, in the case of micelles, critical micelle concentration (CMC). While the sharpness is governed by m, the amphiphile concentration at which the change of behavior occurs is determined by α.

Fig. 3.13 Mole fractions free monomers and aggregates versus total monomer mole fraction for different parameter combinations

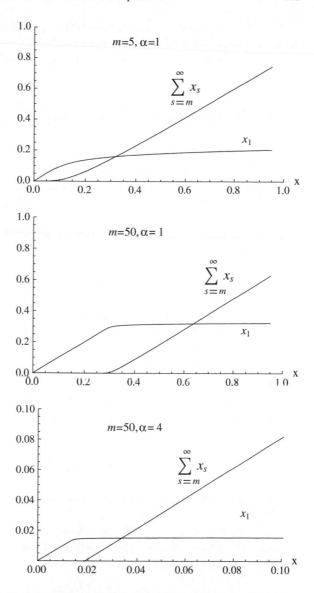

We note that the existence of a CMC is not tied to the specific form of Eq. (3.89). For instance, assuming that monomers may form minimum aggregates only, i.e. only the $s = m$-term in the sum in Eq. (3.91) is present, still yields a CMC. The true size distribution, x_s, in fact is a complicated function of molecular interactions as well as thermodynamic conditions. One interesting and quite general ingredient ignored here is the aggregate dimensionality, which is discussed in the following.

More information on molecular assemblies (micelles, membranes, etc.) can be found in J. Israelachvili (1992) *Intermolecular & Surface Forces*. Academic Press or in D. F. Evans and H. Wennerström (1994) *The Colloidal Domain*. VCH.

3.3.4   Surface Effects in Condensation

The assumption underlying Eq. (3.90) is that all monomers inside an aggregate are equivalent. For a spherical micelle this is in accord with intuition. But what if we study droplets containing monomers completely embedded in their interior and monomers on their surface? These two are certainly different and Eq. (3.90) no longer holds.

A simple model for α allowing to distinguish between bulk and surface monomers in aggregates is

$$\alpha = \alpha_{bulk} - \delta s^{-1/d}, \tag{3.93}$$

where d is the dimensionality of the aggregates ($d = 1$: linear aggregates; $d = 2$: disk- or layer-like aggregates; $d = 3$: spherical aggregates). α_{bulk} is the same for all monomers inside aggregates and independent of s. The second term, $-\delta s^{-1/d}$, is a surface contribution. We note that a three-dimensional spherical droplet containing s monomers has a volume proportional to s. Thus its radius is proportional to $s^{1/3}$ and its surface is proportional to $s^{2/3}$. Expressed more generally the surface is proportional to $s^{(d-1)/d} = ss^{-1/d}$. This means that in this simple case $\bar{\mu}_s$ can be expressed as

$$\bar{\mu}_s = s\bar{\mu}_{bulk} + RT\delta s^{(d-1/d)}, \tag{3.94}$$

where $\bar{\mu}_{bulk}$, the chemical potential of a monomer in the interior of an aggregate, is s-independent. What are the consequences?

First we study the question whether monomers and/or finite size aggregates can coexist with infinite size aggregates, i.e. the bulk phase. If the answer is yes, then the following must be true:

$$\bar{\mu}_{bulk} = \bar{\mu}_{bulk} + RT\delta s^{-1/d} + \frac{1}{s}RT\ln\frac{x_s}{s}. \tag{3.95}$$

This means

$$x_s = s\exp[-\delta s^{(d-1)/d}]. \tag{3.96}$$

Consequently $\sum_s x_s < 1$ is possible only if $d > 1$. If $d = 1$ we have $x_s \propto s$ and the sum diverges. This means that the total monomer mole fraction x diverges. The inconsistency that this imposes (note: $x \leq 1$) is interpreted as the impossibility of coexistence between monomers and/or finite aggregates with a bulk phase in one dimension!

But there is more to discover here. We concentrate on $d > 1$ and simplify our calculation by requiring the aggregates to be monodisperse, i.e. s is the same for all aggregates. The total free enthalpy therefore is

$$G_{total} = n_1 \mu_1 + \frac{n_s}{s} \mu_s. \tag{3.97}$$

Here n_s denotes moles monomer on average bound in aggregates. This equation describes coexistence between a gas of monomers and aggregate droplets. Using Eq. (3.94) yields

$$G_{total} = n_1 \mu_1 + \frac{n_s}{s} \mu_{s,bulk} + n_s RT \delta s^{-1/d}, \tag{3.98}$$

where $\mu_{s,bulk} = s\bar{\mu}_{bulk} + RT \ln[x_s/s]$. If we vary the mass distribution between monomers and aggregate droplets near equilibrium we find

$$dG_{total} = dn_s \left(-\mu_1 + \frac{1}{s} \mu_{s,bulk} + RT\delta s^{-1/d} \right) = 0. \tag{3.99}$$

Note that $dn_1 = -dn_s$. We may usefully apply this equation using $\partial \mu / \partial P|_T = 1/\rho$ or $d\mu = \rho^{-1} dP$ at constant T. Because the monomers form a gas with density ρ_{gas}, while $\rho_{liq} \gg \rho_{gas}$ is the (liquid) monomer density inside the droplets, we may write

$$-\frac{1}{\rho_{gas}} + \frac{1}{\rho_{liq}} \approx -\frac{1}{\rho_{gas}} \approx -RT\delta \frac{\partial s^{-1/d}}{\partial P}\bigg|_T. \tag{3.100}$$

Replacing ρ_{gas} by $P/(RT)$, the ideal gas law, and $s^{1/d}$ by cr, where r is the droplet radius and c is a constant, we find

$$d \ln P = \delta d \left(\frac{1}{cr} \right) \tag{3.101}$$

or

$$P = P_\infty \exp\left[\frac{\delta}{cr} \right]. \tag{3.102}$$

This equation describes the radius, r, of a droplet at equilibrium when the external pressure is P. P_∞ is the saturation pressure, when the monomer gas coexists with the infinite bulk phase. Figure 3.14 shows a sketch of P versus r according to Eq. (3.102).

Fig. 3.14 Sketch of P versus
r according to Eq. (3.102)

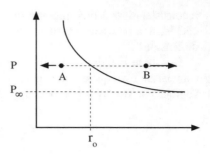

But what happens to droplets which do not have the proper equilibrium radius?
At point A in Fig. 3.14 a droplet will find the external pressure too low for its size,
which is less than the proper equilibrium size r_o, and therefore evaporates monomers. This decreases the radius and the evaporation continues until the droplet
disappears. At point B the droplet is under too high a pressure and additional
monomers from the gas phase condense on its surface. The droplet continues to
grow and finally the limit of a continuous bulk phase is approached. Thus
Eq. (3.102) defines the critical size of a droplet at a given pressure. Below the
critical size droplets disappear, above the critical size they grow without bound. In
turn this means that finite droplets in general are not stable for $d > 1$. In the
following example we reexamine Eq. (3.102) in a specific context for $d = 3$ and
from a slightly different angle.

Example—Critical Droplet Size This example from the microphysics of
clouds combines the above with our previous discussion of relative humidity
in Sect. 3. We consider what is called homogeneous nucleation, i.e. the
condensation of pure vapor, water vapor in this case, into droplets.

Let us assume that a small droplet is created through chance collision of
molecules in the vapor. The subsequent fate of the droplet is decided by the
balance between condensation and evaporation of molecules. This balance we
study in terms of the free enthalpy change dG in a system containing the
droplet inside surrounding bulk gas, i.e.

$$dG_{T,P} = \left(\mu_{H_2O}^{(g)} - \mu_{H_2O}^{(l)} \right) dn_{H_2O}^{(g)} + \gamma dA. \qquad (3.103)$$

The thermodynamic variables temperature T and total (air) pressure P are
constant. The factor multiplying $dn_{H_2O}^{(g)}$ is equal to $RT \ln \varphi$ (cf. Eq. (3.37)),
where φ, defined in Eq. (3.36), is the ratio of the partial water vapor pressure
to the saturation pressure at the same T and P. Finally, the quantity γ is the
surface tension of the droplet and dA is the change of its surface resulting
from condensation or evaporation. Note that $dn_{H_2O}^{(g)} \propto -d\,r^3$ and that

Fig. 3.15 Reduced free enthalpy $G_{T,P}/(RT)$ of a water droplet versus the droplet's radius r at different humidities

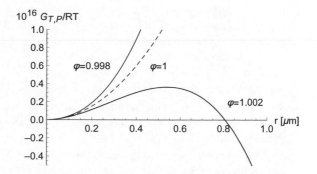

$dA \propto d\,r^2$, where r is the droplet radius. Thus we can express $dG_{T,P}$ in terms of φ, γ, r and dr, i.e.

$$d\frac{G_{T,P}}{RT} = d\left(-\frac{4\pi \ln \varphi}{3\,v_{H_2O}}r^3 + \frac{4\pi \gamma}{RT}r^2\right). \qquad (3.104)$$

The quantity v_{H_2O} is the (partial) molar volume of pure water under the given thermodynamic conditions.

Figure 3.15 shows the expression inside the brackets plotted versus the droplet radius at different φ-values. Here $\gamma = 0.0728$ N/m is the surface tension of water, $T = 293$ K, and $v_{H_2O} = 18$ cm^3. Below and at saturation, i.e. $\varphi \leq 1$, the curves rise monotonously and the equilibrium droplet radius is zero. In the case of supersaturation ($\varphi > 1$), however, the free enthalpy $G_{R,T}$ features a maximum at

$$r_c = \frac{2\gamma\, v_{H_2O}}{RT\, \ln \varphi}. \qquad (3.105)$$

A droplet possessing this radius evaporates following a fluctuation causing an arbitrarily small reduction of r_c. An opposite fluctuation increasing the droplet's size ever so slightly will result in its unlimited growth. Note that this corresponds, for the case of water droplet condensation, to our previous discussion of Eq. (3.102). Solving this equation for $r(\equiv r_c)$ yields $r_c = \delta/(c \ln \varphi)$, where $\varphi = P_{H_2O}/P_{H_2O,\infty}$. The special case of Eq. (3.105) is also known as Kelvin's formula.

Equation (3.102) also tells us that the saturation vapor pressure over a curved (water) surface is greater than over a flat surface. Expanding the exponential to first order Eq. (3.102) becomes

$$P_{H_2O} \approx P_{H_2O,\infty}\left(1 + \frac{\delta}{cr}\right). \qquad (3.106)$$

In the present case of course $\delta/c = 2\gamma v_{H_2O}/(RT)$. This is a curvature correction that enters into most equations in cloud physics.

It is important to note that the above assumption, i.e. "Let us assume that a small droplet is created through chance collision of molecules in the vapor", is nearly impossible to satisfy. A droplet possessing the radius $r_c \approx 0.54$ μm, as shown in Fig. 3.1 requires the collision of around $2 \cdot 10^{10}$ molecules. And even if we increase φ from 1.002 to, for instance, 1.1 this number is still around $2 \cdot 10^5$ molecules. In principle we can keep increasing the supersaturation, but an experimental supersaturation exceeding 1% is rarely observed (see Chap. 9 in Salby 2012).

Equation (3.102) or, equivalently, Eq. (3.105) must be modified when a second component is present in addition to water (solute correction). This means we are talking about the situation depicted in Fig. 3.7. Let's use the same notation as in the figure, i.e. water is indicated by the index A and the second component, the solute, is B. We shall assume that the mole fraction A greatly exceeds the mole fraction B in the liquid phase, i.e. $x_A^{(l)} \gg x_B^{(l)}$. In this case we can apply Raoult's law (3.44) to calculate the saturation pressure of A in the gas phase, P_A, from its value when A is the only component present, P_A^*, i.e. $P_A = P_A^* x_A^{(l)}$. Thus far both pressures in (3.106) refer to pure A, even though we have not given them an asterisk. Hence, if there is B present in the droplet as well as in the gas phase, then we must change Eq. (3.102) or, equivalently, Eq. (3.105) to

$$\frac{P_{H_2O}}{P_{H_2O,\infty}} \approx \exp\left[\frac{2\gamma v_{H_2O}}{RT} \frac{1}{r}\right] \frac{x_{H_2O}(r)}{x_{H_2O}(\infty)}. \tag{3.107}$$

Next we must calculate $x_A^{(l)}(r) \equiv x_{H_2O}(r)$, i.e.

$$x_A^{(l)} = \frac{n_A^{(l)}}{n_A^{(l)} + n_B^{(l)}} = \frac{1}{\left(1 + n_B^{(l)}/n_A^{(l)}\right)} \approx 1 - \frac{n_B^{(l)}}{n_A^{(l)}}. \tag{3.108}$$

Now we use $N_A n_\alpha = (M_\alpha/m_\alpha)$, where $\alpha = A$ or B. N_A is Avogadro's constant, M_α is the total mass of α in the droplet, and m_α is the molecular mass of α. We express the total solvent mass via $M_A \approx c_A 4\pi r^3/3$, where c_A is the bulk mass density of liquid A. Putting everything together we find

$$x_A^{(l)}(r) = 1 - z\frac{M_B}{m_B}\frac{m_A}{4\pi c_A/3}\frac{1}{r^3}. \tag{3.109}$$

The extra factor z accounts for the important cases of salts like *NaCl*, dissociating into two ions which means $z = 2$. Inserting this into (3.107) we obtain our final result

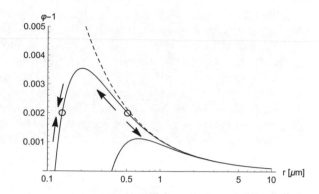

Fig. 3.16 Köhler curves

$$\frac{P_{H_2O}}{P_{H_2O,\infty}} \approx \exp\left[\frac{2\gamma\, v_{H_2O}}{RT}\frac{1}{r}\right]\left(1 - \frac{b}{r^3}\right), \tag{3.110}$$

where b is the factor multiplying $1/r^3$ in Eq. (3.109). This function, i.e. $\varphi - 1 = P_{H_2O}/P_{H_2O,\infty} - 1$, is depicted in Fig. 3.16. The dashed line is the line of the maxima in the Fig. 3.15, i.e. $\varphi(r_c) - 1$, for $\varphi > 1$. The two solid lines are obtained with Eq. (3.110), which includes the concentration effect. The left curve, possessing the higher maximum, is for $M_B = 10^{-16}$ g solute (NaCl in this case and $z = 2$). The right curve is for $M_B = 10^{-15}$ g. Note that a spherical volume containing 10^{-16} g of crystalline NaCl has a radius of about 0.02 μm. This is between 4 to 5 times smaller than the smallest r in Fig. 3.16 for which $M_B = 10^{-16}$ g and $\varphi - 1 > 0$. In particular this confirms that $x_A^{(l)} \gg x_B^{(l)}$ is cloud satisfied.

The one on the right of the two circles in Fig. 3.16 including its two arrows, has the same meaning as the point at r_o in Fig. 3.14. Let us assume the droplet spontaneously increases (decreases) its size at constant $\varphi - 1$. For this increased (decreased) droplet the pressure is too high (low) and it will grow (shrink). The left circle on the solid line, on the other hand, corresponds to a droplet reacting to size fluctuations in exactly the opposite way. When the radius increases (decreases) at constant $\varphi - 1$ the droplet shrinks (grows) back to its original size. Thus, the concentration effect tends to stabilize droplets on the left side of the maximum. This of course is true for both solid lines or for any other so called Köhler curve. Note in particular that the concentration effect reduces the supersaturation necessary for continuous droplet growth. This reduction is enhanced when, at constant M_B, m_B is increased.

We have omitted the range $\varphi - 1 < 0$ in Fig. 3.16. Cloud formation usually takes place by heterogeneous nucleation instead of homogeneous nucleation. Heterogeneous nucleation means that water vapor condenses onto already existing aerosol particles—so called cloud condensation nuclei.

Hygroscopic condensation nuclei may build up significant layers of water (note the relation to our previous discussion of adsorption and Fig. 3.11), thereby increasing the "original droplet", which then requires less supersaturation to grow—especially when the droplet at this point contains dissolved salts, like sodium chloride or ammonium sulfate, as we have seen.

In Chap. 4 we shall resume this discussion venturing into the topic of the growth dynamics of droplets—looking at diffusion as well as collision.

3.3.5   Debye-Hückel Theory

An overall neutral system contains mobile charges. Of special interest in this context are electrolytes, i.e. substances containing free ions. Typically these are ionic solutions, e.g. aqueous solutions of dissociated acids, bases or salts (e.g. $NaCl_{(s)} \rightarrow Na^{1+}_{(aq)} + Cl^{1-}_{(aq)}$). What we want is an approximate description for the electrostatic interaction as part of the chemical potential of the (ionic) charges. In Sect. 2.3.1 we had discussed the relations of the free energy and the free enthalpy to the second law. In the present case electrical work must be included and therefore we have

$$dG|_{T,P} \leq -\delta w_q \tag{3.111}$$

or at equilibrium and expressed as free enthalpy per mole.

$$d\mu_q|_{T,P} = -N_A \delta w_q \overset{(1.15)}{=} N_A dq \phi^{(s)}_{ba}. \tag{3.112}$$

The meaning of last equation on the right in the current context is illustrated in Fig. 3.17. One of the charges, charge q, is shown as thick vertical bar. The charge q is part of a charge density $\rho(r)$, assumed to be radially symmetric and centered on

Fig. 3.17 Spatial distribution of charge

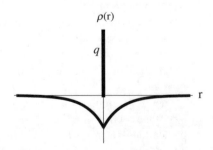

q. An observer on q should notice that the surrounding charge (distribution) preferentially is negative if q is positive and vice versa. However, this q-induced
distribution extends over a finite range only. Beyond a sufficiently large distance r
the central charge q is electrically invisible or screened. The quantity δw_q is the
work done on the system when a infinitesimally small (molar) amount of charge dq
is brought in from infinity (index a) and added to the central charge at zero (index
b). In principle the result is infinite, no matter how small dq is, and therefore useless
to us. But what we really want, is the work due to the screening part of the potential
ϕ_{ba}—indicated by the index (s)—excluding the "bare" potential, q/r, causing the
divergence. It is this screening part of the potential which is the manifestation of the
interaction between the charges in the system.

But how do we calculate $\phi_{ba}^{(s)}$? One equation which comes to mind is Poisson's
equation, i.e.

$$-\vec{\nabla}^2 \phi(\vec{r}) = 4\pi\rho(\vec{r}),\qquad(3.113)$$

where $\phi(\vec{r})$ is the electrostatic potential of a charge density $\rho(\vec{r})$. In the present case

$$\rho(\vec{r}) = q\delta(\vec{r}) + e\sum_i c_i z_i h_i(\vec{r}).\qquad(3.114)$$

The first term on the right is the central charge q at the origin. The second term is
the charge density in a volume element δV located at $\vec{r}$. The factor e is just the
magnitude of the elementary charge. The index i indicates different types of charges
possibly present. c_i is the overall number concentration of these charges, and z_i is
the charging of type i. For instance in the case of $NaCl$ in an aqueous solution there
are Na^+ and Cl^- ions for which $z_{Na} = +1$ and $z_{Cl} = -1$. But there also may be
ions for which $z_i \neq \pm 1$, e.g. CO_3^{2-} with $z_{CO_3} = -2$ or Zn^{2+} with $z_{Zn} = +2$. The
function $h_i(\vec{r}) = h_i(r)$ describes the variation of the i-type screening charge. Notice
that $h_i(\vec{r})$ should vanish as r approaches infinity—but this is all we can say about
$h_i(\vec{r})$ at the moment. We therefore anticipate the following form of $h_i(r)$:

$$h_i(\vec{r}) \approx \exp\left[-\frac{ez_i N_A \phi(\vec{r})}{RT}\right] - 1.\qquad(3.115)$$

The argument of the exponential is the ratio of the electrostatic energy of one
mole of i-charges (at $\vec{r}$) divided by the "thermal energy" RT. This form of $h_i(\vec{r})$ is
by no means exact. It neglects completely structural correlations between charges in
the vicinity of the central charge. It merely considers the effect of the surrounding
charges in the form of a smooth "screening field" as part of the potential $\phi(\vec{r})$, i.e.
no two charges interact directly—each charge interacts with the others through their
collective "screening field". Equation (3.115) relates $h_i(\vec{r})$ as a measure of the
probability for finding a certain charge concentration at distance r from the central
charge to the electrostatic energy of this assembly. The specific form, however, we

can understand only on the basis of microscopic theory as explained in Chap. 5 (notice in particular Eq. (5.150)). Nevertheless, the combination of Eqs. (3.113) to (3.115) yields

$$-\vec{\nabla}^2 \phi(\vec{r}) \approx 4\pi \left(q\delta(\vec{r}) + e\sum_i c_i z_i \exp\left[-\frac{ez_i N_A \phi(\vec{r})}{RT}\right] - 1 \right). \qquad (3.116)$$

This is the desired equation for $\phi(\vec{r})$. However its nonlinearity is inconvenient and thus we go one step further by expanding the exponential, i.e.

$$\exp\left[-\frac{ez_i N_A \phi(\vec{r})}{RT}\right] \approx 1 - \frac{ez_i N_A \phi(\vec{r})}{RT}. \qquad (3.117)$$

This additional approximation is quite in line with our above assumption for the form of $h(\vec{r})$. It requires that the electrostatic energy is much less than the thermal energy and thus that the temperature is "high" (the theory still should be applicable at room temperature though). High temperature also tends to diminish structural correlations. The final equation for $\phi(\vec{r})$ is

$$-\vec{\nabla}^2 \phi(\vec{r}) \approx 4\pi q\delta(\vec{r}) - \frac{8\pi e^2 N_A}{RT} I\phi(\vec{r}). \qquad (3.118)$$

The quantity

$$I = \frac{1}{2}\sum_i c_i z_i^2 = \frac{1}{2}\underbrace{\sum_i v_i z_i^2}_{=p} c \qquad (3.119)$$

is called ionic strength. Here $c_i = v_i c$, where c is the electrolyte concentration and v_i is the number of j-ions per electrolyte molecule.

We solve Eq. (3.118) via Fourier transformation. That is we insert

$$\phi(\vec{r}) = \int_{-\infty}^{\infty} d^3 k \hat{\phi}(\vec{k}) e^{i\vec{k}\cdot\vec{r}} \qquad (3.120)$$

and

$$\delta(\vec{r}) = \frac{1}{(2\pi)^3} \int_{-\infty}^{\infty} d^3 k e^{i\vec{k}\cdot\vec{r}} \qquad (3.121)$$

to obtain

$$\hat{\phi}(\vec{k}) \approx \frac{1}{(2\pi)^3} \frac{4\pi q}{k^2 + \lambda_D^{-2}} \tag{3.122}$$

with

$$\lambda_D = \sqrt{\frac{RT}{8\pi e^2 N_A I}}. \tag{3.123}$$

Insertion of (3.122) into (3.120) and solving the integration yields

$$\phi(\vec{r}) \approx -\frac{q}{r} \left(e^{r/\lambda_D} - e^{-r/\lambda_D} \right). \tag{3.124}$$

Because the first term in brackets grows without bound for $r \to \infty$, we discard this unphysical part of the mathematical solution and thus use

$$\phi(\vec{r}) \approx \frac{q e^{-r/\lambda_D}}{r}. \tag{3.125}$$

When the central charge is approached from a distance, its bare potential, q/r, becomes "visible" when $r \ll \lambda_D$. On the other hand, if $r \gg \lambda_D$ then the potential essentially vanishes, i.e. the central charge is screened. λ_D is the Debye screening length.

How big is λ_D? We first note that the system of our current units requires the replacement of e^2 by $e^2/(4\pi\varepsilon_o\varepsilon_r)$ if we want to use SI-units. Notice that ε_r is the dielectric constant of the background medium containing the charges, e.g. ions in water (water: $\varepsilon_r = 78.3$ at $T = 298$ K and $P = 1$ bar). Thus we have

$$\lambda_D = 1.988 \times 10^{-3} \sqrt{\frac{T\varepsilon_r}{I}} \text{ nm} \quad \text{with} [T] = K \text{ and } [c] = \text{mol}/l. \tag{3.126}$$

For example, in the case of a 0.1 molar aqueous $NaCl$ solution $\lambda_D = 0.96$ nm.

The $\phi^{(s)}$ we want is obtained by subtracting the bare potential q/r from the above $\phi(\vec{r})$, i.e.

$$\phi(\vec{r})^{(s)} \approx \frac{q e^{-r/\lambda_D}}{r} - \frac{q}{r}. \tag{3.127}$$

The potential difference $\phi_{ba}^{(s)}$ is

$$\phi_{ba}^{(s)} = \phi(0)^{(s)} - \phi(\infty)^{(s)} \approx \lim_{r \to 0} \frac{q(1 - r/\lambda_D - \mathcal{O}(r^2)) - q}{r} = -\frac{q}{\lambda_D}. \tag{3.128}$$

Returning now to Eq. (3.112) we can write

$$d\mu_q|_{T,P} \approx -\frac{N_A q}{\lambda_D} dq. \qquad (3.129)$$

The complete μ_q we obtain by increasing dq to the full q, i.e.

$$\mu_q \approx -\int_0^q dq' \frac{N_A q'}{\lambda_D} = -\frac{1}{2} \frac{N_A q^2}{\lambda_D}. \qquad (3.130)$$

This is half the electrostatic energy of two charges $\pm q$ (SI-units: $q^2 \to q^2/(4\pi\varepsilon_o\varepsilon_r)$) at a distance λ_D (multiplied by N_A). Notice also that μ_q is what we must add to the ideal chemical potential of charge q in order to approximately account for its electrostatic interaction with all other charges in the system. In other words, we may write for the (electrostatic) activity coefficient

$$RT \ln \gamma_q \approx -\frac{1}{2} \frac{N_A q^2}{\lambda_D}. \qquad (3.131)$$

Before we proceed with an example, we want to discuss the inclusion of excluded volume. Thus far the ions are point-like. We may include the effect of finite ion size as follows. Due to overall neutrality we require

$$-4\pi q \approx -\frac{1}{\lambda_D^2} \int_b^\infty d^3 r \phi(r) \qquad (3.132)$$

(cf. the second term on the right side of Eq. (3.118)). Here b is the radius of the ion carrying the charge q. Inserting $\phi(r) = A r^{-1} \exp[-r/\lambda_D]$, where A is a constant, we obtain

$$\phi(\vec{r}) \approx \frac{q e^{-(r-b)/\lambda_D}}{r(1+b/\lambda_D)}, \qquad (3.133)$$

instead of Eq. (3.125). The potential difference now becomes

$$\phi_{ba}^{(s)} = \phi(b)^{(s)} - \phi(\infty)^{(s)} \approx \lim_{\delta r \to 0} \left(\frac{q e^{-(r-b)/\lambda_D}}{r(1+b/\lambda_D)} \bigg|_{r=b+\delta r} - \frac{q}{r} \right)$$
$$= -\frac{q}{\lambda_D} \frac{1}{1+b/\lambda_D}, \qquad (3.134)$$

Fig. 3.18 Spherical cell with
a semipermeable wall
containing an electrolyte
solution

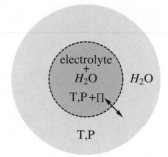

and the charging process yields

$$\mu_q \approx -\frac{1}{2}\frac{N_A q^2}{\lambda_D}\frac{1}{1+b/\lambda_D}.$$

(3.135)

However, for the moment we continue to use Eq. (3.130), i.e. the limit $b \to 0$, and return to Eq. (3.135) when we discuss the phase behavior of simple systems in Chap. 4. Setting $b = 0$ results in the so-called Debye-Hückel limiting law.[9]

Example—Osmotic Pressure in Electrolyte Solutions   In this example we begin by studying osmotic pressure from a somewhat different angle than before. The spherical cell in Fig. 3.18 is submerged inside a water (or solvent) reservoir kept at constant temperature, T, and pressure, P. The water passes freely between the cell, which has a constant volume V, and the reservoir. A suitable mechanism allows to add electrolyte (or solute) to the cell. Contrary to the water the electrolyte, which we assume fully dissociated into its ions, cannot pass the cell's wall. We know from our previous discussion of osmotic pressure that the total pressure inside the cell will rise to $P + \Pi$. The dependence of osmotic pressure, Π, on solute concentration follows via the Gibbs-Duhem Eq. (2.180) applied to the interior of the cell, i.e.

$$VdP \mid_T = \sum_j n_j d\mu_j \mid_T .$$

(3.136)

Here j stands for the different types of ions (or different solute components). The water chemical potential does not appear, because it may adjust to the same value inside and outside the cell. And the outside water chemical potential is constant of course. According to Eqs. (3.72) and (3.73) we may express the change of the j-ion's chemical potential via

---

[9] Peter Debye, Nobel prize in chemistry for his many contributions to the theory of molecular structure and interactions, 1936.

$$d\mu_j = RTd\ln[x_j\gamma_j].\tag{3.137}$$

The desired relation between the osmotic pressure and the electrolyte concentration expressed in moles, n, inside V follows via integration of Eq. (3.136):

$$\Pi = \Pi_{id} + \Pi_{ex} = \frac{RT}{V}\int_0^n \sum_j n_j d\ln x_j + \frac{RT}{V}\int_0^n \sum_j n_j d\ln\gamma_j.\tag{3.138}$$

Using $n_j = v_j n$ the ideal part becomes

$$\Pi_{id} = \frac{RT}{V}\sum_j v_j \int_0^n n' d\ln\left[\frac{n'}{n_{H_2O}(n') + \sum_j v_j n'}\right].\tag{3.139}$$

Note that $d\ln\sum_j v_j = 0$. We recover the van't Hoff equation if we replace $n_{H_2O}(n) + \sum_j v_j n$ by $n_{H_2O}^*$, where $n_{H_2O}^*$ is the water content of the cell at vanishing electrolyte concentration. This approximation requires a negligible solute content, i.e. small electrolyte concentration, and also neglects compressibility effects. Now we can use $d\ln[n_{H_2O}(n) + \sum_j v_j n] \approx d\ln n_{H_2O}^* = 0$ and thus we find

$$\Pi_{id} \approx \Pi_{vH} = \frac{RT}{V}\sum_j v_j n.\tag{3.140}$$

Remembering Eq. (3.131), the Debye-Hückel result for the activity coefficient, we can approximate the excess osmotic pressure, Π_{ex}, via

$$\Pi_{ex} \approx \Pi_{DH} = \frac{RT}{V}\sum_j v_j \int_0^n n' d\left(-\frac{1}{2}\frac{N_A q_j}{\lambda_D}\right).\tag{3.141}$$

With $\lambda_D^{-1} \propto \sqrt{n}$ and

$$\int_0^n n' d\sqrt{n'} = \frac{1}{2}\int_o^n n'^{1/2} dn' = \frac{1}{3}n^{3/2}\tag{3.142}$$

we finally obtain

$$\Pi_{DH} = -\frac{RT}{24\pi N_A \lambda_D^3}.\tag{3.143}$$

Of special interest is the ratio $\phi - 1 \equiv \Pi_{ex}/\Pi_{id} \approx \Pi_{DH}/\Pi_{vH}$. Here ϕ is the osmotic coefficient, i.e.

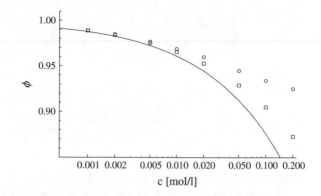

Fig. 3.19 Osmotic coefficient versus electrolyte concentration

$$\phi \approx 1 + \frac{\Pi_{DH}}{\Pi_{vH}} = 1 - \frac{N_A e^2}{6RT\lambda_D}. \tag{3.144}$$

Figure 3.19 shows ϕ versus electrolyte concentration for $AgNO_3$ and $NaCl$ in aqueous solution (the data are from Hamer and Wu 1972)—squares: $AgNO_3$; circles:

$NaCl$; solid line: Eq. (3.144). The limiting law is a good approximation at very small electrolyte concentrations only. However, various approximations like the neglect of finite ion size, ion-ion correlations, and the explicit interaction with the solvent quickly cause deviations with increasing electrolyte concentration.

3.3.6 Gibbs-Helmholtz Equation

In the following we need the Gibbs-Helmholtz equation:

$$\left.\frac{\partial G/T}{\partial T}\right|_P = -\frac{H}{T^2}. \tag{3.145}$$

It may be derived via Eq. (2.117) combined with Eq. (2.123), i.e.

$$G = H - T\left.\frac{\partial G}{\partial T}\right|_P.$$

Dividing both sides by T^2 immediately yields Eq. (2.117). It is useful to rewrite (3.145) in terms of the chemical potential μ_j in a multicomponent system. Differentiation of the left side of Eq. (3.145) with respect to n_j yields

$$\frac{\partial}{\partial n_j}\frac{\partial G/T}{\partial T}\bigg|_{P,n_i}\bigg|_{T,P,n_{i(\neq j)}} = \frac{\partial}{\partial T}\frac{1}{T}\frac{\partial}{\partial n_j}G\bigg|_{T,P,n_{i(\neq j)}}\bigg|_{P,n_i} = \frac{\partial \mu_j/T}{\partial T}\bigg|_{P,n_i}$$

and thus

$$\frac{\partial \mu_j(T,P,n_i)/T}{\partial T}\bigg|_{P,n_i} = -\frac{h_j}{T^2}, \tag{3.146}$$

where h_j is the partial molar enthalpy of component j.

We note that equations analogous to (3.145) and (3.146) hold for the free energy, i.e.

$$\frac{\partial F/T}{\partial T}\bigg|_V = -\frac{E}{T^2} \tag{3.147}$$

and

$$\frac{\partial \mu_j(T,V,n_i)/T}{\partial T}\bigg|_{V,n_i} = -\frac{e_j}{T^2}, \tag{3.148}$$

where e_j is the partial molar energy of component j.

Example—Saha Equation   During the cosmic evolution a phase called recombination occurred. Neutral hydrogen and helium was formed, when the temperature had dropped to about 3000 K (In his book *Cosmology* (Oxford University Press, 2008) Steven Weinberg (Nobel Prize in physics for his contributions to the unification of fundamental interactions, 1979) points out that recombination may be misleading because no neutral atoms had ever existed until this point. But this is the usual term and in addition recombination is still occurring today in the atmospheres of stars.). Even though we cannot provide a complete discussion of this process, which can be found in the aforementioned reference, we still want to get a feel for why it is associated with such a distinct temperature.

Here we study the reaction

$$p + e \rightleftharpoons 1s.$$

p and e stand for one proton and one electron, respectively, while $1s$ denotes the atomic hydrogen ground state. In analogy to the example on p. 109 we write

$$\frac{x_{1s}}{x_p x_e} = PK(T, P^o)$$

assuming ideal gas behavior. However, what we want to calculate is the fraction of ionized hydrogen

$$X = \frac{x_p}{x_p + x_{1s}}.$$

We may combine the last two equations into one, i.e.

$$X(1 + SX) = 1, \qquad (3.149)$$

where $S = (\rho_p + \rho_{1s})PK(T, P^o)/\rho$. Note that $x_p = x_e$ and $\rho_i = \rho x_i$, where ρ is the total number density of massive particles in the universe at this time. Equation (3.149) is the Saha equation (Meghnad Saha, 1893–1956, Indian astrophysicist).

What we really want is $X = X(T)$ and thus we need the explicit temperature dependence of $K(T, P^o)$. The latter quantity is given by

$$K(T, P^o) = \frac{1}{P^o} \exp\left[\frac{1}{RT}\left(\mu_p(T, P^o) + \mu_e(T, P^o) - \mu_{1s}(T, P^o)\right)\right].$$

Using Eq. (3.146) we have

$$\frac{\mu_i(T, P^o)}{T} = \frac{\mu_i(T^o, P^o)}{T^o} - \int_{T^o}^{T} dT' \frac{h_i(T')}{T'^2}.$$

The partial molar enthalpy is $h_i = e_i + RT$, where the internal energy is $e_i = e_i^{(o)} + 3RT/2$. We also use $e_{1s}^{(o)} - e_p^{(o)} - e_e^{(o)} = -13.6 \, eV N_A$, where the right side is the ionization energy for one mole of 1s hydrogen. Overall we obtain

$$\frac{\mu_i(T, P^o)}{RT} = \frac{\mu_i(T^o, P^o)}{RT^o} + \frac{e_i^{(o)}}{R}\left(\frac{1}{T} - \frac{1}{T^o}\right) + \frac{5}{2}\ln\frac{T^o}{T},$$

and thus

$$S = S_o(\rho_p + \rho_{1s})(T/1K)^{-3/2}\exp[158,000K/T],$$

where we have used the ideal gas law to replace P and $13.6 \, eV = 1.58 \cdot 10^5 K$. Here S_o is a number depending on the reference state (T^o, P^o), which thermodynamics does not reveal.

This factor we shall obtain later from Statistical Mechanics, where we learn that the chemical potential of an ideal system of point-like particles is $\mu_i = RT \ln[\rho_i \Lambda_{T,i}^3] + e_i^{(o)}$. Here

$$\Lambda_{T,i} = \sqrt{\frac{2\pi\hbar^2}{m_i k_B T}}$$

is the so called thermal wavelength and $e_i^{(o)}$ is an internal contribution to the particle's chemical potential in the above sense. Setting the masses of the proton and the 1s hydrogen equal, i.e. $m_p = m_{1s}$, we find

$$S_o = 4.14 \cdot 10^{-22} \text{ m}^3.$$

Finally we need to know $\rho_p + \rho_{1s}$. This quantity is given by

$$\rho_p + \rho_{1s} \approx (1 - Y) \frac{\Omega_{m,b}\rho_{c,0}}{m_p c^2} (T/T_0)^3.$$

The quantity $\Omega_{m,b}\rho_{v,0}(T/T_0)^3$ is the baryonic mass density at the time when the radiation temperature is T (cf. the example "The expanding universe and its temperature" in Sect. 2.2) and $m_p c^2$, the proton mass multiplied by the speed of light squared, is the rest energy of a proton. In the aforementioned example it is stated that $\Omega_{m,b} \approx 0.05$ and $\rho_{c,0} \approx 8 \cdot 10^{-10}$J m^{-3}. The current temperature of the background radiation is $T_o \approx 2.7K$. The factor $1 - Y$, where Y is the primordial helium fraction, is roughly $1 - 0.25 = 0.75$. It follows from the theory of primordial nucleosynthesis (a topic which we cannot cover here). With this we have

$$S \approx 4 \cdot 10^{-24}(T/1K)^{3/2} \exp[158{,}000K/T]. \tag{3.150}$$

The result is shown in Fig. 3.20. It is worth noting that neither the position nor the shape of the step do significantly depend on the exact values of $\rho_p + \rho_{1s}$ or S_o (the reader is encouraged to check this).

Weinberg points out that the calculation thus far gives the correct order of magnitude of the temperature of the steep decline in fractional ionization, but it is not correct in detail. However, the in depth discussion is complicated and the interested reader is referred to the above reference.

Fig. 3.20 Fraction of ionized
hydrogen versus temperature

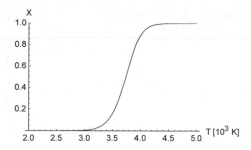

3.3.7   Boiling-Point Elevation

Figure 3.21 sketches a hypothetical crossing at constant pressure from a liquid
phase into the gas in a one-component system consisting of the substance A. The
temperature at which this happens is T_b. From what we know already we may guess
the general form of the attendant chemical potential—in a narrow temperature range
around T_b. This guess is shown in the upper portion of Fig. 3.22. The solid line
depicts the chemical potential along our path. The dotted lines are extensions of the
liquid and gas chemical potentials, respectively, to where they are not stable any-
more. Essentially this picture is based on the inequality (3.68).

   Now suppose we do add a small amount x_B of a second component B to the
liquid. According to Eq. (3.45) we find

$$\delta\mu_A^{(l)} = \mu_{A*}^{(l)} - \mu_A^{(l)} = -RT\ln x_A^{(l)} \approx RTx_B^{(l)}. \tag{3.151}$$

   As always * indicates the pure A and $x_A + x_B = 1$. In addition $x_A \gg x_B$. Equation
(3.151) predicts a downward shift of the liquid chemical potential of A, which is
shown as long-dashed line in Fig. 3.22. We note that we are interested only in the
immediate vicinity of the boiling temperature T_b. Therefore this line and the corresponding solid line are parallel to good approximation. Furthermore we assume

Fig. 3.21 Hypothetical
crossing of the saturation line

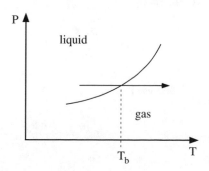

Fig. 3.22 Sketch illustrating
boiling-point elevation

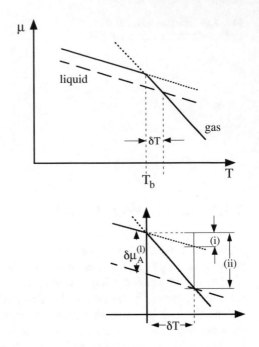

that the amount of B in the gas phase is negligible or causes only a negligible shift of the gas phase chemical potential. Therefore we find that the intersection of the chemical potentials of A in the liquid phase and the gas phase has shifted to a higher temperature $T_b + \delta T$.

Using the Gibbs-Helmholtz equation we may relate $\delta \mu_A^{(l)}$ at T_b to the boiling-point elevation δT. Geometrically $\delta \mu_A^{(l)}$ is given by

$$\delta \mu_A^{(l)} = (ii) - (i), \tag{3.152}$$

where (ii) and (i) are defined in the bottom portion of Fig. 3.22. By simple trigonometry

$$(ii) = -\left.\frac{\partial \mu_{A^*}^{(g)}}{\partial T}\right|_{T_b} \delta T \qquad \text{and} \qquad (i) = -\left.\frac{\partial \mu_{A^*}^{(l)}}{\partial T}\right|_{T_b} \delta T. \tag{3.153}$$

Thus we have

$$\delta \mu_A^{(l)} = -\left.\frac{\partial (\mu_{A^*}^{(g)} - \mu_{A^*}^{(l)})}{\partial T}\right|_{T_b} \delta T \equiv -\left.\frac{\partial \Delta \mu_{A^*}}{\partial T}\right|_{T_b} \delta T. \tag{3.154}$$

Table 3.2 Latent heats of vaporization and melting for various compounds

Compound	$\Delta_{vap}H$ [J/g]	$\Delta_{melt}H$ [J/g]
Ice	2838 ($T = 273.15$K)	333.6 ($T = 273.15$K)
Water	2258 ($T = 373.12$K)	
N_2	199 ($T = 77.35$K)	25.3 ($T = 63.15$K)
O_2	213 ($T = 90.2$K)	13.7 ($T = 54.36$K)
Octane	364 ($T = 298$K)	182 ($T = 273.5$K)

The Gibbs-Helmholtz equation enters via

$$\delta\mu_A^{(l)} = -\left.\frac{\partial\Delta\mu_{A^*}}{\partial T}\right|_{T_b}\delta T = -\frac{\partial}{\partial T}\left(T\frac{\Delta\mu_{A^*}}{T}\right)_{T_b}\delta T$$

$$= -\left(\frac{\Delta\mu_{A^*}(T_b)}{T_b} + T\left.\frac{\partial\Delta\mu_{A^*}/T}{\partial T}\right|_{T_b}\right)\delta T \qquad (3.155)$$

$$\delta\mu_A^{(l)} \overset{(3.137)}{=} \frac{\Delta_{vap}h}{T_b}\delta T.$$

Notice that $\Delta\mu_{A^*}(T_b) = 0$ (chemical equilibrium!) and $\Delta_{vap}h$ is the molar enthalpy change upon crossing from pure liquid A to pure gaseous A, i.e. the enthalpy of vaporization of pure A. The enthalpy change during a phase transition also is called latent heat—here latent heat of vaporization. Table 3.2 compiles latent heats of vaporization and melting for a number of substances. We remark in this context that the heat content of a substance build up without changing phase, $\Delta H = m\int_{T_1}^{T_2} dT C_p(T)$, where m is the mass of the substance, is called sensible heat.

Combining (3.155) with Eq. (3.151) we finally arrive at

$$\delta T \approx \frac{RT_b^2}{\Delta_{vap}h}x_B^{(l)}. \qquad (3.156)$$

If we look up $\Delta_{vap}h$ for the transition of water to steam at $1bar$, e.g. from Table 3.2, we obtain $\delta(T/K) \approx 28.5x_B^{(l)}$, i.e. for amounts B in accord with our above approximations the shift is quite small.

3.3.8 Freezing-Point Depression

Here the above solution containing mostly A and little B is in equilibrium with the solid A, where again the B-content is negligible. Analogously to Fig. 3.21 we may draw the sketch shown in Fig. 3.23. Apparently this time the transition temperature,

Fig. 3.23 Sketch illustrating
freezing-point depression

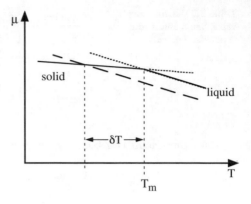

which is the melting temperature T_m, is reduced by the addition of B. A completely
analogous calculation yields

$$\delta T \approx -\frac{RT_m^2}{\Delta_{melt}h}x_B^{(l)}, \tag{3.157}$$

where $\Delta_{melt}h$ is the molar melting enthalpy at given pressure. For water at $1 bar$ we
obtain the freezing-point depression $\delta(T/K) \approx -103x_B^{(l)}$. Figure 3.24 shows this
relation (solid line) in comparison to data points for D-Fructose (crosses) and
Silvernitrate ($AgNO_3$) (open squares) taken from HCP. Notice that in the case of
$AgNO_3$ the mole fraction refers to mole ions, i.e. Ag^+ and NO_3^- taken individually.
The dashed line, apparently an improved description of the $AgNO_3$-data, is discussed in the next section.

Remark—cooling This is a good place to address the following question.
Figure 3.25 depicts a glass of water with a floating ice cube. Their respective
weights are 200 and 20 g and their momentary temperatures are 20 and $0 \,^\circ$C. What
will the temperature of the contents of the glass be after the ice has melted?

Fig. 3.24 Theoretical
predictions of freezing point
depression compared to
experimental data

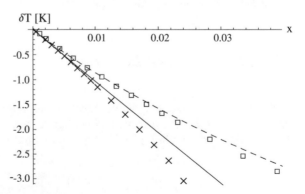

Fig. 3.25 A glass of liquid
water with a floating ice cube

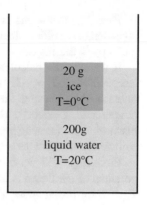

We assume, as so often, no transfer of heat between the contents of the glass and
its surroundings including the glass itself. Then according to the first law we have

$$\Delta E + \Delta w = E_{220g} - E_{200g} - E_{20g} + P(V_{220g} - V_{200g} - V_{20g}) = 0, \qquad (3.158)$$

i.e. the overall enthalpy change is zero:

$$H_{220g} - H_{200g} - H_{20g} = 0. \qquad (3.159)$$

Here the indices refer to the initial water and ice by their respective weights as
well as to the final liquid by its weight.

Neglecting for the moment the melting enthalpy of the ice, i.e. the enthalpy
change when the ice is converted into liquid at $T = 0\,°C$, we obtain the final
temperature T via

$$220gC_PT - 200gC_PT_{20\,°C} - 20gC_PT_{0\,°C} = 0. \qquad (3.160)$$

C_P is the isobaric heat capacity (we use its value at $T = 0\,°C$ which is $4.22\,J/(gK)$),
also assumed to be constant in the relevant temperature range. The resulting final
temperature after the ice has melted is $T = 18\,°C$. But this is incorrect, because we
have not yet included the enthalpy change during melting. There is a cost associated
with the breaking down of the ice structure—most notably the reduction in
hydrogen bonding. The price is paid in the form of heat extracted from the content
of the glass. This melting enthalpy is tabulated for numerous substances in HCP,
where we find $\Delta_{melt}H = 6.01\,kJ/mol$ for water at ambient pressure and $0\,°C$.
Including this contribution yields

$$(20\,g/18g)\Delta_{melt}H + 220\,gC_PT - 200\,gC_PT_{20\,°C} - 20\,gC_PT_{0\,°C} = 0. \qquad (3.161)$$

Here $18g$ is the molar weight of water. The new result is $T = 11\,°C$. This is considerably colder than the previous temperature. Apparently the transition enthalpy is the major contribution[10]!

If we redo our calculation with more ice, let's say 100 g, we obtain even lower temperatures—$T = -13\,°C$ in this case. We immediately object that this is unreasonable, because the whole content of the glass freezes before reaching this temperature. Correct[11]! Nevertheless it brings up an idea connecting our present discussion to freezing point depression. We need to depress the freezing point sufficiently in order to reach such low temperatures.

From Fig. 3.23 we can see that melting above T_m results because the chemical potential of the liquid is lower. Therefore melting continues until the temperature is T_m.[12] If for instance we add salt to our liquid water in the above example, we may, depending on the amount we add, lower the temperature far below the freezing temperature of pure water.

In addition we may take advantage of a second and third enthalpy change associated with a possible phase change of the substance we add or the mixing process itself. If we select the proper substances then we can get quite low temperatures in this fashion—around $-100\,°C$! Of course, the disadvantage is that this method allows to maintain low temperatures over short periods of time only.

Remark The above reasoning is very much simplified. It is incorrect to conclude that increasing the amount of solute (e.g., salt) allows to continuously depress the freezing point. For instance in the case of $NaCl$ we can only get down to about $-21.1\,°C$. At lower temperatures ice and solid salt coexist. What this means we discuss in the next chapter, where we study simple phase diagrams (in particular liquid-solid coexistence in binary systems (cf. Sect. 4.3.3)).

3.3.9 The Osmotic Coefficient Revisited

Boiling point elevation and freezing point depression can be tied to the osmotic coefficient, ϕ, and are practical means for its measurement. We start with the Gibbs-Duhem equation at constant pressure and temperature:

---

[10] This also is something to keep in mind when buying a new washing machine. A higher spin speed is usually better, because of the decreased residual water content in the laundry. This water must be evaporated in the dryer, and the enthalpy of evaporation again is considerable. On the other hand, a modern condenser dryer often is capable of reclaiming some of the invested energy upon condensation.

[11] More precisely, the process comes to a halt at coexistence of ice and liquid water.

[12] Under "pool conditions" the sun transfers heat to the contents of our glass and the melting continues until the ice is gone.

$$-n_A d\mu_A\Big|_{T,P} = \sum_j n_j d\mu_j\Big|_{T,P}. \tag{3.162}$$

Here j stand for different solute components. This equation expresses an infinitesimal change of the A-chemical potential via corresponding changes of the j-chemical potentials. The right side of this equation is comparable to the right side of Eq. (3.136). The difference is that in Eq. (3.162) the pressure is constant while in (3.136) it is not.

However, in the liquid phase, where the compressibility is small, we may to very good approximation equate the two right sides and consequently we arrive at

$$\delta\mu_A\Big|_{T,P} \approx -\frac{V\Pi_{id}}{n_A}\phi. \tag{3.163}$$

Here $\delta\mu_A$ is the chemical potential change due to increasing the solute concentration from zero to some final concentration. The sign is just the opposite of the definition in Eq. (3.151) and thus

$$\phi \approx \frac{n_A^{(l)}}{\sum_j n_j}\frac{\Delta_{vap}h}{RT_b^2}\delta T. \tag{3.164}$$

This is the desired equation for the osmotic coefficient in terms of the boiling-point elevation, where we have inserted the van't Hoff equation for Π_{id}. An analogous relation follows for the osmotic coefficient in terms of the freezing point depression:

$$\phi \approx -\frac{n_A^{(l)}}{\sum_j n_j}\frac{\Delta_{melt}h}{RT_m^2}\delta T. \tag{3.165}$$

The dashed line in Fig. 3.24 is obtained if the osmotic coefficient is calculated via Debye-Hückel theory according to Eq. (3.144). However, the reader should be aware that $AgNO_3$ is an example for which the limiting law works particularly well.

3.3.10   Measuring Surface Tension

Consider the interfacial area $A = xy$ in Fig. 3.26 (left surface). Increasing A to $A + dA = (x + dx)(y + dy)$ (right surface) requires the reversible work $dw = \gamma dA$. Here γ is the interface tension. We may also express dw in terms of the pressure difference on the two sides of the surface ΔP multiplied by the volume change $dV = Adz$, i.e. $dw = \Delta PAdz$. Hence

Fig. 3.26 A Mechanical equilibrium for a curved surface

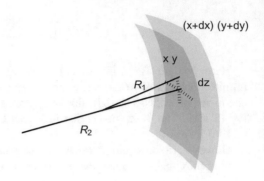

$$\Delta P A dz = \gamma dA. \tag{3.166}$$

This transforms into a very useful equation if we can rewrite dA as a function of dz. Expressing dA first in terms of dx and dy yields

$$dA = xy\left(\frac{x+dx}{x}\right)\left(\frac{y+dy}{y}\right) - xy. \tag{3.167}$$

Now note that the dotted cross on the inner surface in Fig. 3.26 is a local xy-coordinate system. Its x-axis points towards the reader and the positive direction of y-axis is up. The dotted section of the x-axis is swept out by an infinitesimal angular rotation of R_1 relative to its origin, whereas the dotted piece of the y-axis is obtained by an analogous sweep of R_2. The origin of R_2 (generally) is different from the origin of R_1. R_1 and R_2 are the principal radii of curvature of the interface at this location. Using the theorem of rays we can rewrite dA as

$$dA = xy\left(\frac{R_1+dz}{R_1}\right)\left(\frac{R_2+dz}{R_2}\right) - xy = xydz\left(\frac{1}{R_1} + \frac{1}{R_2}\right). \tag{3.168}$$

Combination of the Eqs. (3.166) and (3.168) yields the Young-Laplace equation

$$\boxed{\Delta P = \gamma\left(\frac{1}{R_1} + \frac{1}{R_2}\right).} \tag{3.169}$$

Example—Capillary Adhesion We practice using the Young-Laplace equation by calculating the force between two smooth plates squashing a drop of liquid between them as shown in Fig. 3.27. Eq. (3.169) is applied at the position marked by the black dot. One of the two radii of curvature is R, which also is the radius of the liquid drop if we look down on it through one of the plate's surfaces. The magnitude of the other radius of curvature, let's

Fig. 3.27 Capillary adhesion

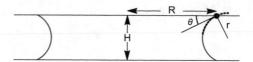

call this one r, is the radius of the circle whose partial circumference is indicated by the dashed line. The angle between the dashed line and the upper plate at the black dot is θ, the contact angle of the liquid on the plate. Simple geometry yields

$$\tan \theta = \frac{\sqrt{r^2 - (H/2)^2}}{H/2}, \tag{3.170}$$

which can be solved for r, i.e.

$$r = \frac{H}{2 \cos \theta}. \tag{3.171}$$

If, in addition, we assume $R \gg r$ then Eq. (3.169) becomes

$$\Delta P = -\frac{2\gamma \cos \theta}{H}. \tag{3.172}$$

The extra minus sign is due to the fact that the center of the second circle lies outside the liquid. In the opposite case it would have been positive. Note the similarity of this equation to Eq. (2.136) derived in the context of the capillary rise problem in Chap. 2. We can convert ΔP into a force between the plates, i.e.

$$F = -\pi R^2 \frac{2\gamma \cos \theta}{H}, \tag{3.173}$$

Here the minus sign indicates attraction between the plates (as long as $\theta < \pi/2$). For example, if the liquid is water ($\gamma = 0.0728$ N/m), (assuming) $\theta = 0$, $R = 1$ cm and $H = 5$ μm then $F \sim -10$ N.

The Young-Laplace equation can be used also to calculate the shape of a drop suspended from a dosing capillary as depicted in Fig. 3.28 (the capillary is omitted). Here the pressure difference is a function of z, i.e.

$$\Delta P = \Delta P(0) - \Delta c g z. \tag{3.174}$$

Note that the origin of the z axis is at the bottom of the drop. Δc is the mass density difference between the liquid inside the drop and the medium surrounding the drop, which may be air or another liquid. g is the acceleration of gravity. The

Fig. 3.28 Pendant drop

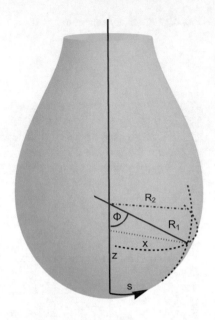

two (principal) radii of curvature at $z = 0$ are identical due to the axial symmetry of the problem, i.e. $R_1 = R_2 \equiv R$ (In the figure the two radii are shown for a different point on the drops surface where they are different.). Inserting this into Eq. (3.169) yields

$$\Delta P(0) = \frac{2\gamma}{R}. \tag{3.175}$$

Substituting (3.175) back into Eq. (3.169), this time for an arbitrary z, yields

$$\frac{1}{R_1} + \frac{\sin\Phi}{x} = \frac{2}{R} - \frac{\Delta c g z}{\gamma}. \tag{3.176}$$

Note that $x = R_2 \sin\Phi$. Note also that the pendant drop is a figure of revolution and that R_2 therefore must originate on the z-axis. The same is not true for R_1 (cf. Fig. 3.28). It is useful to introduce a parameter variable s, which is the contour length along the drop's surface starting at the bottom ($z = 0$) and going up (see again Fig. 3.28). This leads to the following set of three coupled first order differential equations:

$$\frac{d\Phi}{ds} = -\frac{\sin\Phi}{x} + \frac{2}{R} - \frac{\Delta c g z}{\gamma} \frac{dx}{ds} = \cos\Phi \frac{dz}{ds} = \sin\Phi, \tag{3.177}$$

where we have used $R_1\, d\Phi = ds$, $dx = R_2 \cos\Phi\, d\Phi = \cos\Phi\, ds$, and $d(R_2 - z) = d(R_2 \cos\Phi)$, i.e. $dz = \sin\Phi\, ds$.

Fig. 3.29 Pendant drop contours

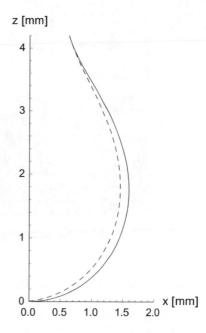

In practice the drop shape is recorded by a camera and the Eqs. (3.177) are solved to fit the theoretical contour to the recorded contour using R and γ as adjustable parameters. The initial conditions are $0 = x(s = 0) = z$ $(s = 0) = \Phi(s = 0)$. Figure 3.29 shows two numerical solutions, i.e. the first solution (solid line) is for $R = 1.5$ mm and γ is the surface tension of water, whereas the second solution (dashed line) is for the same value of R but a 10% smaller surface tension. The first solution can be calculated via the following *Mathematica* code:

$R = 0.0015; \gamma = 0.0728;$

$t = \text{ND Solve}[\{\phi'[s] == -\text{Sin}[\phi[s]]/x[s] + 2/R - (9810/\gamma)z[s], x'[s] == \text{Cos}[\phi[s]],$

$z'[s] == \text{Sin}[\phi[s]], x[0] == z[0] == \phi[0] == 0.000001\}, \{\phi, x, z\}, \{s, 0.006\}]$

$\text{ParametricPlot}[\text{Evaluate}[\{x[s] * 1000, z[s] * 1000\}/.t], \{s, 0, 0.006\},$

$\text{PlotRange} \rightarrow \{\{0, 2\}, \{0, 4.2\}\}, \text{AxesLabel} \rightarrow \{``x \text{ [mm]}", ``z \text{ [mm]}"\}]$

The second solution is obtained by changing the γ-value (In addition, in the figure this contour is shifted horizontally so that the two contours match up at the top.). This method is the so called pendant drop method. It can be used to obtain

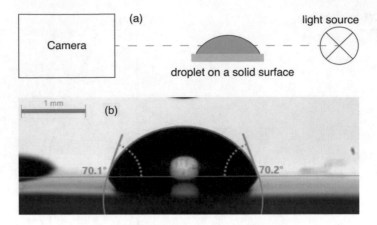

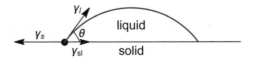

Fig. 3.30 a Schematic of the sessile drop measurement. **b** Picture of a liquid drop on a solid surface

either the surface tension, if the surrounding medium is air, or the interface tension, if the surrounding medium is another liquid. Another common technique for measuring surface tensions is the sessile drop method depicted in Fig. 3.30a. A small drop ($\approx 1 - 2\,\mu l$) of a liquid is deposited on a clean smooth solid surface. After the droplet has settled into its equilibrium shape a camera snaps a picture (cf. Fig. 3.30b). This picture is then used to measure the contact angle, i.e. the perpendicular angle between the solid surface and the tangent to the droplet's contour at its base. There are two such angles on either side of the drop. Usually they differ slightly due to surface irregularities, a possible tilt of the substrate surface, etc.

The theoretical analysis of this measurement is based on the following three equations, the first of which is Young's equation:

$$\boxed{\gamma_s = \gamma_{sl} + \gamma_l \cos \theta}. \tag{3.178}$$

The sketch in Fig. 3.31 depicts the droplet and the contact angle, θ, in relation to the tensions in Young's equation. Here the index s indicates the solid surface and l is the liquid. γ_s is the surface tension of the solid-gas interface and γ_l is the analogous quantity for the liquid instead of the solid. γ_{sl} is the interface tension of the solid-liquid interface. Notice that the unit of γ is energy/length^2 or force/length.

Fig. 3.31 Standard setup for Young's equation

The interpretation of Eq. (3.178) in terms of a force equilibrium is quite obvious. The change in surface free energy at constant temperature when we put the drop down on the surface is

$$\Delta G^{(s)} = (\gamma_{sl} - \gamma_l - \gamma_s)\Delta A. \qquad (3.179)$$

A second, albeit empirical, approximation of the same quantity is

$$\Delta G^{(s)} = -2\left(\sqrt{\gamma_l^d \gamma_s^d} + \sqrt{\gamma_l^p \gamma_s^p}\right)\Delta A. \qquad (3.180)$$

Here γ_i^d is the dispersive part of the surface tension γ_i and γ_i^p is its polar part, i.e.

$$\gamma_i = \gamma_i^d + \gamma_i^p. \qquad (3.181)$$

Combination of the Eqs. (3.178), (3.179) and (3.180) yields

$$\sqrt{\gamma_l^d \gamma_s^d} + \sqrt{\gamma_l^p \gamma_s^p} = (\gamma_l^d + \gamma_l^p)\frac{\cos\theta + 1}{2}. \qquad (3.182)$$

The distinction of "dispersive" versus "polar" is not always easy and never clear-cut. Two examples may suffice at this point. Let's consider the liquid phase of an n-alkane. Its methylene groups do not exhibit a particular separation of charges (partial charges) and its polarizibility is small. In this case $\gamma_i^p \approx 0$. The (attractive) dispersive inter and intra-molecular interactions, which are at work here and essentially govern γ_s^d, are due to interacting quantum fluctuations of the electron distribution around the nuclei. These fluctuations of course are universally present. Water, our second example, is very different from any n-alkane. In the gas phase it possess a, compared to its size, large dipole moment of about 1.85 D. In the condensed phase, due to polarization, it has an even larger (average) dipole moment. Thus, liquid water is the liquid of choice when a large γ_i^p is needed.

While an accurate distinction between γ_i^d and γ_i^p can be difficult on the level of microscopic interactions, it is quite straightforward in the following analysis. Dividing Eq. (3.182) by $\sqrt{\gamma_l^d}$ yields the equation of a straight line, $y = mx + b$, with

$$y = \frac{\gamma_l}{\sqrt{\gamma_l^d}}\frac{\cos\theta + 1}{2} \quad \text{and} \quad x = \sqrt{\frac{\gamma_l^p}{\gamma_l^d}}. \qquad (3.183)$$

If the experiment is repeated with at least two test liquids, i.e. liquids whose surface tension components are known, then the y-intercept b and the slope m will yield the dispersive part and the polar part of the solid's surface tension:

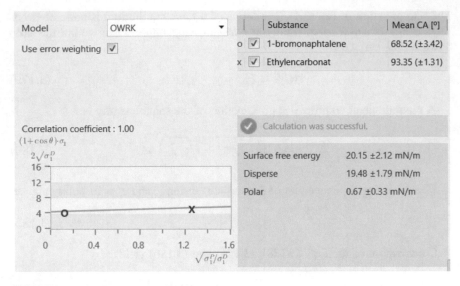

Fig. 3.32 Standard application of Eq. (3.182)

$$\gamma_s^d = b^2 \quad \text{and} \quad \gamma_s^p = m^2. \tag{3.184}$$

Figure 3.32 shows a fairly typical example taken from a customer application report (KRÜSS GmbH, 2020). Only two liquids are used, but some care is taken that these liquids cover a significant x-range. The temperature is 60 °C. The values of the substrate's surface free energy per area, the substrate is a polymer in this case, and its components are shown in the box in the lower right corner. The measured contact angles (CA) are shown in the box above. The quality of the result obviously benefits from a wide range of x-values and a "large" number of test liquids. Table 3.3 lists γ_s, γ_s^d and γ_s^p for a number of test liquids. Among these liquids water

Table 3.3 Surface tension of some test liquids (taken from a KRÜSS GmbH customer application report)

Test liquid	γ [mN/m]	γ^d [mN/m]	γ^p [mN/m]
Water (25°C)	72.8	21.8	51.0
Ethylene glycol (20°C)	47.7	30.9	16.8
1-bromonaphthalene (20°C)	44.6	44.6	0.0
1-bromonaphthalene (60°C)	43.0	42.4	0.6
1-bromonaphthalene (80°C)	42.1	41.4	0.7
1-bromonaphthalene (120°C)	40.6	39.6	1.0
ethylene carbonate (60°C)	51.9	20.2	31.7
ethylene carbonate (80°C)	50.3	20.6	29.7
ethylene carbonate (120°C)	47.5	21.1	26.4

at room temperature has the largest x-value (≈ 1.5), whereas 1-bromonaphthalene has the smallest x-value ($= 0$). Note the temperature dependence of the surface tension.

The development of Eq. (3.180) into (3.182) is called the OWRK-method in the literature. Here OWRK stands for D. Owens and R. Wendt (1969), W. Rabel (1971) and D. H. Kaeble (1970). However, much of the OWRK-method is described in earlier work by F. M. Fowkes (1964). Motivated by the form of certain key formulas in the theory of intermolecular and surface forces, Fowkes introduces the geometric mean approximation (3.180) for the dispersive parts of the surface tensions, i.e. he also arrives at Eq. (3.182) albeit with $\gamma_s^p = 0$. In his paper he considers the entire range of interfaces occurring between gases, liquids and solids. He also obtains the value for $\gamma_{H_2O}^d$ in Table 3.3 based on the combination of Eqs. (179) and (3.180). In his case the index l stands for water and the index s stands for a series of hydrocarbons with negligible polar parts of their surface tensions. Since γ_{H_2O} is known (it can be measured by a number of methods), Fowkes' procedure also fixes $\gamma_{H_2O}^p$. Similarly one can obtain γ_l^d and γ_p^d for other (test) liquids.

In principle we can divide Eq. (3.182) by $\sqrt{\gamma_s^d}\gamma_l$ instead of $\sqrt{\gamma_l^d}$. The result is another straight line equation in which $b = \sqrt{\gamma_l^d}/\gamma_l$ and $m = \sqrt{\gamma_l^p}/\gamma_l$. Thus we can obtain the unknown components of a liquid's surface tension by measuring its θ on at least two "test surfaces".

Eq. (3.182) may be cast into another useful form by defining

$$\gamma_l^d = R\cos\phi \quad \text{and} \quad \gamma_l^p = R\sin\phi. \tag{3.185}$$

Inserting this into Eq. (3.182) and solving for R yields:

$$R \equiv R(\phi, \theta) = \left(\frac{\sqrt{\gamma_s^d}\cos\phi + \sqrt{\gamma_s^p}\sin\phi}{\cos\phi + \sin\phi}\right)^2 \frac{4}{(\cos\theta + 1)^2}. \tag{3.186}$$

For every fixed value of θ we now can obtain γ_l^d and γ_l^p by letting ϕ vary from 0 to $\pi/2$. Figure 3.33 shows two examples of so called wetting envelopes. In the case

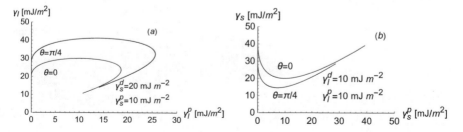

Fig. 3.33 Examples of wetting envelopes

(a) γ_s^d and γ_s^p are assigned fixed values and Eqs. (3.185) and (3.186) are used to plot γ_l vs. γ_l^p for two values of θ. If $\theta = 0$ wetting is complete. Panel (b) shows the analogous result if Eq. (3.185) is replaced by

$$\gamma_s^d = R' \cos \phi' \quad \text{and} \quad \gamma_s^p = R' \sin \phi'. \tag{3.187}$$

We obtain R' if we insert (3.187) once again into (3.182):

$$R'(\phi', \theta) = \left(\frac{\gamma_l^d + \gamma_l^p}{\sqrt{\gamma_l^d} \cos \phi' + \sqrt{\gamma_l^d} \sin \phi'} \right)^2 \frac{(\cos \theta + 1)^2}{4}. \tag{3.188}$$

In panel (b) of Fig. 3.33 γ_l^d and γ_l^p are assigned fixed values and the Eqs. (3.187) and (3.188) are used to plot γ_s versus γ_s^p for two values of θ. The information content is the same in both cases, because it is always Eq. (3.182) which is solved.

There are quite a few techniques and methods available for measuring contact angles or surface and interface tensions. Surface tensions may be extracted from the shape of drops in various configurations (hanging (pendant drop), rotating, deposited (sessile drop), ...). As we have mentioned, interface tensions are obtained for instance when the drops are embedded in another liquid. Here we cannot discuss all of these measuring techniques in the detail they deserve. Instead we want to conclude this section by briefly describing two selected methods in addition to the above pendant and sessile drop methods. The reason for including the first method is its frequent use, which is comparable to the pendant and sessile drop methods, and the main reason for including the second method is the illustration of a problem. The first method is the Wilhelmy plate method which can be employed to determine the surface tension of a liquid, the interface tension between two liquids (if air in the discussion below is replaced by a second liquid), and the contact angle between a liquid and a solid. What is measured here is the force F acting on a thin smooth plate touching or partially submerged in a liquid as shown in Fig. 3.34. This force is

Fig. 3.34 Illustration of the Wilhelmy plate method

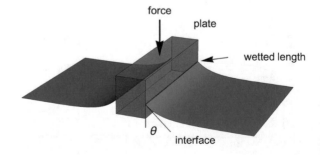

$$F = \gamma_l L \cos\theta - V\Delta c\, g. \tag{3.189}$$

Here L is the perimeter of the plate, i.e. the length of the entire plate-liquid contact line, V is the volume of the displaced liquid, Δc is the density difference between liquid and air, and g is the acceleration of gravity. Note that the second term is the buoyancy of the submerged portion of the plate.

Commonly a platinum plate is used if the surface tension of an unknown liquid is the quantity of interest. Platinum possesses a high surface free energy which results in complete or total wetting, i.e. $\theta = 0$. Another common procedural step of the Wilhelmy method is a slow submersion followed by the likewise slow retraction of the plate. The corresponding dynamic contact angles, θ_a (advancing θ) and θ_r (receding θ) are different and their difference is called contact angle hysteresis.

But the non-uniqueness of the contact angle θ in Young's equation is not limited to dynamic measurements. It is observed in static situations due to the non-ideality of the surfaces involved. Consider for example a liquid drop trapped inside a narrow glass capillary. Now turn the capillary, which is open on both ends, so that the force of gravity acting on the liquid is parallel to the capillary. (Usually) the drop remains stuck inside the capillary forming an upper and a lower meniscus possessing different contact angles θ_{top} and θ_{bottom}. Generalizing our previous formula for the height h of a liquid column inside a capillary, i.e Eq. (2.136), to this case, we find

$$\frac{2\gamma}{R}\left(\cos\theta_{top} - \cos\theta_{bottom}\right) = c\,g\,h. \tag{3.190}$$

Here h is the length of the liquid column trapped inside the capillary, from which it would escape if $\cos\theta_{top} - \cos\theta_{bottom} = 0$ However, usually this is not the case and it takes some extra work, e.g. blowing into the capillary, to remove the trapped liquid. More specifically, for the liquid column to remain motionless (equilibrium), $\theta_{top} > \theta_r$ and $\theta_{bottom} < \theta_a$. Here θ_r and θ_a are the limiting contact angles below or beyond which a solid-liquid contact line, or in this case the liquid column, begins to recede or advance spontaneously. Roughly, on a "good" surface the hysteresis $\theta_a - \theta_r$ is small ($<5°$). On rough or dirty surfaces it can be much larger. The experimentally observed advancing contact angle is usually considered to be closest to the contact angle θ in the Young equation.

A substrate may not possess a "smooth" surface but may be a powder or granulate instead. An important example are fillers consisting of nanoparticles (<100 nm in diameter), usually carbon black or silica, which are used for rubber reinforcement in the tire industry. The dispersion of the particles within the polymer matrix is governed by the attendant interfacial free energies, which means that it is of significant importance to reliably measure surface tensions of powders or granulates. One can try to prepare thin layers of these particles on otherwise smooth substrates (e.g. adhesive tape) and measure contact angles with for instance Wilhelmy's method. But this will produce a pronounced contact angle hysteresis and results are hardly reproducible. A method especially designed for porous substrates was developed by E. W. Washburn (1921). Here the rate of penetration of a test liquid into a compressed powder cake is monitored, i.e. the depth of the liquid front l as a function of time t is recorded. The contact angle θ then follows via

$$l^2 = \frac{r\, t\, \gamma_l \cos\theta}{2\eta} \, . \tag{3.191}$$

The quantity η is the liquid's viscosity and r represents the pore radius. In the static limit of this experiment the wetting liquid penetrates upward vertically through the powder cake until the capillary pressure balances the liquid's weight. Essentially the theoretical description is again Eq. (2.136) (capillary rise example), where in this case r once again represents the pore radius. Off course, the latter is not very well defined and therefore usually replaced by an effective capillary radius r_{eff} defined in terms of the volume fraction of solid ϕ, the density of the solid material c and the specific surface area per gram of solid A:

$$r_{eff} = \frac{2(1-\phi)}{\phi\, c\, A} \, . \tag{3.192}$$

For more details of this, the Wilhelmy and others methods the reader is referred to Yuan and Lee (2013). The theory of surface and interface tension is covered in de Gennes et al. (2004).

Remark 1   The temperature dependence of the surface tension, which we do not discuss here, nevertheless is of great technical importance. Since in most applications a linear approximation is sufficiently accurate, many reports of measured data do include the gradient $d\gamma/dT$ in the range of relevant temperatures T. At the gas-liquid critical point of a substance its surface tension γ vanishes. For temperatures T not too different from the critical temperature T_c one can show that the scaling relation $\gamma \propto (1 - T/T_c)^\mu$ is satisfied, where μ is a critical exponent. The original value $\mu = 1$ (Eötvös' law) was modified empirically to $\mu = 11/9$ by E. A. Guggenheim (1945) and finally calculated on the basis of the hyperscaling relation $\mu = (d-1)\nu$, where d is the space dimension and ν is the critical exponent of the temperature scaling of the order parameter fluctuation correlation length (J. S Rowlinson, B. Widom, Molecular Theory of Capillarity (Oxford University Press, Oxford, 1989)). In three dimensions $\nu \approx 0.63$ and therefore $\mu \approx 1.26$.

Remark 2   nA simple form describing the dependence of the surface tension of polymer liquids on molecular weight M is $\gamma = \gamma_\infty + cM^{-x}$. Here γ_∞ and c are constants. The exponent x usually is close to 2/3 for low molecular weights but crosses over to 1 at high molecular weights (Thompson et al. 2008).

Chapter 4
Simple Phase Diagrams

4.1 Van Der Waals Theory

4.1.1 The Van Der Waals Equation of State

The van der Waals[1] theory assumes a molecular structure of matter, where matter means gases or liquids. The interaction between molecules requires modification of the ideal gas law:

$$\underbrace{P}_{P + a(n/V)^2} \underbrace{V}_{V-nb} = nRT,$$

i.e.

$$\boxed{P = \frac{nRT}{V-nb} - a\left(\frac{n}{V}\right)^2}. \tag{4.1}$$

Molecules, or atoms in the case of noble gases, at close proximity tend to repel. The attending volume reduction is $-nb$, where b is a parameter accounting for the exclusion of (molar) volume that a particle imposes on the other particles in V. At large distances particles attract, which in turn reduces the pressure. The particular form of this pressure correction, i.e. $a(n/V)^2$, may be motivated as follows. The number of particle pairs in a system consisting of N particles is $N(N-1)/2 \approx N^2/2$. Expressed in moles this leads to the factor n^2. The attraction is limited to "not too large" particle-to-particle separation. We assume that two particles feel attracted if they are in the same volume element ΔV. The probability that two particular particles are found within ΔV simultaneously is proportional to $(\Delta V/V)^2$. Assuming this to be true for all possible pairs leads to an overall number

---

[1] Johannes Diderik van der Waals, Nobel prize in physics for his work on the phase behavior of gases and liquids, 1910.

© Springer Nature Switzerland AG 2022   149
R. Hentschke, *Thermodynamics*, Undergraduate Lecture Notes in Physics,
https://doi.org/10.1007/978-3-030-93879-6_4

of attracted molecules proportional to $(n/V)^2$. The resulting Eq. (4.1) is the van der Waals equation of state for gases and liquids. The (positive) parameters a and b are characteristic for the specific material. The van der Waals equation of state is by no means accurate, but its combination of simplicity and utility is outstanding.

The parameters a and b may be estimated by measuring the pressure as function of temperature at low densities. The result may then be approximated using the following low density expansion of Eq. (4.1):

$$P = RT \sum_{i=1}^{\infty} B_i(T; a, b) \left(\frac{n}{V}\right)^i, \tag{4.2}$$

where

$$B_1(T; a, b) = 1 \tag{4.3}$$

$$B_2(T; a, b) = b - \frac{a}{RT} \tag{4.4}$$

$$B_3(T; a, b) = b^2 \tag{4.5}$$
$$\vdots$$

are so called virial coefficients. In practice one determines $B_2^{(exp)}(T)$ by fitting low order polynomials to experimental pressure isotherms at low densities. The resulting $B_2^{(exp)}(T)$ is then plotted versus temperature. Now a and b may be obtained by fitting Eq. (4.4) to these data points.

If we introduce the following reduced quantities, p, t, and v, via

$$P = P_c p \tag{4.6}$$

$$T = T_c t \tag{4.7}$$

$$V = V_c v, \tag{4.8}$$

where

$$P_c = \frac{1}{27} \frac{a}{b^2} \tag{4.9}$$

$$RT_c = \frac{8}{27} \frac{a}{b} \tag{4.10}$$

$$V_c = 3nb, \tag{4.11}$$

we may rewrite the van der Waals Eq. (4.1) into

$$p = \frac{8t}{3v - 1} - \frac{3}{v^2}. \tag{4.12}$$

This is the so called universal van der Waals equation. It is universal in the sense that it does no longer depend on the material parameters a and b. Notice that the ideal gas law in these units is

$$p_{id.gas} = \frac{8t}{3v}. \tag{4.13}$$

4.1.2   Gas-Liquid Phase Transition

The upper portion of Fig. 4.1 shows plots of the universal van der Waals equation for three different values of t (solid lines). Of course Eq. (4.12) always deviates from the ideal gas law at low v. In fact we did not plot the pressure for v-values below the singularity at $v = 1/3$, because there the molecules overlap. But we also notice that the universal van der Waals equation exhibits strange behavior if $t < 1.0$. There is a v-range in which the pressure rises even though the volume increases. Here we find an isothermal compressibility $\kappa_T < 0$—in clear violation of the mechanical stability condition in (3.16)! Had we plotted Eq. (4.12) for even smaller t-values, we would have obtained negative pressures in addition. All in all, for certain v and t, the van der Waals equation does describe states which cannot be equilibrium states. It turns out however that we can fix this problem, and at the same time we may describe a new phenomenon—the phase transformation between gas and liquid.

To understand how the model may be fixed we look at the free energy obtained via integration of the pressure

$$F(V) = F_o - \int_{V_o}^{V} dV P \tag{4.14}$$

(cf. Eq. (2.109). The bottom part of Fig. 4.1 shows the result obtained using the universal van der Waals equation (with $F_o = 3$) for the same three temperatures as above. The $t = 0.9$-curve, which violates mechanical stability according to the attendant pressure isotherm, is sketched in somewhat exaggerated fashion in Fig. 4.2.

We notice that the system represented by the filled black circle may lower its free energy by decomposing into regions in which the free energy is f_l or f_g. In between the free energy is

Fig. 4.1 Van der Waals
pressure and free energy
versus volume at three
different temperatures

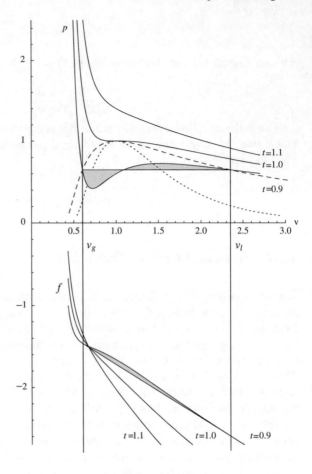

Fig. 4.2 Reduction of the
van der Waals free energy via
phase separation below t = 1

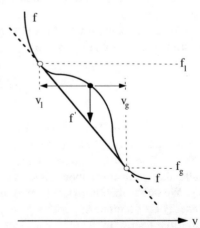

$$f' = f_l x + f_g (1 - x) \quad \text{with} \quad x = \frac{v - v_g}{v_l - v_g}. \tag{4.15}$$

Notice that this is the lowest free energy the system can achieve via decomposition into regions with high density, denoted v_l, and regions with low density, denoted v_g. The respective volume fractions of the two different regions are assigned by the parameter x (note: $v = x v_l + (1 - x) v_g$) according to the value of v. In other words, for volumes $v_l < v < v_g$ the homogeneous system is unstable relative to the decomposed or inhomogeneous system.

Imagine we move along an isotherm $t < 1$ starting from a large volume $v > v_g$. We are in a homogeneous so called gas phase. Upon decreasing v we are entering the range $v_g > v > v_l$. Here, depending on the value of v, we observe a "mixture" of regions having a homogeneous density n/v_g or n/v_l. As we approach v_l the volume fraction of the latter regions increases to unity. If $v_l \geq v$ we are again inside a homogeneous system—the liquid phase. This augmented van der Waals theory therefore predicts a phase change from gas to liquid and vice versa.

Notice also that for $v_g \geq v \geq v_l$

$$p' = -\frac{\partial f'}{\partial v}\bigg|_T = -\frac{f_g - f_l}{v_g - v_l} = \text{constant}$$

(the straight line in Fig. 4.2 is the common tangent to f at v_l and v_g) and

$$f_g - f_l = p' v_l - p' v_g \quad \text{or} \quad \underbrace{f_g + p' v_g}_{=n\mu_g} = \underbrace{f_l + p' v_l}_{=n\mu_l},$$

which means that mechanical and chemical stability are satisfied.

In turn we may calculate v_l and v_g via the conditions $p(t, v_l) = p(t, v_g)$ and $\mu(t, v_l) = \mu(t, v_g)$ based on the universal van der Waals equation itself, i.e.

$$p(t, v_l) = p(t, v_g) \tag{4.16}$$

and

$$-\int_{v_o}^{v_l} dv \, p(t, v) + p(t, v_l) v_l = -\int_{v_o}^{v_g} dv \, p(t, v) + p(t, v_g) v_g \tag{4.17}$$

using $n\mu = f + pv$. The numerically obtained values $v_l(t)$ and $v_g(t)$ are shown as a dashed line, the binodal line, in the upper part of Fig. 4.1. The area beneath the binodal line is the gas-liquid coexistence region. Notice that no solutions exist if $t > 1$, i.e. no gas-liquid phase transition is encountered above $t = 1$. In addition to the binodal line there is a dotted line, the spinodal line, which indicates the (mechanical) stability limit. This means that the isothermal compressibility, κ_T, is negative below this line.

Fig. 4.3 Pressure versus chemical potential below t = 1

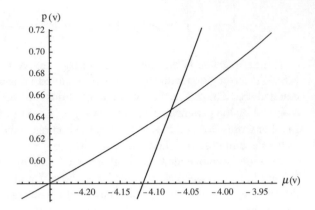

We remark that, among other methods, the simultaneous numerical solution of Eqs. (4.16) and (4.17) may be programmed as "graphical" search for the intersection of pressure and chemical potential in a plot of $\mu(v)$ versus $p(v)$. An example is shown in Fig. 4.3 for $t = 0.9$.

The largest t-value for which a solution is obtained is $t = 1$. Here one finds $v_l = v_g$. The corresponding values of the unreduced pressure, temperature, and volume are P_c, T_c, and V_c given by the Eqs. (4.6)–(4.8). We can find this so called critical point directly by simultaneous solution of $P_c = P(T_c, V_c)$, $dP/dV|_{T_c,V_c} = 0$, and $d^2P/dV^2|_{T_c,V_c} = 0$ (the second and third equations are due to the constant pressure in the coexistence region).

We may rewrite Eq. (4.17) using $p'(t) = p(t, v_l) = p(t, v_g)$ as

$$\int_{v_l}^{v_g} dv\, p(t, v) - p'(t)(v_g - v_l) = 0. \tag{4.18}$$

The left side of this equation is the sum of the two shaded areas, one positive and one negative, in the upper graph in Fig. 4.1. The equation states that the two areas between the van der Waals pressure $p(t, v)$, and the constant pressure, $p'(t)$, which replaces it between v_g and v_l, are equal. Therefore v_g and v_l may also be found graphically via this equal area or Maxwell construction . We remark that between v_g and v_l the van der Waals pressure isotherms are said to exhibit a van der Waals loop.

Figure 4.4 shows the possibly simplest of all phase diagrams in the t-v- and in the t-p-plane. The solid line in the upper diagram is the same as the dashed line, i.e. the phase coexistence curve, in Fig. 4.1. The dotted line is the spinodal. The lower graph shows the phase boundary between gas and liquid in the pressure-temperature plane. Notice that here no coexistence region appears because the pressure is constant throughout this region (at constant t). The crosses are vapor pressure data for water taken from HCP.

The next figure, Fig. 4.5, shows three isobars above, at, and below the critical pressure.

Fig. 4.4 Phase diagrams in the t-v- and in the t-p-plane

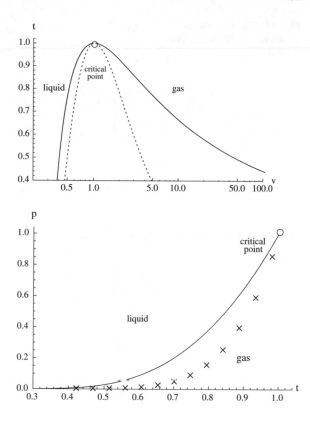

Fig. 4.5 Isobars in the vicinity of the critical pressure

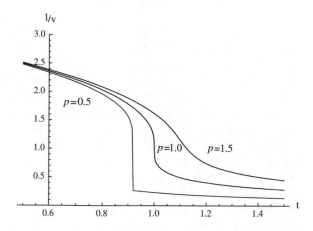

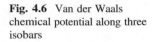

Fig. 4.6 Van der Waals chemical potential along three isobars

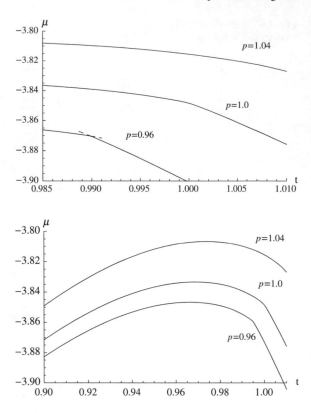

Figure 4.6 shows the van der Waals chemical potential along three isobars close to the critical point (top) and over an expanded temperature range (bottom), i.e. we compute μ along three horizontal lines in the lower panel of Fig. 4.4 just below, at, and just above the critical pressure. For $p = 0.96$ we cut across the gas-liquid phase transition. Notice that the dashed lines indicate the continuation of the liquid (low t) and gas (high t) chemical potentials into their respective metastable region, i.e. the region between spinodal and binodal line. This justifies our sketches of the chemical potential in Fig. 3.22 and in principle also in Fig. 3.23. At $p = 1$ the chemical potential still exhibits a kink, whereas above its slope changes smoothly.

The general quality of the van der Waals equation is nicely demonstrated in Fig. 4.7.[2] The figure shows coexistence data for seven different substances plotted in units of their critical parameters. The data, in almost all cases, indeed fall onto a universal curve. This behavior is called *law of corresponding states*. The universal van der Waals equation certainly is not an exact description, but considering its simplicity the agreement with the experimental data is quite remarkable!

---

[2] The data shown here are taken from the book by Stanley (1971); the original source is Guggenheim (1945).

Fig. 4.7 Law of corresponding states demonstrated for various compounds in comparison to the van der Waals prediction

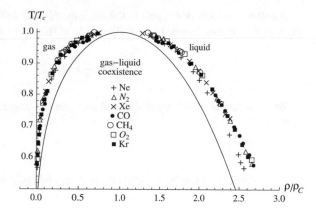

Fig. 4.8 Reduced second virial coefficient according to the van der Waals equation in comparison to experimental data

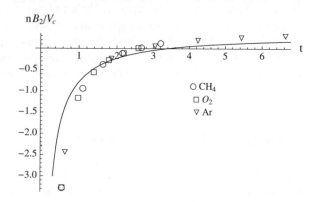

We briefly return to the second virial coefficient, $B_2(T; a, b)$, in Eq. (4.4). Figure 4.8 illustrates the comparison between the van der Waals prediction (solid line), i.e.

$$\frac{nB_2}{V_c} = \frac{1}{3}\left(1 - \frac{27}{8t}\right),\tag{4.19}$$

and experimental data from HTTD (Appendix C). The agreement with the experimental data is qualitatively correct. We note however, that the form of B_3 in Eq. (4.5) is an oversimplification. The third virial coefficient, B_3, is not independent of temperature as this equation suggests. Notice that $B_2(T) = 0$ defines the so called Boyle temperature, T_{Bolyle}. According to the van der Waals theory

$$T_{Boyle} = \frac{27}{8}T_c.\tag{4.20}$$

Another quantity of interest is the compressibility factor at criticality, for which the van der Waals theory predicts that it is a simple universal number:

$$\frac{P_c V_c}{nRT_c} = \frac{3}{8} = 0.375. \tag{4.21}$$

In the cases of argon, methane, and oxygen the experimental values are close to 0.29.

4.1.3 Other Results of the Van Der Waals Theory

The gas-liquid phase behavior described by the van der Waals theory is considered "simple". In this sense it is a reference distinguishing "simple" from "complex". We emphasize that this refers to the qualitative description rather than to the quantitative prediction of fluid properties. Other phenomenological equations of state may be better in this respect, but it is the physical insight which here is important to us. Because of this we compute a number of other thermophysical quantities in terms of t, p, and/or v.

Isobaric Thermal Expansion Coefficient

Figure 4.9 shows the temperature dependence of the isobaric thermal expansion coefficient, α_p, below, at, and above the critical pressure. The dashed line is the ideal gas result. Below the critical pressure a jump occurs when the gas-liquid saturation line (cf. Fig. 4.4; right panel) is crossed. At the critical point we observe a divergence. Above the critical point a maximum marks the smooth "continuation" of the gas-liquid saturation line, which sometimes is called Widom line.

Fig. 4.9 Temperature dependence of the isobaric thermal expansion coefficient for different pressures

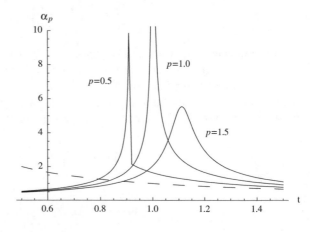

Fig. 4.10 Volume
dependence of the isothermal
compressibility in the vicinity
of the critical temperature

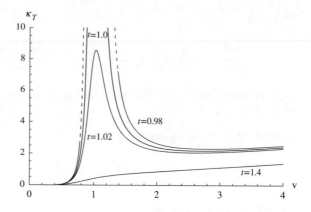

Isothermal Compressibility

Figure 4.10 shows the volume dependence of the isothermal compressibility, κ_T,
below, at, and above the critical temperature. The dashed lines are the continuation
of κ_T into the metastable region. Notice that κ_T diverges when $\partial p/\partial v|_T = 0$. This
condition defines the stability limit, $\kappa_T \geq 0$, i.e. inside the gap between the dashed
lines the van der Waals equation gives negative κ_T. Notice that κ_T also diverges at
the critical temperature. Above the critical temperature κ_T exhibits a maximum near
the critical volume, which diminishes as the temperature increases. In general the
compressibility increases as v increases—the gas is less dense and easier to
compress.

Isochoric Heat Capacity

Having discussed α_P and κ_T the next obvious function to look at is the isochoric
heat capacity, C_V. It turns out that within the van der Waals theory all we obtain is

$$C_V^{vdW} = C_V^{vdW}(T),\tag{4.22}$$

i.e. C_V^{vdW} is a function of temperature only.

We show this via

$$\underbrace{\frac{\partial}{\partial T}\frac{\partial F}{\partial V}\Big|_T\Big|_V}_{=-\frac{\partial p}{\partial T}\big|_V} = \underbrace{\frac{\partial}{\partial V}\frac{\partial F}{\partial T}\Big|_V\Big|_T}_{=-\frac{\partial S}{\partial V}\big|_T}.$$

Therefore

$$\frac{\partial^2 P}{\partial T^2}\Big|_V = \frac{\partial}{\partial T}\frac{\partial S}{\partial V}\Big|_T\Big|_V = \frac{\partial}{\partial V}\frac{\partial S}{\partial T}\Big|_V\Big|_T \overset{(2.158)}{=} \frac{\partial}{\partial V}\frac{C_V}{T}\Big|_T = \frac{1}{T}\frac{\partial C_V}{\partial V}\Big|_T.$$

This proves Eq. (4.22), because according to the van der Waals equation $\partial^2 P/\partial T^2|_V = 0$.

Remark The above equation, i.e.

$$\frac{\partial C_V}{\partial V}\Big|_T = T\frac{\partial^2 P}{\partial T^2}\Big|_V, \tag{4.23}$$

may be integrated to yield

$$C_V - C_{V,ideal} \approx -nR\left[2T\frac{\partial B_2(T)}{\partial T} + T^2\frac{\partial^2 B_2(T)}{\partial T^2}\right]\frac{n}{V}, \tag{4.24}$$

which is a low density approximation to C_V. Of course this correction vanishes if the second virial coefficient, $B_2(T)$, of the van der Waals theory (cf. Eq. (4.4)) is used.

Inversion Temperature

In Sect. 2.2 we had discussed the Joule-Thomson coefficient. Based on the universal van der Waals Eq. (4.12) we want to calculate the inversion line in the t-p-plane. Eq. (2.106) is inconvenient, because we have to express v in terms of t and p. However, using Eq. (A.2) we may write

$$\frac{\partial v}{\partial t}\Big|_p = -\frac{\partial v}{\partial p}\Big|_t\frac{\partial p}{\partial t}\Big|_v = -\frac{\frac{\partial p}{\partial t}\Big|_v}{\frac{\partial p}{\partial v}\Big|_t}.$$

The inversion line now is the solution of

$$0 = t\frac{\partial p}{\partial t}\Big|_v\Big/\frac{\partial p}{\partial v}\Big|_t + v,$$

which we may find analytically:

$$p = 12(\sqrt{t} - \sqrt{3/4})(\sqrt{27/4} - \sqrt{t}). \tag{4.25}$$

The inversion temperatures at $p = 0$ therefore are $t_{min} = 3/4$ and $t_{max} = 27/4$. Equation (4.25) is shown in Fig. 4.11. The curve encloses the area in the t-p-plane, where the Joule-Thomson coefficient, μ_{JT}, is positive (cooling). Outside this area

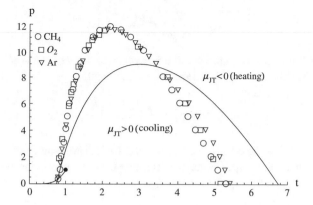

Fig. 4.11 Inversion temperature according to Eq. (4.25) including experimental data

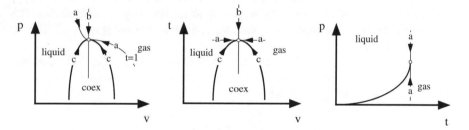

Fig. 4.12 Different paths approaching the gas-liquid critical point

μ_{JT} is negative corresponding to heating. Experimental data for methane, oxygen, and argon from Fig. 4.12 in Hendricks et al. (1972) *Joule-Thomson inversion curves and related coefficients for several simple fluids* NASA Technical Note D-6807 are included for comparison. Qualitatively the van der Waals predictions are correct. But the quantitative quality is quite poor for *t-p*-conditions far from the critical point (the gas-liquid saturation line (cf. Fig. 4.4; right panel) terminating in the critical point (circle) is included), which we use to tie our theory to reality.

It is interesting to work out $\partial v / \partial t |_p$ based on the virial expansion (4.2), because this allows a better understanding of the Joule-Thomson effect on the basis of molecular interaction. To leading order we find

$$\frac{\partial v}{\partial t}\bigg|_p = \frac{v}{t} + t\frac{\partial(b_2/t)}{\partial t}, \tag{4.26}$$

where $b_2 = nB_2(T)/V_c$. Consequently the Joule-Thomson coefficient in this approximation is

$$\mu_{JT} = \frac{V_c}{C_P} \left(t^2 \frac{\partial (b_2/t)}{\partial t} \right). \tag{4.27}$$

This equation is general, i.e. we have not yet used the van der Waals equation of state. If we do this, i.e. we insert Eqs. (4.4), the result is

$$V_c^{-1} C_P \mu_{JT} = \frac{1}{3} \left(\frac{27}{4t} - 1 \right). \tag{4.28}$$

We recognize that the equation describes the Joule-Thomson coefficient at $P = 0$ (or small P) near the upper inversion temperature. In particular we verify that $\mu_{JT} > 0$ for $t < t_{max}$ and $\mu_{JT} < 0$ for $t > t_{max}$.

Van Der Waals Critical Exponents
Close to the gas-liquid critical point one can show that the quantities $\pm \delta p$, $\pm \delta v$, and $\pm \delta t$, which are small deviations from the critical point in terms of the variables p, v, and t, are simply related to each other as well as to thermodynamic functions like the isobaric thermal expansion coefficient, α_p, and the isothermal compressibility, κ_t (as well as all others!).[3]
Again we focus on the universal van der Waals equation, i.e. Eq. (4.12). Along the critical isotherm, path (a) in Fig. 4.12, we set $p = 1 + \delta p$ and $v = 1 - \delta v$. Inserting this into (4.12) we obtain

$$\delta p = -\frac{3}{2} \delta v^3 + \mathcal{O}(\delta v^4).$$

Ignoring constant factors and additional terms, like the corrections to the leading behavior, we write instead

$$\delta p \sim \pm \delta v^3 \quad \text{where} \quad \pm : v = 1 \mp \delta v. \tag{4.29}$$

Approaching the critical point from above, $t = 1 + \delta t$, along the critical isochor, $v = 1$, i.e. path (b) in Fig. 4.12, we find

$$\delta p \sim \delta t. \tag{4.30}$$

Another special line is the coexistence curve, shown as the dashed line in Fig. 4.1. On the sketches in Fig. 4.12 the path along the coexistence curve is labeled (c). We insert $t = 1 - \delta t$ and $v = 1 \pm \delta v$ into the universal van der Waals equation

---

[3] The small quantities δp, δv, and δt are all positive.

and expand the result in powers of δv. Finally we assume $\delta v \equiv \delta v_\pm = c_\pm \delta t^\beta$.[4] Note that δt and δv are positive. The result is

$$p_\pm = 1 - 4\delta t \mp \frac{3}{2}c^3 \pm \delta t^{3\beta} \pm 6c_\pm \delta t^{1+\beta} + \ldots.$$

Here ... stands for higher order terms $\mathcal{O}(\delta t^{4\beta})$ and $\mathcal{O}(\delta t^{2+\beta})$, and $+$ and $-$ stand for the gas and the liquid side of the coexistence curve respectively. The stability condition (4.16) requires $p_- = p_+$. Setting $c_+ = -c_-$ fulfills this equality but does not yield the desired result because $v_- = v_+$. The only other solution requires $3\beta = 1 + \beta$ and $\mp \frac{3}{2}c_\pm^3 \pm 6c_\pm = 0$. Consequently we obtain

$$\beta = \frac{1}{2} \quad \text{and} \quad c_\pm = 2, \tag{4.31}$$

i.e. the leading relation between δv and δt along the coexistence curve is

$$\delta v \sim \delta t^{1/2}. \tag{4.32}$$

We may use this to work out the dependence of the isothermal compressibility, $\kappa_t = -(1/v)\partial v/\partial p|_t$, near the critical point and along the coexistence curve. Again we insert $t = 1 - \delta t$ and $v = 1 \pm \delta v$ into the universal van der Waals equation and expand the result in powers of δv. We then work out the derivative $\partial p/\partial v|_t$ and insert the above result (4.32). This yields

$$\kappa_t^{-1} \sim \delta t. \tag{4.33}$$

Using the general thermodynamic relation $\frac{\partial v}{\partial t}\big|_p = -\frac{\partial p}{\partial t}\big|_v / \frac{\partial p}{\partial v}\big|_t$ (cf. Eq. (A.3)) we obtain the δt-dependence of the isobaric expansion coefficient, $\alpha_p = (1/v)\partial v/\partial t|_p$, near the critical temperature and also along the coexistence curve. Because $\partial p/\partial t|_v$ to leading order contributes a constant only, we obtain, as above for κ_t,

$$\alpha_p \sim \delta t^{-1}. \tag{4.34}$$

We return briefly to the critical isochore, $v = 1$, and compute the δt-dependence of κ_t^{-1}, when we approach the critical point via this path. Working out $\partial p/\partial v|_t$ and setting $v = 1$ as well as $t = 1 + \delta t$ yields

$$\kappa_t^{-1} \sim \delta t \tag{4.35}$$

---

[4] This is the leading term in a power series expansion of δv in δt. Notice that the coexistence curve is not symmetric with respect to reflection across the critical isochore—except very close to the critical point (cf. below).

Table 4.1 Selected critical exponents and their vdW-values

Exponent	Definition	Conditions	vdW-value
α'	$C_V \sim \delta T^{\alpha'}$	$V = V_c \ (T < T_c)$	0
α	$C_V \sim \delta T^{\alpha}$	$V = V_c \ (T > T_c)$	0
β	$\rho_l - \rho_g \sim \delta T^{\beta}$	coex curve	$\frac{1}{2}$
γ	$\kappa_T^{-1} \sim \delta T^{\gamma}$	coex curve $(T < T_c)$	1
γ'	$\kappa_T^{-1} \sim \delta T^{\gamma'}$	$V = V_c \ (T > T_c)$	1
δ	$\delta P \sim \pm (\rho_l - \rho_g)^{\delta}$	$T = T_c \left(\begin{array}{l} + \ : \rho > \rho_c \\ - \ : \rho < \rho_c \end{array} \right)$	3

as before. Only the prefactors (called scaling amplitudes) are different, i.e. $\kappa_t^{-1} \approx 12\delta t$ on the path along the coexistence curve and $\kappa_t^{-1} \approx 6\delta t$ along the isochore. Again we find the relation (4.34) for the same reason as before, i.e. $\partial p / \partial t \big|_v$ to leading order contributes a constant only.

The reader may want to work out the divergences of α_p in Fig. 4.9 and κ_t in Fig. 4.10 to leading order in δt and δv, respectively. The result is $\alpha_p \sim \delta t^{-2/3}$ along the critical isobar and $\kappa_t \sim \delta v^{-2}$ along the critical isotherm.

The exponents in the above power laws are called critical exponents. Table 4.1 compiles a selected number of them together with their definition, thermodynamic conditions, and van der Waals values. Here $\rho_l - \rho_g$ is the density difference across the coexistence curve.[5] This quantity is called order parameter. Notice that by construction the order parameter vanishes above T_c. In addition, $\delta T = |T - T_c|$ and $\delta P = |P - P_c|$. Notice also that we have not yet talked about the heat capacity exponent α.[6] The prime indicates the same critical exponent below T_c. The van der Waals theory yields the same values for the two exponents listed here, i.e. $\alpha = \alpha'$ and $\gamma = \gamma'$. But the van der Waals values are not correct! Even though the correct exponent values turn out to be nevertheless the same below and above T_c, we again adhere to the (safe) standard notation, which distinguishes the two conditions. α_P does not appear in this list, because, as we have seen, its exponents are also γ and γ' at the indicated conditions.

---

[5] What is the relation between $\rho_l - \rho_g$ and $\pm \delta v$? Setting $\rho_l = \rho_c + \delta \rho_l$ and $\rho_g = \rho_c + \delta \rho_g$ and using $|\delta \rho / \rho| = |\delta v / v|$ yields $\rho_l - \rho_g \propto \delta v_- + \delta v_+$.

[6] The present critical exponent notation is standard. We want to adhere to it, even though certain letters are used for other quantities also.

4.2 Beyond Van Der Waals Theory

Thermodynamic Scaling

The preceding discussion of van der Waals critical exponents has highlighted the power law-relations connecting thermodynamic functions close to the gas-liquid critical point. This led to the idea to express the latter as generalized homogeneous functions.[7] That is if $f = f(x, y)$ is a generalized homogeneous function then

$$f(x, y) = \lambda^{-1} f(\lambda^p x, \lambda^q y), \tag{4.36}$$

where p and q are parameters. This is best explained via a simple example. We choose

$$f(x, y) = \frac{A}{x^2} + By^3,$$

where A and B are constants. Applying the right side of Eq. (4.36) yields

$$\lambda^{-1} f(\lambda^p x, \lambda^q y) = \frac{A}{\lambda^{2p+1} x^2} + B\lambda^{3q-1} y^3 \underset{=}{\scriptstyle p=-1/2, q=1/3} \frac{A}{x^2} + By^3 = f(x, y).$$

This also works for $f(x, y) = Ax^{-2}y^3$. However it does not work for

$$f(x, y) = \frac{A}{1 + x^2} + By^3,$$

because

$$\lambda^{-1} f(\lambda^p x, \lambda^q y) = \frac{A}{\lambda + \lambda^{2p+1} x^2} + B\lambda^{3q-1} y^3 \underset{=}{\scriptstyle p=-1/2, q=1/3} \frac{A}{\lambda + x^2} + By^3.$$

Therefore we do not expect that this idea applies to thermodynamic functions in general, except close to the critical point, where they may be expressed in terms of powers of δT and δP.

But what can we learn from Eq. (4.36)? We apply this equation to the free energy, i.e.

$$f_\infty(\delta t, \delta v) = \lambda^{-1} f_\infty(\lambda^p \delta t, \lambda^q \delta v). \tag{4.37}$$

---

[7] See Widom (1965).

The index ∞ is a reminder that we stay close to the critical point and we consider the leading part of the free energy in the sense discussed above. The exponent p should not be confused with the reduced pressure used in the universal van der Waals equation. With the particular choices (a) $\lambda = \delta v^{-1/q}$ and (b) $\lambda = \delta t^{-1/p}$ we transform Eq. (4.37) into

$$
\begin{aligned}
(a): \quad & f_\infty(\delta t, \delta v) = \delta v^{1/q} f_\infty(\delta v^{-p/q} \delta t, 1) \\
(b): \quad & f_\infty(\delta t, \delta v) = \delta t^{1/p} f_\infty(1, \delta t^{-q/p} \delta v).
\end{aligned}
\tag{4.38}
$$

We want to use this to work out $C_V = -T \partial^2 F / \partial T^2 |_V$ and $\kappa_T^{-1} = -V \partial^2 F / \partial V^2 |_T$, i.e.

$$
\begin{aligned}
\text{using (a):} \quad & C_V = \delta v^{1/q - 2p/q} \tilde{C}_V(\delta v^{-p/q} \delta t, 1) \\
\text{using (b):} \quad & \kappa_T^{-1} = \delta t^{1/p - 2q/p} \tilde{\kappa}_T(1, \delta t^{-q/p} \delta v).
\end{aligned}
\tag{4.39}
$$

Along the coexistence curve we had defined the exponent β via $\delta v \sim \delta t^\beta$. We use this to eliminate δv from the Eqs. (4.39). Together with $C_V \sim \delta t^{-\alpha}$ and $\kappa_T \sim \delta t^{-\gamma}$ we obtain the following equations relating the exponents:

$$
-\alpha = \beta\left(\frac{1}{q} - 2\frac{p}{q}\right), \quad -\gamma = -\frac{1}{p} + 2\frac{q}{p}, \quad \beta = \frac{q}{p}.
\tag{4.40}
$$

The third equation ensures that the scaling functions $\tilde{C}_V$ and $\tilde{\kappa}_T$ are "well behaved" at the critical point, i.e. $\delta v^{-p/q} \delta t$ approaches a constant at the critical point. We may eliminate p and q from these equations, which yields the critical exponent relation

$$
\alpha + 2\beta + \gamma = 2.
\tag{4.41}
$$

The van der Waals values from Table 4.1 obviously satisfies this relation. Conversely we can use this relation to justify $\alpha = 0$!

Figure 4.13 shows C_V measured for sulfurhexafluoride (SF_6) along the critical isochore copied with permission from Haupt and Straub (1999) (SF_6: $T_c = 318.7$ K, $P_c = 37.6$ bar, $V_c = 200$ cm^3/mol).[8] The value of the critical exponent α determined from these data is $\alpha = 0.1105^{+0.025}_{-0.027}$. The currently accepted theoretical value is $\alpha = 0.110 \pm 0.003$ (Sengers and Shanks 2009). For β and γ the accepted theoretical values are $\beta = 0.326 \pm 0.002$ and $\gamma = 1.239 \pm 0.002$ in agreement with experiments and with the exponent relation (4.41). Note that these exponents do not depend on the molecular details of the fluid systems. Theoretical arguments show that the critical exponent values depend on space dimension, (order parameter)

---

[8] On the critical isochore we have $f_\infty(\delta t, \delta v = 0) = \delta t^{1/p} f_\infty(1, 0)$ and therefore $C_V \sim \delta t^{1/p - 2} \sim \delta t^{-\alpha}$. That is the exponent is the same as on the coexistence curve.

Fig. 4.13 Temperature dependence of the isochoric heat capacity of sulfurhexafluoride near the critical point

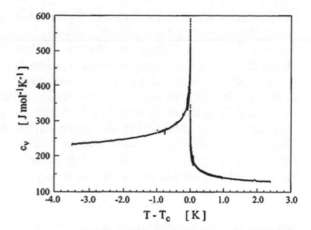

symmetry, and range of interaction ("short" versus "long") but not on the details of molecular interaction. This allows to define so called universality classes of fluids (or near critical physical systems in general) with identical exponent values.[9]

It is easy to derive a critical exponent relation involving the exponent δ. Applying $P = -\partial F/\partial V|_T$ to (b) in (4.38) yields

$$P = \delta t^{1/p-q/p}\tilde{P}(1, \delta t^{-q/p}\delta v). \quad (4.42)$$

Again with $\delta v \sim \delta t^\beta$ along the coexistence curve we find $\beta\delta = 1/p - q/p$ or

$$\gamma = \beta(\delta - 1). \quad (4.43)$$

And again this relation is fulfilled by the van der Waals exponent values in Table 4.1. It is worth emphasizing that exponent relations like (4.41) and (4.43) and others mainly serve to unify the picture, i.e. the number of independent critical exponents is greatly reduced.

Already we have mentioned that the van der Waals exponents are incorrect. This incorrectness is not "just" due to the modest quantitative predictive power of the van der Waals approach. Here the latter misses the underlying physical picture completely.

Every thermodynamic quantity fluctuates around an average value. Usually the fluctuations can be ignored entirely. This is what the van der Waals model does too. However, close to the critical point fluctuations become increasingly important and dominate over the average values.[10] There are many models which in this respect are

---

[9] It is a hypothesis that all fluids belong to one and the same universality class. A discussion of this hypothesis may be found in the above article by Sengers and Shanks.

[10] More precisely what happens is that the local fluctuations influence each other over large distances. These distances are measured in terms of the fluctuation correlation length which diverges at the critical point.

like the van der Waals model (Kadanoff 2009). When these models describe critical points, they all yield the same critical exponents—the so called *mean fieldcritical exponents*.[11] This is striking, because the models look quite different indeed. Nevertheless, near their respective critical points they all posses the same "symmetry". Only after a method, the so called *renormalization group*, was invented how to properly deal with the dominating fluctuations in the critical region, was it possible to actually calculate the correct values for the critical exponents—and they still obey the above relations together with others derived via thermodynamic scaling.[12] This is because of the great generality of the thermodynamic laws and, in addition, the fact that the power law behavior of thermodynamic functions near criticality turns out to be an integral part of the new theory as well.

4.2.1   The Clapeyron Equation

Along the gas-liquid transition line in Fig. 4.14 we always have at "1" and "2"

$$\mu_l(1) = \mu_g(1) \quad \text{and} \quad \mu_l(2) = \mu_g(2) \quad \text{or} \quad \mu_l(2) - \mu_l(1) = \mu_g(2) - \mu_g(1).$$

If "1" and "2" are infinitesimally close we may write

$$d\mu_l = -s_l dT + v_l dP = -s_g dT + v_g dP = d\mu_g. \tag{4.44}$$

Here lower case letters indicate molar quantities. Consequently we find

$$\left.\frac{dP}{dT}\right|_{coex} = \frac{s_g - s_l}{v_g - v_l}, \tag{4.45}$$

where dP/dT is the slope of the gas-liquid transition line in Fig. 4.14. This equation is more general of course, because it applies not only to the transition from gas to liquid and vice versa but to the transition between any two phases we choose to call I and II. We remark that a transition with a non-zero latent heat, i.e. $T\Delta s \neq 0$, we call a first order phase transition.[13] Thus we have

$$\left.\frac{dP}{dT}\right|_{coex} = \frac{s_{II} - s_I}{v_{II} - v_I} \overset{(2.119),(2.177)}{=} \frac{1}{T}\frac{h_{II} - h_I}{v_{II} - v_I}. \tag{4.46}$$

---

[11] In models for magnetic systems δP is replaced by the corresponding magnetic field variable and δv is replaced by the magnetization. The compressibility is therefore replaced by the magnetic susceptibility.

[12] A nice reference including historical developments is Fisher (1998).

[13] Transitions without such discontinuity, e.g. at the gas-liquid critical point, are called continuous or (generally) second order.

Fig. 4.14 Thermodynamic paths on either side of the saturation line

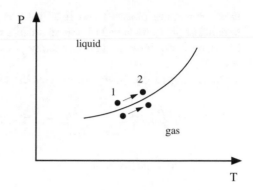

This is the Clapeyron equation.

Example—Enthalpy of Vaporization for Water Here we calculate the enthalpy of vaporization, $\Delta_{vap}h$, for water from the saturation pressure data shown in Fig. 4.4. Using Eq. (4.46) we may write

$$\left.\frac{dP}{dT}\right|_{coex} \approx \frac{1}{T}\frac{\Delta_{vap}h}{v_{gas}}. \tag{4.47}$$

Compared to the molar volume of the gas we may neglect the liquid volume. If in addition v_{gas} is expressed via the ideal gas law Eq. (4.47) becomes

$$d\ln P \approx \frac{\Delta_{vap}h}{R}\frac{dT}{T^2} \tag{4.48}$$

or, after integration,

$$\ln\frac{P}{P_o} \approx -\frac{\Delta_{vap}h}{R}\left(\frac{1}{T} - \frac{1}{T_o}\right). \tag{4.49}$$

Using the values $P = 4.246$ kPa at $T = 303.15$ K and $P = 0.6113$ kPa at $T = 273.15$ K from the aforementioned figure, we obtain $\Delta_{vap}h = 44.5$ kJ/mol in very good accord with $\Delta_{vap}h = 45.05$ kJ/mol at $T = 273.15$ K or $\Delta_{vap}h = 44.0$ kJ/mol at $T = 298.15$ K taken from HCP.

One important application of the Clapeyron equation is the following. Whereas the van der Waals theory only describes the transition between gas and liquid, we know that already a one component system may exhibit other phases—like the solid state. A sketch of the situation is shown in Fig. 4.15. There are transition lines (solid lines) separating phases I and II as well as II and III. The two lines may come together at (?) to form what is called a triple point.[14] How would this look like, i.e. how would we draw the line separating phases I and III? Can the dashed lines be correct?

According to Eq. (4.46) we have at the triple point

$$T_t \frac{dP}{dT} = \frac{\Delta h_{I \to II}}{\Delta v_{I \to II}} = \frac{\Delta h_{II \to III} + \Delta h_{III \to I}}{\Delta v_{II \to III} + \Delta v_{III \to I}} = \frac{\Delta h_{I \to III}}{\Delta v_{I \to III}} \left(\frac{1 - \frac{\Delta h_{II \to III}}{\Delta h_{I \to III}}}{1 - \frac{\Delta v_{II \to III}}{\Delta v_{I \to III}}} \right).$$

The second equality follows via

$$\Delta h_{I \to II} + \Delta h_{II \to III} + \Delta h_{III \to I} = 0$$
$$\Delta v_{I \to II} + \Delta v_{II \to III} + \Delta v_{III \to I} = 0,$$

corresponding to a path enclosing the triple point in infinitesimal proximity (Both functions are state functions!). Because, according to our assumption, the slopes of the coexistence lines I–II and I–III are identical, we must require $(1 - ...)/(1 - ...) = 1$ and thus

$$\frac{\Delta h_{I \to III}}{\Delta v_{I \to III}} = \frac{\Delta h_{II \to III}}{\Delta v_{II \to III}}.$$

This means, according to Eq. (4.46), that the slopes of the two solid lines in Fig. 4.15 should coincide close to the triple point. This is not a satisfactory result. We conclude that the slopes of all three lines must be different near the triple point.[15] This leaves us with the two alternatives depicted in Fig. 4.16. In alternative (a) the broken lines correspond to the continuation of the coexistence lines between phases I and II and phases I and III. In particular the shaded area is a region in which phase I is unstable with respect to II. On the other hand in the same area phase II is less stable than III. According to the solid lines, however, phase I is the most stable, which clearly is inconsistent. Thus we discard alternative (a). Alternative (b) does not suffer from this problem and is the correct one. We conclude that the continuation of the coexistence line between any two of the phases must lie inside the third phase.

With this information we may now sketch out the phase diagram of a simple one component system shown in Fig. 4.17. There are three projections of course. The T-P-projection is what we just have talked about. Here G means gas, F means liquid and K means solid. According to the van der Waals theory the gas-liquid coexistence line should terminate in a critical point (C). We do not posses any knowledge

---

[14] According to the phase rule this is the most complicated case in a one-component system.

[15] Note that this conclusion based on the Clapeyron equation does not hold in cases when there are transitions involved without discontinuities Δh or Δv.

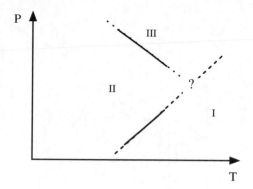

Fig. 4.15 Phase coexistence lines in the T-P-plane

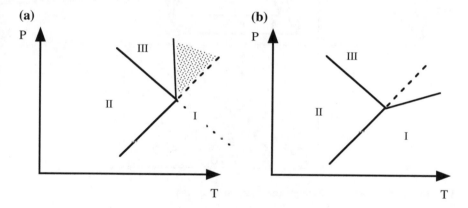

Fig. 4.16 Alternative phase diagrams near a triple point

of whether or not the liquid-solid line terminates similarly.[16] All three lines meet in the triple point. The remaining projections contain areas of phase coexistence due to the volume discontinuity at the transitions.

It is worth noting that even a one-component system exhibits a much more complicated phase diagram, i.e. here we always concentrate on partial phase diagrams. Fig. 4.18 shows an extended but still partial phase diagram for water (data sources: lower graph—HCP; upper graph—Martin Chaplin (http://www.lsbu.ac.uk/water/phase.html). Even though things already get complicated, the rules we have established thus far for phase diagrams are always satisfied. The special

---

[16] However, gas and liquid differ in no essential aspect of order or symmetry. This clearly sets them apart from the crystal. We can choose a path in the T-P-plane leading us from the gas to the liquid phase without crossing a phase boundary, i.e. very smoothly. Based on this concept we would not expect to find a liquid-solid critical point.

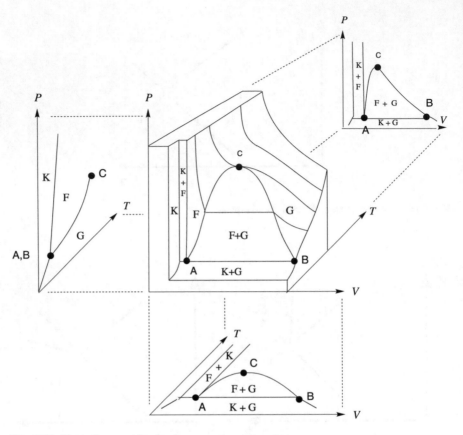

Fig. 4.17 Phase diagram of a simple one-component system

temperatures are the freezing temperature at 1 bar, T_f, the boiling temperature at 1 bar, T_b, the critical temperature, T_c (together with the critical pressure, P_c), as well as the triple point temperature, T_t (together with the triple point pressure, P_t). Roman numerals in the upper graph distinguish different high pressure ice phases.

> ***Example—Moist Air Parcel Lapse Rate*** Here we use the Clapeyron equation to generalize our calculation of the temperature profile of the troposphere in Sect. 2.2 to the case when the air parcel contains water vapor. Let's briefly recap the adiabatic expansion of a dry air parcel. According to energy conservation, i.e. the first law of thermodynamics, we have
>
> $$dE = \delta q + \delta w = -PdV. \qquad (4.50)$$
>
> Notice that $\delta q = 0$, i.e. there is no net exchange of heat between the air parcel and its environment, and the (reversible) volume work is given by $dw = -PdV$.

Fig. 4.18 Partial phase diagram for water

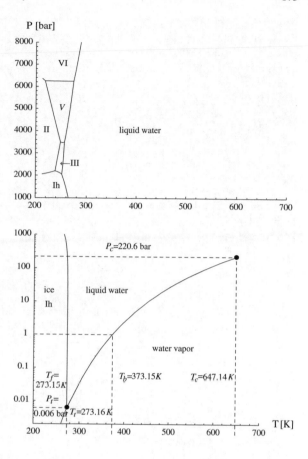

Fig. 4.19 Droplet growth through condensation according to Eq. (4.82). Solid line: full equation; dashed line: the growth when l is set to one.

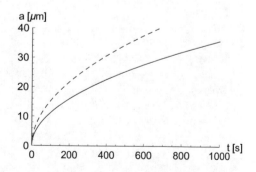

The internal energy $E = E(T, V)$ is a function of temperature, T, and volume, V. In this context, however, V is inconvenient and we decide to use the enthalpy, $H = H(T, P)$, where P, the pressure, is a more natural variable. Hence

$$dH = d(E + PV) = dE + PdV + VdP = VdP. \qquad (4.51)$$

In addition, since $H = H(T, P)$,

$$dH = C_P dT + \left. \frac{\partial H}{\partial P} \right|_T dP. \qquad (4.52)$$

Because in the following we shall exclusively work with ideal gases, i.e. the ideal gas law $PV = nRT$ applies, we have $\partial H / \partial P|_T = 0$. Equating the right hand sides of Eqs. (4.51) and (4.52) yields

$$\frac{dP}{dT} = \frac{C_P}{V}. \qquad (4.53)$$

If the air parcel is a cube of height dh we measure a pressure difference, dP, from bottom to top, which is due to the weight of the air parcel itself, i.e.

$$dP = -c \, g \, dh. \qquad (4.54)$$

Here $c = M/V = MP/(nRT)$ is the mass density of the air parcel, where M is the mass, V the volume, and n the number of moles of gas in the parcel. The quantity g is the acceleration of gravity. The combination of Eqs. (4.53) and (4.54) yields the so called dry lapse rate

$$\frac{dT}{dh} = -\frac{m_{mol} g}{c_P}, \qquad (4.55)$$

where m_{mol} and c_P are the molar mass and the molar isobaric heat capacity of the air parcel, respectively.

Let's make a quick estimate. Air mostly contains nitrogen and oxygen. Both are diatomic molecules possessing 3 translational degrees of freedom plus 2 rotational degrees of freedom. At the temperatures of interest here, these five degrees of freedom contribute $2.5 \, nR$, where R is the gas constant, to the heat capacity. Thus $c_P = (2.5 + 1)R \approx 29$ J/K. Notice that vibrations are frozen out. The molar mass of (dry air) is $m_{mol} = 0.21 \cdot 32 + 0.78 \cdot 28 \approx 0.029$ kg. This means that the dry lapse rate is roughly

$$\frac{dT}{dh} \approx -10 \text{ K/km}, \qquad (4.56)$$

Thus far we have not paid much attention to the moisture content of the air parcel. The mass of 1 m^3 of air at sea level is roughly 1 kg. At $T = 20\,°C$ and 40% humidity this much air contains about 7 g of water. Even though the specific heat capacity of water is about twice that of the above mixture of nitrogen and oxygen, the overall effect on the lapse rate is negligible.

Surprisingly, this becomes different the moment when the rising air parcel crosses the saturation line of water, i.e. the water partial pressure in the air, $P_{H_2O}(T)$, becomes equal to the saturation pressure at this temperature. Fig. 3.6 hows two dashed lines corresponding to adiabatic curves of air at two different humidities. We now want to know what happens to moist air expanding adiabatically, when the water vapor in the air crosses its saturation line—where of course it must condense. The equation that changes is Eq. (4.51), which now contains a second contribution, i.e.

$$dH = VdP - \Delta_{vap}h\, dn_{H_2O}. \tag{4.57}$$

The additional contribution is the enthalpy of condensation, which is the negative of the enthalpy of vaporization, $\Delta_{vap}h$, per mol multiplied by the differential molar amount of condensed water, dn_{H_2O}. Our task is to work out the differential

$$dn_{H_2O} = \left.\frac{\partial n_{H_2O}}{\partial T}\right|_P dT + \left.\frac{\partial n_{H_2O}}{\partial P}\right|_T dP. \tag{4.58}$$

We note that we can express the amount of water vapor in the parcel via

$$n_{H_2O}(T, P) = n\frac{P_{H_2O}(T)}{P}, \tag{4.59}$$

where n is the total number of moles of gas (in the parcel). Note again, P_{H_2O} is the water vapor partial pressure. Thus

$$\begin{aligned}
dn_{H_2O} &= \frac{n}{P}\left.\frac{\partial P_{H_2O}}{\partial T}\right|_P dT + nP_{H_2O}(T)\left.\frac{\partial 1/P}{\partial P}\right|_T dP \\
&= \frac{n}{P}\left.\frac{\partial P_{H_2O}}{\partial T}\right|_P dT - \frac{n_{H_2O}}{P} dP.
\end{aligned} \tag{4.60}$$

The derivative $\partial P_{H_2O}(T)/\partial T|_P$ is tricky. What is happening here? Well, the water vapor cannot exist across from the saturation line. On that side only liquid water is stable. If the water vapor wants to rise, it must do it following the saturation line. So instead of $\partial P_{H_2O}(T)/\partial T|_P$ we work out $\partial P_{H_2O}(T)/\partial T|_{coex}$. Since the total pressure does hardly change anyway, we do not expect trouble. Using the Clapeyron equation means

$$\left.\frac{\partial P_{H_2O}}{\partial T}\right|_{coex} = \frac{1}{T}\frac{\Delta_{vap}h}{\Delta v_{H_2O}}, \tag{4.61}$$

where $\Delta v_{H_2O} \approx RT/P_{H_2O}$. We therefore obtain

$$dn_{H_2O} = \frac{n_{H_2O}\Delta_{vap}h}{RT^2}dT - \frac{n_{H_2O}}{P}dP. \tag{4.62}$$

Inserting this result into Eq. (4.57) yields

$$dH = VdP - \frac{n_{H_2O}\Delta_{vap}h^2}{RT^2}dT + \frac{n_{H_2O}\Delta_{vap}h}{P}dP = C_P dT \tag{4.63}$$

and thus

$$\frac{dP}{dT} = \frac{C_P + \frac{n_{H_2O}(T,P)\Delta_{vap}h^2}{RT^2}}{V + \frac{n_{H_2O}(T,P)\Delta_{vap}h}{P}}. \tag{4.64}$$

Before we take the last step, which is to express dP in terms of dh, we want to pause and discuss this expression. Note that if there is no water at all, i.e. $n_{H_2O} = 0$ then the result coincides with Eq. (4.53). In this case the slope is that of the dry adiabatic curve. If on the other hand the water moisture completely dominates, which corresponds to neglecting C_P as well as V, then the slope is the slope in the Clapeyron equation, i.e. the slope of the saturation line. The slope of the moist adiabatic curve, beyond the saturation line, as given by Eq. (4.64), is in between the two. How can we understand this based on a physical picture?

In the ideal gas a water molecule has the average energy, e_g, which is the kinetic energy, k_g, determined by its temperature. When the water molecules are in the liquid state then they have the average energy $e_l = k_l - \epsilon$. Here $-\epsilon < 0$ is their potential energy. As long as the water molecules have not traded energy with other parts of the system $e_g = e_l$ and the kinetic energy in the liquid state therefore is $k_l = k_g + \epsilon$. This means that at this moment, the water is effectively hotter than the surrounding gas. How big is ϵ? In ice each water molecule participates in four hydrogen bonds. In the liquid state this number is reduced - which we ignore here. Thus, the potential energy per water molecule is $4 \times (1/2)$ hydrogen bonds $\approx 2 \times 25$ kJ/mol (incidentally this is quite close to the enthalpy of vaporization of water). The liquid water drops formed at the saturation line now must equilibrate with the remaining gas— mostly N_2 and O_2. If we assume that 5 g of water transfer their extra energy to 1000 g of dry air, then this leads to a temperature increase ΔT in the air given by

$$\frac{5g}{18g} 50\text{kJ/mol} = \frac{1000g}{29g} \underbrace{3.5R}_{=c_P} \Delta T. \tag{4.65}$$

The result is $\Delta T \approx 14$ K. This is crude. Nevertheless, it illustrates the point that the droplets formed at the saturation line contain sufficient excess energy, which originates from the (negative) cohesion energy between the molecules —even though the total mass of the droplets is a lot less than the total mass of the other gas components.

In addition there is the effect that upon reaching the saturation line the water vapor must be compressed in the first place. The attendant work, i.e. $P_{H_2O}\Delta V$, must be done by the other gas components. This in turn means that they must do double-duty. The gas continues to invest work into its adiabatic expansion but to a lesser extend, because at the same time it also invests work into the compression of the water vapor. The overall effect, however, is small compared to the aforementioned one. If 5 g water vapor is compressed at a partial pressure $P_{H_2O} = 0.01$ bar (cf. Fig. 3.6) across the saturation line the attendant work is $(5/18)RT \approx 0.6$ kJ/mol. The above cohesion energy on the other hand is $(5/18)\,50$ kJ/mol ≈ 14 kJ/mol. It is worth noting that the condensation of water vapor does not occur at one particular height. As the air becomes dryer due to condensation, the condensation shifts down the saturation line (cf. Fig. 3.6).

We obtain the lapse rate of moist air beyond the saturation line combining Eqs. (4.64) and (4.54), i.e.

$$\frac{dT}{dh} = -\frac{m_{mol}g}{c_P} \frac{1 + x_{H_2O}\frac{\Delta_{vap}h}{RT}}{1 + x_{H_2O}\left(\frac{\Delta_{vap}h}{RT}\right)^2 \frac{R}{c_P}}. \tag{4.66}$$

Here $x_{H_2O} = n_{H_2O}/n$ is the mole fraction water in the air parcel.

Let's estimate the difference between the dry lapse rate as described by Eq. (4.55) and the lapse rate described by Eq. (4.66). The difference is the additional factor in Eq. (4.66). Because this factor does depend on thermodynamic conditions, we decide to do our calculation for $T \approx 5$ to 10 °C (cf. Fig. 3.6). The attendant enthalpy of vaporization is roughly 45 kJ/mol and thus $\Delta_{vap}h/(RT) \approx 19$. In addition $c_P/R \approx 3.5$. For 10 g of water per 1 kg of air we have $x_{H_2O} \approx 0.016$. This yields

$$\frac{dT}{dh} \approx 0.5 \left.\frac{dT}{dh}\right|_{dry}, \tag{4.67}$$

the lapse rate upon crossing of the saturation line, which means just above the cloud base, reduced by roughly a factor of two. If the water content is reduced to 1 g per 1 kg of air the factor 0.5 is replaced by 0.9.

Example—Droplet Growth   In this example, in which the Clapeyron equation again plays a central role, we resume our discussion of droplet formation and stability which started in Chap. 3. In the following we discuss the growth dynamics of droplets via condensation and collision.

Let's begin with condensation. We assume that our droplet of radius $a(t)$ grows at the center of a coordinate system embedded in a uniform, time-independent vapor phase. The rate at which the mass $m(r)$ inside an sphere of radius $r \geq a$ changes is given by the continuity equation

$$\frac{dm(r)}{dt} = - \int_{A(r)} d\vec{A} \cdot \vec{j}(r). \tag{4.68}$$

The right hand side is an integral over the flux of molecules $\vec{j}(r)$ through a spherical surface $A(r) = 4\pi r^2$. Note that $d\vec{A} = \vec{e}_r dA$, where $\vec{e}_r$ is a radial unit vector. Note also that the left hand side is not really a function of r, because the droplet is embedded inside a uniform, time-independent vapor phase, which means that the mass change occurs on the surface ($r = a$) of the droplet only. Hence

$$\frac{dm(a)}{dt} = -4\pi r^2 \vec{e}_r \vec{j}(r). \tag{4.69}$$

Now we express the flux $\vec{j}(r)$, using Fick's law, $\vec{j}(r) = -D\vec{\nabla}c(r)$, in terms of a hypothetical vapor mass density gradient $\vec{\nabla}c(r) = \vec{e}_r \partial_r c(r)$, i.e.

$$\frac{dm(a)}{dt} = 4\pi D r^2 \frac{dc(r)}{dr}. \tag{4.70}$$

Here D is the (constant) diffusion constant in the vapor phase surrounding the droplet. The differential equation (4.70) can be solved via separation of variables, i.e.

$$\frac{dm(a)}{dt} \int_a^\infty \frac{dr}{r^2} = 4\pi D \int_{c(a)}^{c(\infty)} dc \tag{4.71}$$

or

$$\frac{dm(a)}{dt} = 4\pi D a(c(\infty) - c(a)). \tag{4.72}$$

We finalize this expression by substituting $m(a) = c^{(l)} 4\pi a^3/3$ and $c(\infty) \equiv c^{(g)}$, the bulk vapor phase density, which yields

$$\frac{d}{dt}a^2 = 2D\frac{c^{(g)} - c(a)}{c^{(l)}}. \tag{4.73}$$

Assuming that the mass density at $r = a$ does not change with time we see that $a^2(t) = a^2(0) + b\,t$, where b is a constant. But note that at this point we do not know $c(a)$.

There is another quantity whose transport must be considered. Net adsorption of molecules from the gas phase onto the droplet's surface increases the local temperature due to a negative heat of vaporization $-\Delta_{vap}h$. This causes a heat flux into the surrounding gas phase. In analogy to Eq. (4.68) we now have

$$-\Delta_{vap}h\,\frac{dm(a)}{dt} = -\int_{A(r)} d\vec{A}\cdot\vec{j}_Q(r). \tag{4.74}$$

The index Q indicates that $\vec{j}_Q$ is a heat current. We follow the same path as before. Replacing Fick's law by Fourier's law, $\vec{j}_Q(r) = -\lambda\vec{\nabla}T(r)$, where λ is the thermal conductivity of the bulk vapor, we obtain

$$-\Delta_{vap}h\,\frac{dm(a)}{dt} = 4\pi\lambda a(T(\infty) - T(a)). \tag{4.75}$$

Here we do not know $T(a)$. However, despite this lack of information, it is possible to obtain an approximate analytic equation for $a(t)$. We can use the Clapeyron equation in the form of Eq. (4.48), i.e.

$$d\ln P \approx d\ln(c\,T) \approx \frac{\Delta_{vap}h}{RT}\frac{dT}{T}, \tag{4.76}$$

where c is the gas mass density (note that here we can replace the number density by the mass density), or

$$\frac{dc}{c} \approx \left(\frac{\Delta_{vap}h}{RT} - 1\right)\frac{dT}{T} \tag{4.77}$$

or

$$\frac{c_2 - c_1}{c_1} \approx \left(\frac{\Delta_{vap}h}{RT_1} - 1\right)\frac{T_2 - T_1}{T_1}. \tag{4.78}$$

It is important to keep in mind that $dc \equiv c_2 - c_1$ and $dT \equiv T_2 - T_1$ are small differences along the saturation line. By replacing T_1 with $T(\infty)$ and T_2 with $T(a)$, and using (4.75), Eq. (4.78) becomes

$$\frac{c_s(a) - c_s(\infty)}{c_s(\infty)} \approx (l - 1) \frac{\Delta_{vap} h \, dm/dt}{4\pi\lambda a T(\infty)}, \tag{4.79}$$

where $l \equiv \Delta_{vap} h / (RT(\infty). \ c_s(\infty)$ is the saturation vapor density at $T(\infty)$. Likewise, $c_s(a)$ is the saturation vapor density at $T(a)$. In order to make use of Eq. (4.72) we expand $c(\infty) - c(a)$, i.e.

$$c(\infty) - c(a) = c(\infty) - c_s(\infty) + c_s(\infty) - c(a)$$
$$= c_s(\infty) \left((S - 1) - \frac{c(a) - c_s(\infty)}{c_s(\infty)} \right), \tag{4.80}$$

where $S = c(\infty)/c_s(\infty)$. The only difference between the left hand side of Eq. (4.79) and the second term in the brackets on the right hand side of Eq. (4.80) is $c_s(a)$ in the former versus $c(a)$ in the latter. Both densities are at the same temperature, however, and we shall assume that their difference is much smaller than the difference of either one to $c_s(\infty)$. Hence we may combine Eqs. (4.72), (4.79) and (4.80) into

$$\frac{d}{dt} a^2 \approx \frac{2(S - 1)}{\frac{c^{(l)}}{D c_s^{(g)}} + (l - 1) \frac{\Delta_{vap} h \, c^{(l)}}{\lambda T^{(g)}}}. \tag{4.81}$$

Note that $c(\infty)$ and $c_s(\infty) \equiv c_s^{(g)}$ are also densities at the same temperature $T(\infty) \equiv T^{(g)}$ and thus $S = P(\infty)/P_s(\infty) = \varphi$. Integration of (4.81) yields

$$a^2(t) \approx a^2(0) + \frac{2(S - 1)}{\frac{c^{(l)}}{D c_s^{(g)}} + (l - 1) \frac{\Delta_{vap} h \, c^{(l)}}{\lambda T^{(g)}}} t. \tag{4.82}$$

An example is shown in Fig. 4.19, the numerical values of the various quantities are $a(0) = 0.1$ μm, $S = 1.01$, $\Delta_{vap} h = 45$ kJ/mol, $T^{(g)} = 273$ K, $c^{(l)} = 5.55 \cdot 10^4$ mol/m^3, $c_s^{(g)} = 0.27$ mol/m^3, $D = 0.24$ cm^2/s for H_2O in air, and $\lambda = \lambda_{air} = 0.024$ W/(mK). The dashed curve illustrates the difference when the latent heat term in the denominator vanishes ($l = 1$).

In this derivation we have omitted the curvature and concentration corrections discussed in the previous part of this example, i.e. the dependence of the saturation pressure above of the droplet on its size and on the concentration of a possible solute.

Another point worth noting is the narrowing of the droplet size distribution as time increases. To see this we subtract $a_1(t)^2$ from $a_2(t)^2$, where the subscripts 1 and 2 here refer to two separate droplets. We obtain

$$a_2^2(t) - a_1^2(t) = a_2^2(0) - a_1^2(0) \quad \text{or} \quad (a_2(t) - a_1(t)))\bar{a}(t) = \left(a_2^2(0) - a_1^2(0)\right)/2,$$

$$(4.83)$$

where $\bar{a}(t) = (a_2(t) + a_1(t))/2$. The right hand side does not depend on t and therefore the left hand side should not depend on t either. Since $\bar{a}(t)$ is growing, we conclude that the difference $a_2(t) - a_1(t)$ becomes smaller and eventually tends to zero, i.e. the above growth mechanism leads to a monodisperse droplet size distribution for long times.

Growth by condensation is too slow to eventually lead to raindrops in the range of several millimeters. But like all masses droplets fall in the gravitational field. Under ideal conditions and when the droplets are not too large the only force which they experience is Stoke's frictional force $f_S = 6\pi \eta a v$, where η is the surrounding air's viscosity and v is the droplets velocity. The droplet's acceleration is given by

$$\ddot{x}(t) = g - \frac{M_b}{M} g - \frac{6\pi \eta a}{M} \cdot x(t). \qquad (4.84)$$

Here $x(t)$ is the distance a droplet has fallen during the time t, M is the mass of the droplet, g is the gravitational acceleration, and $M_b g$ is the droplet's buoyancy. Since M_b, the mass of the air displaced by the droplet, is small compared to M, we neglect this force and the velocity of the droplet becomes

$$v(t) = v_\infty \left(1 - e^{-t/\tau}\right), \qquad (4.85)$$

where $v_\infty = g\tau$ is the terminal velocity and $\tau = M/(6\pi \eta a)$. For example: let $a = 5$ μm and $\eta = \eta_{air} = 18.6$ μPa at 300 K. Using $c_{H_2O} = 1$ g/cm^3 we obtain $\tau \approx 0.03$ s and $v_\infty \approx 0.3$ cm/s. If $a = 50$ μm instead, we find $\tau \approx 3$ s and $v_\infty \approx 29$ cm/s). For still larger a our formula for v_∞ overestimate the actual terminal velocity (f_S should be replaced by a drag force proportional to $(av)^2$), e.g. droplets with $a = 500$ (2500) μm possess an actual terminal velocity of about 400 (910) cm/s.

A falling droplet may "collect" other droplets in its path and thereby accumulate mass according to

$$\frac{dm}{dt} \approx \pi a^2 E c_{moist} v_\infty, \qquad (4.86)$$

increasing its radius by

$$\frac{da}{dt} \approx \frac{E c_{moist}}{4 c_{H_2O}} v_\infty. \qquad (4.87)$$

The idea is that during a short time dt the falling droplet collects a fraction E of the moisture, in the form of smaller droplets, in a cylindrical column of radius a and length $v_\infty dt$. This length of course is rough, because we have replaced the velocity of the collector droplet relative to the velocities of the collected droplets simply by v_∞. The quantity E is the collection efficiency. Some of the droplets in the path of the collector just get pushed aside and around the falling collector and some just fall apart in the collision, etc. However, we want to conclude our discussion of clouds and droplets at this point and refer the interested reader to Salby (2012) for an in-depth discussion.

Example—Superconductor Thermodynamics   Figure 4.20 shows a phase diagram in the $T - H$-plane, i.e. in this system the magnetic H-field assumes the role of P. The two phases are named s and n. We easily can work out a version of the Clapeyron equation in this case. Our starting point is Eq. (4.44), i.e.

$$-s_s dT - \frac{v}{4\pi} \boldsymbol{B}_s \cdot d\boldsymbol{H} = -s_n dT - \frac{v}{4\pi} \boldsymbol{B}_n \cdot d\boldsymbol{H}. \tag{4.88}$$

Here we have used the mapping from $(P, -V)$ to $(\boldsymbol{H}, v\boldsymbol{B}/(4\pi$ described on p. 26 in the context of the discussion of Eq. (1.51). Notice that now v is a constant molar volume of the material to which the phase diagram in Fig. 4.19 applies. Analogous to (4.45) we find in the present case

$$\frac{v}{4\pi} (\boldsymbol{B}_n - \boldsymbol{B}_s) \cdot \frac{d\boldsymbol{H}}{dT}\bigg|_{coex} = s_s - s_n. \tag{4.89}$$

Our phase diagram is meant to apply to a type I superconductor. The letters s and n label the superconducting and the normal conducting phases, respectively. In the s-region we have therefore $\boldsymbol{B}_s = 0$. If in addition we use the linear relation $\boldsymbol{B}_n = \mu_r \boldsymbol{H}$, where μ_r is the magnetic permeability, then Eq. (4.89) becomes

$$\frac{v}{8\pi} \mu_r \frac{dH^2}{dT}\bigg|_{coex} = s_s - s_n. \tag{4.90}$$

Based on this equation and some additional information we may work out the coexistence line. The additional information consists of the empirical approximations to the molar heat capacities in superconducting and normal conducting phases at low temperatures, i.e. $c_s = aT^3$ and $c_n = bT^3 + \gamma T$ (an example may be found in Chap. 33 of Ashcroft and Mermin (1976). The quantities a, b, and γ are constants, which may by obtained via suitable

Fig. 4.20 A phase diagram in the T-H-plane

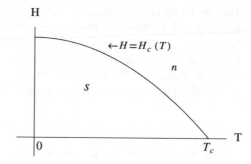

experimental data. By simply integrating the thermodynamic relation $c = T \partial s / \partial T|_H$ from zero temperature to T we obtain $s_s(T) = s_s(0) + (a/3)T^3$ and $s_n(T) = s_n(0) + (b/3)T^3 + \gamma T$ (It may be puzzling that c_n is used below T_c. Here the normal phase can be produced by a weak magnetic field destroying the superconducting state with little effect on the heat capacity.). If we invoke what is called the third law of thermodynamics, i.e. $s(0) = 0$, which we discuss in Chap. 5, we can integrate Eq. (4.90). The result, after some algebra, is

$$H_c(T) = H_c(0)\left(1 - \frac{T^2}{T_c^2}\right),$$ (4.91)

where

$$H_c(0) = \sqrt{\frac{2\pi\gamma}{\nu\mu_r}}T_c \quad \text{and} \quad T_c = \sqrt{\frac{3\gamma}{a-b}}.$$ (4.92)

Notice that the latent heat, $T\Delta s$, vanishes in the two end-points of the transition line. Even though thermodynamics by itself does not explain superconductivity, it does allow additional predictions provided that certain input is available. However, if this input does consist of approximations then the additional predictions will be approximate as well.

Phase Separation in the RPM

Arguably the most important insight provided by van der Waals theory is the role of intermolecular interaction on gas-liquid phase separation. The latter requires short-ranged repulsion as well as attraction. Here we discuss an overall neutral system consisting of charges $+q$ and $-q$ possessing hard core excluded volume.

This model system is termed the restricted primitive model (RPM). Corresponding experimental systems are molten salts (Weiss 2010; Pitzer 1990). Does such a system posses a critical point analogous to the gas-liquid critical point in the van der Waals theory? A priori there is no easy answer, because there is no obvious net attractive interaction as in the latter theory.[17]

The free enthalpy of our model system is

$$G = n_+ \mu_+ + n_- \mu_- . \tag{4.93}$$

The indices refer to the two charge types. We approximate the chemical potentials via

$$\mu_\pm \approx \mu_{o,\pm} + RT \ln \frac{n_\pm N_A}{V} - \frac{1}{2} \frac{N_A q^2}{\lambda_D} \frac{1}{1 + b/\lambda_D} . \tag{4.94}$$

The first term describes contributions to the chemical potentials not dependent on ion concentration. The concentration dependence here enters through an ideal gas term, the second term in Eq. (4.94), and via the interaction between the ions in the framework of the Debye-Hückel theory when the ions posses the radius b (cf. Eq. (3.135)), the third term in Eq. (4.94). Does this yield a critical point? We insert Eq. (4.94) into (4.93) and the result into Eq. (2.195). The dotted line in the left panel in Fig. 4.4 is the spinodal line obtained in the van der Waals theory. On the spinodal line the compressibility diverges and thus $\partial G/\partial V \, |_{T,n_1,n_2,...} = 0$ (cf. Eq. (2.195)). Straightforward differentiation of our present G, using $\lambda_D \propto V^{1/2}$, leads to

$$\frac{1}{4T^*} \frac{x}{(1+x)^2} = 1 \tag{4.95}$$

employing this condition. Here $T^* = bRT/(N_A q^2)$ and $x = b/\lambda_D$. If we insert $x = 1 + \delta x$ into (4.95), we obtain to leading order in the small quantity δx

$$1 - \delta x^2 = 16T^*. \tag{4.96}$$

We recognize that there are always two solutions $x = 1 \pm \delta x$ with the same T^*. The two solutions coincide if $x = x_c = 1$ corresponding to $T^* = T_c^* = 1/16$. The conclusion is that the RPM possesses a gas-liquid spinodal curve, here worked out in the vicinity of the critical point at

$$T_c = \frac{1}{16} \frac{N_A q^2}{Rb} \quad \text{and} \quad \rho_c = 2c_c = \frac{1}{64\pi b^3} . \tag{4.97}$$

---

[17] One may argue that the immediate neighborhood of $+q$ on average contains an excess of $-q$ and that this leads to the net attraction. But this is a truly complicated system and such arguments should always be backed up by calculation.

Note that presently $q^2 = e^2$, and ρ is the total ion number concentration. Notice also that the replacement $q^2 \rightarrow q^2/(4\pi\varepsilon_o\varepsilon_r)$ yields T_c in SI-units.

The pressure follows via integration of

$$\left.\frac{\partial P}{\partial V}\right|_{T,\pm n} = \frac{1}{V}\left.\frac{\partial G}{\partial V}\right|_{T,\pm n}, \tag{4.98}$$

i.e.

$$\frac{PV}{(n_+ + n_-)RT} = 1 + \frac{N_A q^2}{RTb}\left(\frac{\ln[1+x]}{x^2} - \frac{1}{2x}\frac{2+x}{1+x}\right). \tag{4.99}$$

The resulting critical compressibility factor is

$$\frac{P_c}{\rho_c RT_c} = 16\ln 2 - 11 \approx 0.09035. \tag{4.100}$$

The above references list experimental critical parameters. In particular we may compare the compressibility factor (4.100), because it is a pure number. It turns out that the above model does not make quantitative predictions—except in selected cases. But for us it provides a valuable exercise. In fact the model is not wrong—it is incomplete. It turns out that association of ions into aggregates is the most important ingredient for a more accurate description of phase separation in molten salts and related ionic systems. A detailed account of this can be found in Levin and Fisher (1996).

Electric Field Induced Critical Point Shift

The following discussion of the electric field induced shift of the critical temperature, density, and pressure is not so much motivated by its practical importance but rather by the rich content of conceptual and technical aspects making this a valuable exercise.

In the preceding section we have used Eq. (2.195), i.e. the divergence of the compressibility, to locate the critical point. Here we consider an ordinary dielectric liquid (dielectric constant ε_r) in an electric field E, where E is the (macroscopic) average electrical field in the liquid. We also may want to apply Eq. (2.195) to deduce the electric field effect on the location of the critical point. We must know, however, whether to work out the partial derivative at constant constant D or at constant E. We can find the answer via the following inequality

$$\frac{\partial^2 f}{\partial D^2}\frac{\partial^2 f}{\partial \rho^2} - \left(\frac{\partial^2 f}{\partial D \partial \rho}\right)^2 > 0. \tag{4.101}$$

Here $f = f(T, \rho, D)$ is the free energy density, depending on (constant) temperature, T, particle density, ρ, and the magnitude of the displacement field, D

(appropriate in an isotropic medium).[18] Inequality (4.101) expresses the requirement that f is convex in terms of ρ and D. A sign change signals a "dent" in f causing phase separation (cf. the 1D situation depicted in Fig. 4.2). Thus, the replacement of > 0 by $= 0$ in (4.101) yields the critical point, i.e

$$\frac{1}{4\pi}\left(\frac{\partial E}{\partial D}\Big|_{T,\rho} \frac{\partial \mu}{\partial \rho}\Big|_{T,D} - \frac{\partial \mu}{\partial D}\Big|_{T,\rho} \underbrace{\frac{\partial E}{\partial \rho}\Big|_{T,D}}_{\overset{(A.3)}{=} -\frac{\partial E}{\partial D}\Big|_{T,\rho} \frac{\partial D}{\partial \rho}\Big|_{T,E}} \right) = 0. \tag{4.102}$$

Making use of Eq. (A.1) we obtain the desired result valid at the critical point, i.e.[19]

$$\frac{\partial \mu}{\partial \rho}\Big|_{T_c,E} = 0 \tag{4.103}$$

We apply this formula to the free energy density

$$f = f_o + \frac{1}{4\pi} E \cdot D. \tag{4.104}$$

Here f_o is the part of the free energy density which does not depend on the field explicitly. The attendant chemical potential is

$$\mu = \mu_o + \frac{1}{4\pi} \frac{\partial}{\partial \rho} ED\Big|_{T,D}. \tag{4.105}$$

Differentiation at constant E and using $D = \varepsilon_r(\rho)E$ yields

$$\begin{aligned}
\frac{\partial \mu}{\partial \rho}\Big|_{T,E} &= \frac{\partial \mu_o}{\partial \rho}\Big|_T + \frac{1}{4\pi} \frac{\partial}{\partial \rho} \frac{\partial}{\partial \rho} ED\Big|_{T,D}\Big|_{T,E} \\
&= \frac{\partial \mu_o}{\partial \rho}\Big|_T - \frac{1}{8\pi} E^2 \frac{\partial^2 \varepsilon_r(\rho)}{\partial \rho^2}\Big|_{T,E}.
\end{aligned} \tag{4.106}$$

We can express $d\mu_o$ in terms of dT and dP as usual

$$d\mu_o = -\frac{S}{\rho} dT + \frac{1}{\rho} dP \tag{4.107}$$

---

[18] On p. 67 we had found $F = F(T, D)$. The density (volume) dependence was ignored, because it did not play a significant role in the example. In addition we use $D = \epsilon_r E$, and thus D and E are along the same direction. Moreover we can use the magnitudes instead of D and E.

[19] On p. 25 we had discussed situations in which one can obtain new thermodynamic relations via replacement of P through, for instance, the electric field strength, E. In the present case we could have applied this to the chemical stability condition in (3.16) to immediately obtain Eq. (4.103).

and thus

$$\frac{\partial \mu_o}{\partial \rho}\bigg|_T = \frac{1}{\rho}\frac{\partial P}{\partial \rho}\bigg|_{T,E=0}.$$

(4.108)

We put everything together by combining Eqs. (4.103), (4.106), and (4.108), i.e.

$$0 = \frac{1}{\rho_c}\frac{\partial P(\rho_c)}{\partial \rho}\bigg|_{T_c,E=0} - \frac{1}{8\pi}E^2\frac{\partial^2 \varepsilon_r(\rho_c)}{\partial \rho^2}\bigg|_{T_c,E}.$$

(4.109)

Note that we first take the derivatives with respect to ρ under the indicated constraints and subsequently evaluate the result at ρ_c. Notice also that the first term does not vanish even though we have learned that $0 = \partial P/\partial V|_{T_c} = -(\rho/V)\partial P/\partial \rho|_{T_c}$ in the context of van der Waals theory. This is because the critical point we study now is for a certain field strength $E \neq 0$. However, we may expand the first term as follows

$$\begin{aligned}
\frac{1}{\rho_c}\frac{\partial P(\rho_c)}{\partial \rho}\bigg|_{T_c,E=0} &= \frac{1}{\rho_{c,o}+\delta\rho}\frac{\partial P(T_{c,o}+\delta T, \rho_{c,o}+\delta\rho)}{\partial \rho}\bigg|_{E=0} \\
&\approx \frac{1}{\rho_{c,o}}\left(1 - \frac{\delta\rho}{\rho_{c,o}}\right)\bigg[\underbrace{\frac{\partial P(T_{c,o},\rho_{c,o})}{\partial \rho}\bigg|_{E=0}}_{=0} \\
&\quad + \underbrace{\frac{\partial^2 P(T_{c,o},\rho_{c,o})}{\partial \rho^2}\bigg|_{E=0}}_{=0}\delta\rho + \frac{\partial^2 P(T_{c,o},\rho_{c,o})}{\partial T \partial \rho}\bigg|_{E=0}\delta T\bigg] \\
&\approx \frac{1}{\rho_{c,o}}\frac{\partial^2 P(T_{c,o},\rho_{c,o})}{\partial T \partial \rho}\bigg|_{E=0}\delta T.
\end{aligned}$$

(4.110)

The index c, o indicates that this quantity is taken at the critical point of the same system in absence of the electric field ($T_c = T_{c,o} + \delta T$; $\rho_c = \rho_{c,o} + \delta\rho$). The first two terms in the square brackets are zero, because they are evaluated for vanishing field strength at the attendant critical point. Analogously we have

$$-\frac{1}{8\pi}E^2\frac{\partial^2 \varepsilon_r(\rho_c)}{\partial \rho^2}\bigg|_{T_c,E} \approx \frac{1}{8\pi}E^2\frac{\partial^2 \varepsilon_r(T_{c,o},\rho_{c,o})}{\partial \rho^2}\bigg|_{E=0}.$$

(4.111)

Notice that E also is a small quantity, so that the right side is the leading term of the expansion. Combination the last two equations yields

$$\delta T \approx \frac{1}{8\pi}\rho_{c,o}E^2\frac{\partial^2 \varepsilon_r(T_{c,o},\rho_{c,o})}{\partial \rho^2}\bigg|_{E=0}\bigg/\frac{\partial^2 P(T_{c,o},\rho_{c,o})}{\partial T \partial \rho}\bigg|_{E=0}.$$

(4.112)

This is the leading contribution to the field induced temperature shift for small field strength. In order to obtain $\delta\rho$ to the same order we must work from $\partial^2\mu/\partial\rho^2|_{T_c,E}=0$. The result is

$$\delta\rho \approx \frac{\left[K(\varepsilon_r)+\frac{1}{\rho_{c,o}}\right]\frac{\partial^2 P(T_{c,o},\rho_{c,o})}{\partial T\partial\rho}\bigg|_{E=0}-\frac{\partial^3 P(T_{c,o},\rho_{c,o})}{\partial T\partial\rho^2}\bigg|_{E=0}}{\frac{\partial^3 P(T_{c,o},\rho_{c,o})}{\partial\rho^3}\bigg|_{E=0}}\delta T, \tag{4.113}$$

where $K(\varepsilon_r)=\frac{\partial^3\varepsilon_r(T_{c,o},\rho_{c,o})}{\partial\rho^3}\bigg|_{E=0}\bigg/\frac{\partial^2\varepsilon_r(T_{c,o},\rho_{c,o})}{\partial\rho^2}\bigg|_{E=0}$. The shift of the critical pressure is simply

$$\delta P \approx \frac{\partial P(T_{c,o},\rho_{c,o})}{\partial T}\bigg|_{E=0}\delta T. \tag{4.114}$$

Notice that the shifts all are quadratic in the field strength.[20]

4.3   Low Molecular Weight Mixtures

4.3.1   A Simple Phenomenological Model for Liquid-Liquid Coexistence

The van der Waals approach is applicable to gas-liquid phase separation in a one-component system. Another type of phase separation is observed in binary mixtures. Depending on thermodynamic conditions the components may be miscible or not. A simple model describing this is based on the following molar free enthalpy approximation

$$g = x_A^{(l)}g_A + x_B^{(l)}g_B + x_A^{(l)}\ln x_A^{(l)} + x_B^{(l)}\ln x_B^{(l)} + \chi x_A^{(l)}x_B^{(l)}. \tag{4.115}$$

Here $g_A = \mu_A^*(l)/RT$ and $g_B = \mu_B^*(l)/RT$ are the reduced molar free enthalpies of two pure liquid components A and B. Mixing A and B gives rise to the mixing free enthalpy described by the ln-terms. Note that the mole fractions are $x_A^{(l)}$ and

---

[20] We remark that the pressure derivatives can be estimated using the van der Waals equation of state $(\partial^2 P/\partial T\partial\rho|_c = 6Z_c, \ \partial^3 P/\partial T\partial\rho^2|_c = 6Z_c/\rho_c; \ \partial^3 P/\partial\rho^3|_c = 9Z_c/\rho_c$, where $Z_c = 3/8$ is the critical compressibility factor. A sufficiently accurate estimate of the dielectric constant derivatives is more difficult. Considering a permanent point dipole, $\boldsymbol{\mu}$, in a spherical cavity inside a continuous dielectric medium characterized by a dielectric constant, ε_r, Onsager (L. Onsager *Electric moments of molecules in liquids*. J. Am. Chem. Soc. **58**, 1486 (1936); Nobel prize in chemistry for his work on irreversible thermodynamics, 1968) has derived the following simple approximation $(\varepsilon_r - 1)(2\varepsilon_r + 1)/\varepsilon_r = 4\pi\mu^2 N_A\rho/(RT)$, which may in principle be used for this purpose. We leave this to the interested reader.

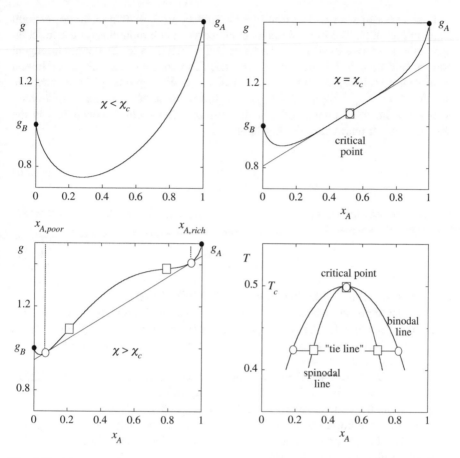

Fig. 4.21 Schematic of the x_A-dependence of g for different χ-values. The *curves* shown here are for $\chi = 1$ (*upper left*), $\chi = 2$ (*upper right*), and $\chi = 3$ (*lower left*) using $g_A = 1.5$ and $g_B = 1.0$. *Lower right*: T-x_A-phase diagram of our model of a binary mixture, where we assume that $T = 1/\chi$

$x_B^{(l)} = 1 - x_A^{(l)}$. We had obtained this contribution earlier, cf. Eq. (3.30), for mixtures of ideal gases. Here we consider liquids. Nevertheless we still assume ideal behavior. The last term is new. It introduces an additional interaction free enthalpy proportional to the two mole fractions. The quantity χ is a parameter in this theory.

Figure 4.21 shows g for different values of the χ-parameter. If χ is less than a critical value then g is a convex function of $x_A^{(l)}$ (or $x_B^{(l)}$). This situation is analogous to the free energy in the van der Waals theory for temperatures above the critical temperature. If $\chi = \chi_c$ then the curvature of g at $x_A^{(l)} = 1/2$ becomes zero. For still larger values of χ a "bump" develops - again analogous to the free energy in the van der Waals theory for temperatures less than the critical temperature. Driven by the second law the system now lowers its free enthalpy by separating into two types of regions, which over time will coagulate into two large domains, one depleted of A

and one enriched with A. The resulting phase diagram is shown in the lower right panel in Fig. 4.21. The binodal line is obtained via a common tangent construction applied to the free enthalpy (cf. the lower left panel) akin to the common tangent construction used in the van der Waals case. The common tangent is the lowest possible free enthalpy in between $x_{A,poor}$ and $x_{A,rich}$. For a given x_A in this range the quantity $(x_A - x_{A,poor})/(x_{A,rich} - x_{A,poor})$ is the fraction of A in the $A, rich$-phase relative to the total amount of A in the system. The second special line is the spinodal line. It marks the stability limit

$$\left. \frac{\partial^2 g}{\partial x_A^2} \right|_{T,P} = 0 \tag{4.116}$$

(cf. the third stability condition in (3.16)). Both lines meet at the critical point where

$$\left. \frac{\partial^3 g}{\partial x_A^3} \right|_{T,P} = 0, \tag{4.117}$$

because at the critical point the curvature obviously changes sign.

Note that temperature here enters via the assumed proportionality $\chi \propto 1/T$. This assumption accounts for the observation that phase separation usually occurs upon lowering temperature. Nevertheless this is purely empirical and more complex descriptions of χ can be found.

Figure 4.22 shows experimental liquid-liquid equilibria data for the binary mixtures water/phenol (solid squares) and methanol/hexane (solid circles).[21] Here x_1 is the mole fraction of water and methanol, respectively. Notice that while both systems show the basic behavior predicted by our theory, only the second system also exhibits the symmetry around $x = 0.5$. Nevertheless, the solid lines are "theoretical" results, which where obtained using

$$\chi = \frac{c_0 + c_1 x}{T} + c_2. \tag{4.118}$$

Here c_0, c_1, and c_2 are constants, which are adjusted so that the theory matches the data points. In particular the c_1-term breaks the symmetry around $x = 0.5$. While it is quite common to introduce such expressions for χ, it is not easy to provide reasonable physical explanations of the individual terms. In addition, the "best fit" usually does not correspond to a unique set of values for c_0, c_1, and c_2. We return to this in the context of polymer mixtures.

---

[21] Data from HTTD.

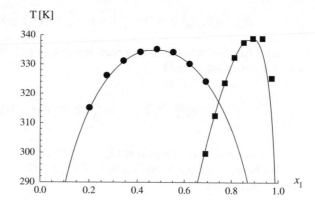

Fig. 4.22 Liquid-liquid equilibria data for the binary mixtures water/phenol (solid squares) and methanol/hexane (solid circles)

4.3.2 Gas-Liquid Coexistence in a Binary System

Can this type of phase separation be used to physically separate components A and B? In principle yes—but not entirely and usually not as a practical means. Let us look at the binary mixture from another angle. The above model did not include possible distribution of components A and B in phases corresponding to different states of matter, e.g. gas, liquid or solid. Here we want to study a situation when gas and liquid coexist containing both A and B. This is depicted in Fig. 4.23—which we encountered before (cf. Fig. 3.7).

In equilibrium we have $\mu_A^{(g)} = \mu_A^{(l)}$ and $\mu_B^{(g)} = \mu_B^{(l)}$ (cf. Eq. (3.38)). Thus we may also write $d\mu_A^{(g)} = d\mu_A^{(l)}$ and $d\mu_B^{(g)} = d\mu_B^{(l)}$. Concentrating on component A and using Eqs. (3.27) and (3.45) we have

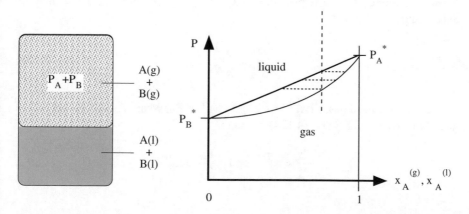

Fig. 4.23 Gas-liquid coexistence in a binary system

$$d(\mu_A^{(g)}(T, P_A^*) + RT \ln x_A^{(g)}) = d(\mu_A^{(l)}(T, P_A^*) + RT \ln x_A^{(l)}), \qquad (4.119)$$

i.e. from the start we assume that both the gas as well as the liquid are ideal. This may be reshuffled to yield

$$d\mu_A^{(g)}(T, P_A^*) - d\mu_A^{(l)}(T, P_A^*) = RTd \ln \frac{x_A^{(l)}}{x_A^{(g)}}. \qquad (4.120)$$

Note that we work at constant temperature.

Combination of the Gibbs-Duhem Eq. (2.180) at constant temperature with Eq. (2.182) yields

$$\left.\frac{\partial \mu_i}{\partial P}\right|_T = v_i = \left.\frac{\partial V}{\partial n_i}\right|_{T,P,n_j(\neq i)}. \qquad (4.121)$$

Here $d\mu_A^{(g)}(T, P_A^*)$ and $d\mu_A^{(l)}(T, P_A^*)$ refer to the pure component A, i.e. v_A^* is the molar volume of A in the gaseous and liquid states, respectively. In particular we may neglect $v_A^{*(l)}$ in comparison to $v_A^{*(g)}$. Therefore Eq. (4.120) becomes

$$v_A^{*(g)} dP \approx RTd \ln \frac{x_A^{(l)}}{x_A^{(g)}}. \qquad (4.122)$$

Integration, after insertion of the ideal gas law, i.e. $v_A^{*(g)} = RT/P$, yields

$$\frac{P}{P_A^*} \approx \frac{x_A^{(l)}}{x_A^{(g)}}, \qquad (4.123)$$

where the reference state is pure A and $x_A^{(l)}/x_A^{(g)} = 1$. Of course A and B may be interchanged and thus

$$\frac{P}{P_B^*} \approx \frac{x_B^{(l)}}{x_B^{(g)}}. \qquad (4.124)$$

Because for ideal gases $Px_A^{(g)} = P_A$ and $Px_B^{(g)} = P_B$, where P_A and P_B are partial pressures, Eqs. (4.123) and (4.124) become

$$\frac{P_A}{P_A^*} \approx x_A^{(l)} \quad \text{and} \quad \frac{P_B}{P_B^*} \approx x_B^{(l)}. \qquad (4.125)$$

In addition we may use Dalton's law, $P = P_A + P_B$, which allows to express P entirely through $x_{A/B}^{(l)}$ or $x_{A/B}^{(g)}$ (Salby 2012). The first relation is

$$P \approx x_A^{(l)}(P_A^* - P_B^*) + P_B^*, \tag{4.126}$$

where we use $x_A^{(l)} + x_B^{(l)} = 1$. Next we replace $x_A^{(l)}$ with $x_A^{(g)}$ via Eq. (4.123) obtaining

$$P \approx \frac{P_B^*}{1 - (1 - P_B^*/P_A^*)x_A^{(g)}}. \tag{4.127}$$

Equations (4.126) and (4.127) both are shown in the right panel of Fig. 4.23. The straight line is Eq. (4.126), whereas the curved line is Eq. (4.127). Their meaning is as follows. The dashed vertical line corresponds to a fixed x_A. Above its intersection with Eq. (4.126) we are in the liquid state of the mixture. Below its intersection with Eq. (4.127) the mixture is a homogeneous gas. In between however liquid and gas do coexist, and the mole fractions $x_A^{(l)}$ and $x_A^{(g)}$ are given by the intersections of the horizontal dashed lines (depending on P) with Eqs. (4.126) and (4.127).

There is a simple law connecting the total amount of liquid, $n(l)$, and the total amount of gas, $n(g)$, with x_A, $x_A^{(l)}$, and $x_A^{(g)}$—the lever rule. Note that (i) $nx_A = n(l)x_A^{(l)} + n(g)x_A^{(g)}$, where $x_A^{(l)} = n_A(l)/n(l)$ and $x_A^{(g)} = n_A(g)/n(g)$, and (ii) $nx_A = n(l)x_A + n(g)x_A$. Combining (i) and (ii) yields the lever rule:

$$\frac{x_A^{(g)} - x_A}{x_A - x_A^{(l)}} = \frac{n(l)}{n(g)}. \tag{4.128}$$

Notice that x_A is indeed bracketed by $x_A^{(g)}$ and $x_A^{(g)}$ as we had assumed.

Experimental isothermal gas-liquid equilibria are shown in Fig. 4.24. The left panel shows the system 1-chlorbutane/toluene at $T = 298.16K$. This system is well described by the above Eqs. (4.126) and (4.127) shown as solid lines. But there are other systems, like water/ethanol at $T = 323.15K$ shown on the right, which are not as ideal.

One may wonder about the difference between our two treatments of binary mixtures. The first one basically is a model composed of the ideal free enthalpy of mixing supplemented by a temperature dependent phenomenological "interaction" free enthalpy. Here the mixture may phase separate into regions of different component concentration depending on T. The second approach assumes a (first order) transition between different states of matter and describes the distribution of components A and B between phases corresponding to those different states (gas/liquid etc.) in terms of pressure. Aside from the assumed coexistence of phases ideality is used throughout. In reality a combination of both approaches may be necessary. However, it is worth noting in this context that a complete theory for the full phase diagram of a real system (or material) does not exist.

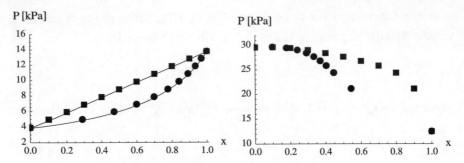

Fig. 4.24 *Left* 1-chlorbutane/toluane; *right* water/ethanol

4.3.3   Solid-Liquid Coexistence in a Binary System

Solubility

Based on the simple model expressed in Eq. (4.115) we want to study the situation depicted in Fig. 4.25. The figure shows a solution of B in A and a pile of solid B on the bottom. This can happen for instance if we try to dissolve too much sugar or salt in water. The following is a rough calculation of the maximum mole fraction, $x_B(T)$, which we can dissolve at a given temperature, T.

Neglecting the χ-parameter in Eq. (4.115) we may write for the chemical potential of B in A:

$$\mu_B(l) = \mu_B^*(l) + RT \ln x_B. \tag{4.129}$$

If $\mu_B^*(s)$ is the chemical potential of the pure solid B, then we have at coexistence $\mu_B^*(s) = \mu_B(l)$, i.e.

$$\mu_B^*(s) = \mu_B^*(l) + RT \ln x_B. \tag{4.130}$$

This can be rewritten into

Fig. 4.25 A solution of B in A including solid B at the bottom

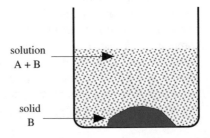

$$\ln x_B = -\frac{\Delta\mu^*(T)}{RT} + \frac{\Delta\mu^*(T_m)}{RT_m}, \tag{4.131}$$

where $\Delta\mu^*(T) = \mu_B^*(l) - \mu_B^*(s)$. Notice that T_m is the equilibrium melting temperature of pure B and thus $\Delta\mu^*(T_m) = 0$. Why have we added this term? To see this we write $\Delta\mu = \Delta h - T\Delta s$, where Δh and Δs are molar enthalpy and entropy changes. Assuming (!) that $\Delta h^*(T) \approx \Delta h^*(T_m) \equiv \Delta_{melt,B}h$ and $\Delta s^*(T) \approx \Delta s^*(T_m)$, i.e. these quantities depend only weakly on T, we immediately find

$$\ln x_B \approx -\frac{\Delta_{melt,B}h}{R}\left(\frac{1}{T} - \frac{1}{T_m}\right). \tag{4.132}$$

The added term has essentially eliminated the entropy and all we need to know is the transition enthalpy (here the melting enthalpy) of pure B as well as its melting temperature.

Notice that the index B in Eq. (4.132) can be replaced by an index i, where $i = A, B$. This means that the roles of A and B may be interchanged. Fig. 4.26 shows what we get for a mixture of tin (Sn) and lead (Pb). For tin we have $\Delta_{melt,Sn}h = 7.17$ kJ/mol and $T_m = 231.9\,°C$ (HCP). In the case of lead $\Delta_{melt,Pb}h = 4.79$ kJ/mol and $T_m = 327.5\,°C$. Using $x_{Pb} = 1 - x_{Sn}$ we can combine both graphs of T versus x_{Sn} and T versus x_{Pb} according to Eq. (4.132) into one plot. They intersect at $x_{Sn} = 0.485$ and $T_e = 81.6\,°C$. Below the intersection both lines are continued as dashed lines. Above T_e and between the solid lines the mixture is a homogeneous liquid. The solid lines are the solubility limit of Sn in Pb or, above $x_{Sn} = 0.485$, Pb in Sn. T_e, the eutectic temperature , is the lowest temperature at which a mixture of Sn and Pb can exist as a homogeneous liquid. The intersection of the lines marks the so called eutectic point.

Predictions of this simple approach, even though they are helpful for our understanding, are neither quantitative nor complete. The true eutectic point of the

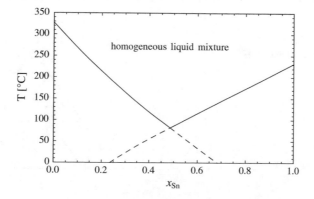

Fig. 4.26 Approximate solubility limits of tin and lead in a binary system

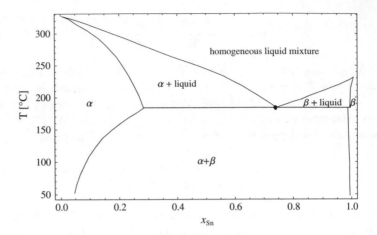

Fig. 4.27 The real phase diagram of the tin/lead-system

Sn/Pb-system is at $x_{Sn} = 0.73$ and $T_e = 183\,°C$. The real phase diagram of this system (at $P = 1bar$) is shown in Fig. 4.27 (based on data in HCP). As in Fig. 4.26 there is a homogeneous liquid mixture with an eutectic point shown as black dot. The regions marked α and β correspond to homogeneous solid phases rich in *Pb* and *Sn*, respectively. The remaining regions are phase coexistence regions.

4.3.4   Ternary Systems

Figure 4.28 explains how to read triangular composition phase diagrams of ternary systems. A ternary system contains the there components *A*, *B*, and *C*. By definition the side lengths of the equilateral triangle *ABC* are equal to one. At the point labeled *Q* the system has the composition x_A, x_B, and x_C. The position of *Q* within the triangle is described via the dashed lines possessing the respective lengths x_A, x_B,

Fig. 4.28 Triangular composition phase diagram of ternary systems

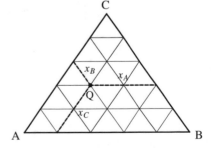

Fig. 4.29 Experimental example of a ternary phase diagram

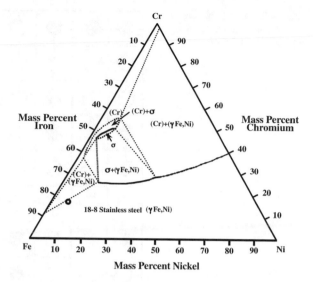

and x_C. The dashed line x_A is parallel to AB, x_B is parallel to BC, and x_C is parallel to AC. This definition satisfies the necessary condition $x_A + x_B + x_C = 1$. The validity of $x_A + x_B + x_C = 1$ is verified easily using the division of the original triangle into a lattice of smaller equilateral triangles. There is no loss of generality due to this meshing, because the coarse mesh in our example can be replaced by one which is arbitrarily fine.

An experimental example is depicted in Fig. 4.29 showing the iron-chromium-nickel ternary phase diagram at 900 °C (adapted from HCP (Fig. 23 of Sect. 12 p. 199, 89th edition)).

Exercise: Determine from Fig. 4.29 the composition of 18-8 stainless steel (open circle).

4.4   Phase Equilibria in Macromolecular Systems

4.4.1   A Lattice Model for Binary Polymer Mixtures

We return to the study of binary mixtures assuming that the two components are linear polymers. A simple but instructive approximation of a linear polymer is a path on a lattice as depicted in Fig. 4.30. Here the lattice is a square lattice and every lattice cell contains one polymer segment. Segments belonging to the same polymer are connected by a solid line. The solid and hollow circles indicate two chemically different types of segments. In the following we consider ν_i polymers of type i with length (or mass) m_i ($i = 1, 2$). This means that all polymers of type i posses the same length, i.e. they are monodisperse. In reality (technical) polymers

Fig. 4.30 Binary mixture of linear polymers represented by paths on a lattice

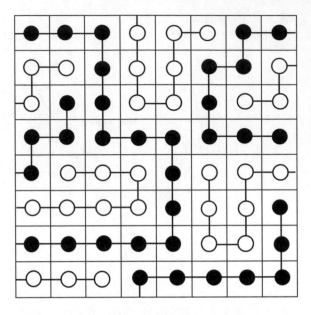

are polydisperse, i.e. they do have a distribution of lengths. Here we avoid this complication. In addition we assume that each of the m_i segments is one monomer of the real polymer for conceptual simplicity. There are $N = N_1 + N_2$ monomers total ($N_i = m_i v_i$) and N is equal to the number of lattice cells. This means that the lattice is fully occupied.

Having specified our model we want to estimate the number of distinct polymer configurations on the lattice:

(i)   We proceed by severing all bonds connecting the monomers in each polymer chain. The individual monomers, which we consider distinguishable at this point, are then placed on a new empty but otherwise identical lattice. There are $N!$ ways to accomplish this.

(ii) Now we ask: What is the probability that in one such configuration all monomers do have the same neighbors they had before in the polymer? We first approximate the probability that a particular monomer is placed in a cell next to its polymer-neighbor monomer via

$$\left(\frac{q-1}{N}\right)^{m_1-1} \quad \text{or} \quad \left(\frac{q-1}{N}\right)^{m_2-1}.$$

Here q is the coordination number of the lattice. This is the number of neighbors each cell has. On a square lattice $q = 4$; on a simple cubic lattice $q = 6$. This means that if we have a polymer partially laid out on the lattice and we put the next monomer, of which we know that it is the neighbor in the polymer, down on the lattice blindfolded, then there are $q - 1$ "good" cells compared to N cells total. Of course we neglect occupancy of the cells by previous monomers—a truly crude approximation. Nevertheless we approximate the above probability as

$$\left(\frac{q-1}{N}\right)^{(m_1-1)v_1}\left(\frac{q-1}{N}\right)^{(m_2-1)v_2}.$$

(iii) This number we multiply with $N!$, the total number of configurations. But we also must divide by the product $v_1!v_2!$, because two polymers of the same type are indistinguishable. All in all we find that the number of distinguishable ways to accommodate the polymers on the lattice, Ω, may be approximated via

$$\Omega \approx \frac{N!}{v_1!v_2!}\left(\frac{q-1}{N}\right)^{(m_1-1)v_1}\left(\frac{q-1}{N}\right)^{(m_2-1)v_2}, \tag{4.133}$$

We can now work out the entropy, which is the configuration entropy, via

$$S = N_A^{-1}R\ln\Omega. \tag{4.134}$$

The justification for this clearly important formula will be given in the next chapter. Here we merely consider its consequences. Using the Stirling formula, i.e.

$$\ln N! \approx N\ln N - N + \ln\sqrt{2\pi N} \approx N\ln N - N \quad \text{(if)} \quad N\text{is large,} \tag{4.135}$$

we obtain

$$\frac{S}{nR} = -\frac{\phi_1}{m_1}\ln\frac{\phi_1}{m_1} - \frac{\phi_2}{m_2}\ln\frac{\phi_2}{m_2} + \left[(1-\frac{1}{m_1})\phi_1 + (1-\frac{1}{m_2})\phi_2\right]\ln\frac{q-1}{e}, \tag{4.136}$$

where $n = N/N_A$ and $\phi_i = N_i/N$.

Before we discuss this, we compute the entropy of mixing given by

$$\frac{\Delta S}{N_A^{-1}R} = -v_1\ln\phi_1 - v_2\ln\phi_2. \tag{4.137}$$

This is the entropy change if we combine two lattices of size N_1 and N_2, each filled with the respective polymers of type 1 and 2, into one lattice of size $N = N_1 + N_2$, i.e.

$$\Delta S = S - S_1 - S_2, \tag{4.138}$$

where $S_i = N_A^{-1} R \ln \Omega_i$ and

$$\Omega_i \approx \frac{N_i!}{v_i!} \left(\frac{q-1}{N_i}\right)^{(m_i-1)v_i}. \tag{4.139}$$

Using again $v_i = N_i/m_i$ and $\phi_i = N_i/N$ Eq. (4.137) becomes

$$\frac{\Delta S}{nR} = -\frac{\phi_1}{m_1} \ln \phi_1 - \frac{\phi_2}{m_2} \ln \phi_2. \tag{4.140}$$

Notice that this equation is quite similar to the free enthalpy of mixing for a $K = 2$-component ideal gas (3.30) (in the case of a fully occupied lattice we have $\phi_i = x_i$). Only the factors $1/m_i$ are new.

Equations (4.136) and (4.140) are the backbone of a method describing thermodynamic properties of macromolecular systems akin to the van der Waals approach to low molecular weight systems . The lattice approach outlined here was pioneered independently by Staverman and van Santen (A. J. Stavermann, J. H. van Santen, Rec. Trav. Chim. **60**, 76 (1941)), Huggins (M. L. Huggins, J. Chem. Phys. **9**, 440 (1941); Ann. NY Acad. Sci. **43**, 1 (1942)) and Flory (Paul John Flory, Nobel prize in chemistry for his work on the physical chemistry of macromolecules, 1974) (P. J. Flory, J. Chem. Phys. **9**, 660 (1941); **10**, 51 (1942)) (cf. R. Koningsveld, L. A. Kleintjens *Fluid phase equilibria.* Acta Polymerica **39**, 341 (1988)).

A Digression—One-Component Gas-Liquid Phase Behavior

Equations (4.136) and (4.140) can be applied to a number of interesting situations. We introduce the replacements $\phi_1 = \phi$, $m_1 = m$, and $\phi_2 = 1 - \phi$. In addition we assume $m_2 = 1$. This corresponds to polymers in a solvent, where the index 2 indicates the solvent. The resulting configuration entropy is

$$\frac{S_{conf}}{nR} = -\frac{\phi}{m} \ln \frac{\phi}{m} - (1-\phi) \ln(1-\phi) + \phi\left(1 - \frac{1}{m}\right) \ln \frac{q-1}{e}. \tag{4.141}$$

If we replace the solvent cells by empty cells, we describe the same type of physical situation described by the van der Waals equation. Here the total volume is $V = bN$, where b, the cell size, also is the monomer size. We may obtain the attendant configurational pressure via

$$P_{conf} = -\frac{\partial}{\partial V}(-TS_{conf})\Big|_T = \frac{RT}{N_A b}\left[-\phi(1 - \frac{1}{m}) - \ln(1 - \phi)\right]. \tag{4.142}$$

Analogous to the van der Waals approach we must add a term accounting for attractive interaction between the monomers. Our choice, in analogy to the van der Waals equation of state, is

$$\frac{N_A bP}{RT} = \frac{N_A bP_{conf}}{RT} - \frac{1}{2}\frac{N_A \epsilon_o}{RT}\phi^2.$$ (4.143)

Here $\epsilon_o > 0$ is a parameter.

The closeness of this and the van der Waals equation of state becomes even more clear if we compute the gas-liquid critical parameters via $\frac{\partial P}{\partial V}|_T = \frac{\partial^2 P}{\partial V^2}|_T = 0$. We find

$$T_c = \frac{N_A \epsilon_o}{R}\frac{m}{(\sqrt{m}+1)^2}$$ (4.144)

$$\phi_c = \frac{1}{\sqrt{m}+1}$$ (4.145)

$$\frac{bP_c}{\epsilon_o} = \frac{\frac{1}{2} - \sqrt{m}\left(1 + \sqrt{m}\ln\left(\frac{\sqrt{m}}{\sqrt{m}+1}\right)\right)}{(\sqrt{m}+1)^2}.$$ (4.146)

In the limit $m = 1$ we therefore have

$$\frac{RT_c}{N_A \epsilon_o} = \frac{1}{4} \qquad \phi_c = \frac{1}{2} \qquad \frac{bP_c}{\epsilon_o} = \frac{2\ln 2 - 1}{8}.$$ (4.147)

Comparison with Eqs. (4.9) to (4.11) yields $N_A \epsilon_o = (32/27)(a_{vdW}/b_{vdW})$ and $N_A b = (3/2)b_{vdW}$.[22] We may work out the relation between critical and Boyle temperature,

$$T_{Boyle} = 4T_c,$$ (4.148)

or the critical compressibility factor

$$\frac{N_A P_c}{RT_c \rho_c} = 2\ln 2 - 1 \approx 0.39.$$ (4.149)

Both values are very close to the same quantities in the van der Waals theory (cf. Eqs. (4.20) and (4.21)).

But we are not interested in a competition with the van der Waals equation. We therefore look at the opposite limit, i.e. very long polymer chains, which is not

---

[22] These relations are not unique. Here we have used T_c and ϕ_c. Instead we can use T_c and P_c or ϕ_c and P_c. The resulting differences are small.

Fig. 4.31 Critical compressibility factor for n-alkanes

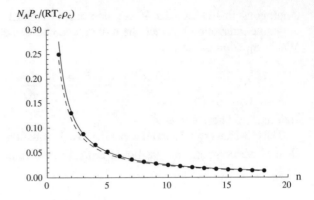

described by the van der Waals equation. In the limit $m \to \infty$ we have to leading order

$$\frac{RT_c}{N_A \epsilon_o} \approx 1 \quad \phi_c \approx \frac{1}{m^{1/2}} \quad \frac{bP_c}{\epsilon_o} \approx \frac{1}{3m^{3/2}}. \tag{4.150}$$

The corresponding leading behavior of the critical compressibility factor is

$$\frac{N_A P_c}{RT_c \rho_c} \approx \frac{1}{3m}. \tag{4.151}$$

Here ρ_c is the number density of the monomer units—not the polymers! Notice that the critical compressibility factor is not a constant independent of the type of molecule as before. Figure 4.31 shows the critical compressibility factor for n-alkanes $(1 \leq n \leq 18)$. The symbols are data from E. D. Nikitin *The critical properties of thermally unstable substances: measurement methods, some results and correlations.* High Temperature **36**, 305 (1998). The mass density was converted to the monomer number density using CH_2 as the monomer unit. This also implies $m = n$. The lines are fits to the data using the full expressions, i.e. Eqs. (4.144)–(4.146), (solid line) and the limiting law, Eq. (4.151), (dashed line). The only fit parameter is a multiplicative constant, i.e. instead of $1/(3m)$ we use $0.24/m$ to match the data for large m.

Notice also that expressing pressure, temperature, and volume or density in terms of their critical values eliminates the material parameters b and ϵ_o, but it does not eliminate m. This means that the resulting equation of state is not universal in the sense that it is different for molecules with different length, i.e. different m. Therefore the law of corresponding states is not obeyed by molecules with different m.

Polymer Mixtures

In Sect. 4.3.1 we had discussed liquid-liquid binodal curves for low molecular weight binary fluid mixtures. Figure 4.32 shows analogous binodal data points for a

Fig. 4.32 Binodal data points and theoretical fits for three binary polymer mixtures

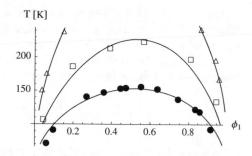

macromolecular fluid mixture determined by observation of the cloud points. The term "cloud point" refers to the turbidity observed upon passing from the homogeneous mixture into the coexistence region, where droplet formation increases the scattering of light. In our theoretical description we assume a fully occupied lattice.

Figure 4.32 shows polystyrene-polybutadiene mixture cloud point data taken from Fig. 4.3 in Roe (1980); PS2-PBD2 (solid circles); PS3-PBD2 (open squares), and PS5-PBD26 (open triangles); $m_{PS2} = 2220$, $m_{PS3} = 3500$, $m_{PS5} = 5200$, $m_{PBD2} = 2350$, $m_{PBD26} = 25000$. Note that ϕ_1 refers to PS. The solid lines are the results of a calculation analogous to the one which produced the solid lines in Fig. 4.22, i.e. we determine the binodal line by the common tangent construction applied to the mixing free enthalpy[23]

$$\frac{\Delta G}{nRT} = \frac{\phi_1}{m_1} \ln \phi_1 + \frac{1 - \phi_1}{m_2} \ln(1 - \phi_1) + \chi \phi_1 (1 - \phi_1). \tag{4.152}$$

Again we use Eq. (4.118) to describe χ. Notice that the χ-term in the literature sometime is denoted as an enthalpic contribution. This is not necessarily true, because for instance $\partial G / \partial T |_P = -S$, and if χ depends on temperature, as it usually does, then the χ-term contributes to the entropy as well. We had already pointed out that the physical interpretation of the χ-term is not straightforward. Here significant insight is needed into the microscopic interaction of polymer systems. A good starting point for the interested reader is the following paper by R. Koningsveld (R. Koningsveld, L. A. Kleintjens *Fluid phase equilibria*. Acta Polymerica **39**, 341 (1988)).

Polymers in Solution

We briefly want to discuss Eq. (4.152) when $m_2 = 1$, i.e.

$$\frac{\Delta G}{nRT} = \frac{\phi}{m} \ln \phi + (1 - \phi) \ln(1 - \phi) + \chi \phi (1 - \phi), \tag{4.153}$$

---

[23] It does not matter whether we apply the common tangent construction to the mixing free enthalpy or to the full free enthalpy.

where $\phi_1 = \phi$. This situation describes a polymer-solvent-system. The phase behavior of this system is in principle described by Fig. 4.21 (bottom-right panel), except of course without the symmetry around $x_A = 0.5$, i.e. $\phi_1 = 0.5$, unless $m_1 = 1$. Setting the coefficient $c_1 = 0$ in Eq. (4.118) we easily work out the critical temperature and the critical packing fraction, i.e.

$$\frac{c_0}{T_c} = \frac{1}{2} - c_2 + \frac{1}{\sqrt{m}} + \frac{1}{2m} \quad \text{and} \quad \phi_c = \frac{1}{\sqrt{m}+1}, \tag{4.154}$$

according to Shultz and Flory (Shultz abd Flory 1952).[24] These critical parameters follow via simultaneous solution of

$$\frac{\partial^2}{\partial \phi^2} \Delta G = 0 \quad \text{and} \quad \frac{\partial^3}{\partial \phi^3} \Delta G = 0 \tag{4.155}$$

(cf. Eqs. (4.116) and (4.117)). The critical solution temperature, T_c, measured for different m may be fitted via (4.154), i.e. T_c^{-1} versus $m^{-1/2} + (2m)^{-1}$, to determine c_0 and c_2 experimentally (for this particular mixture).

Osmotic Pressure in Polymer Solutions

Equation (4.153) may be used to calculate the osmotic pressure of polymers in solution. Again we employ the Gibbs-Duhem equation at constant temperature (3.136):

$$V\Pi_1 = \int_0^{v_1} d\mu_1(v_1') \frac{v_1'}{N_A} = \int_0^{v_1} dv_1 \frac{v_1'}{N_A} \frac{\partial^2 \Delta G}{\partial v_1^2}. \tag{4.156}$$

Notice that $d\mu_1(v_1)$ is due to altering the relative polymer content of the solution, which solely affects the mixing contribution of the free enthalpy. After some work, using $n = N/N_A$, $N = m_1 v_1 + m_2 v_2$, and $\phi_i/m_i = v_i/(nN_A)$, we find

$$\Pi_1 = \frac{RT}{V} n_2 \left[-\phi_1 \left(1 - \frac{1}{m}\right) - \ln(1 - \phi_1) - \chi \phi_1^2 \right], \tag{4.157}$$

where n_2 is the mole fraction solvent. It is instructive to expand the right side for small polymer concentration, i.e.

---

[24] Notice that T_c and ρ_c do agree with the same critical parameters in the case of the previously discussed gas-liquid critical point, cf. Eqs. (4.144) and (4.145), if $c_o = N_A \epsilon_o/(2R)$ and $c_2 = 0$.

$$\Pi_1 = \frac{RT}{V} n_1 + \frac{RT}{V} \frac{1}{n_2} \left(\frac{1}{2} - \chi \right) m^2 n_1^2 + \mathcal{O}(n_1^3), \qquad (4.158)$$

where we also have used $m \gg 1$. We do not want to discuss Eqs. (4.157) and (4.158) in much detail. It turns out that the lattice approach has a number of shortcomings. An insightful discussion of osmotic pressure in polymer solutions, including the present result, can be found in P.-G. de Gennes (1988) *Scaling Concepts in Polymer Physics*. Cornell University Press.

However, it is worth to compare Eq. (4.157) to the equation of state (4.143), obtained via the same combinatorial lattice approach applied to a fluid of small molecules. We recognize that both equations posses the same functional dependence on ϕ and ϕ_1. In Sect. 4.1.1 we had discussed the virial expansion of the van der Waals equation of state. The same expansion may be carried out in the case of the lattice equation of state (4.143). Analogously we may expand the osmotic pressure in powers of the solute concentration. Such an expansion, Eq. (4.158) shows the first two terms in the case of the lattice model discussed here, can be used to describe the deviations between van't Hoff's law and experimental data as in Fig. 3.10.

Somebody may object to this pointing out that the leading correction to van't Hoff's law in the case of electrolyte solutions is proportional to $c^{3/2}$ (cf. Eq. (3.143)), where c is the electrolyte concentration, rather than to c^2. However, expansions like (4.2) or (4.158) are based on assuming short-ranged microscopic interactions. In Statistical Mechanics it is shown how the configuration integral in the partition function (partition functions are introduced in Chap. 5) can be expanded in particle clusters consisting of one, two, three, ... particles at a time—corresponding to integer powers of the density. A particle may be a nobel gas atom or a molecule. The one-particle term results in the ideal gas law. The two-particle term results in its leading correction as described by the second virial coefficient—etc. This cluster or virial expansion is sensible only if the inter-particle interactions are short-ranged. Coulomb interactions, on the other hand, are long-ranged.[25] Even at low concentrations a particle (ion) interacts with numerous other particles—despite the screening which beyond some distance shrouds the presence of the particle at the origin.

---

[25] What is the meaning of short versus long? If two particles at a separation r interact with a potential r^{-n}, then the average potential energy per particle due to this interaction is $e \propto (\rho/2) \int_a^\infty dr r^{d-1-n}$ (cf. p. 78). Here ρ is the particle number density, a is the distance of closest approach (particle diameter), and d is the space dimension. The integral is finite only for $n > d$. Here long-ranged means that this condition is not satisfied. Of course this does not mean that e is infinite if $n \leq d$—it is not. It just means that the microscopic interaction must be dealt with more carefully. In particular it means that cluster expansions of the above type are not possible.

Chapter 5
Microscopic Interactions

5.1 The Canonical Ensemble

It is of course desirable to combine thermodynamics with our knowledge of the structure of matter. In particular we want to calculate Thermodynamic quantities on the basis of microscopic interactions between atoms and molecules or even subatomic particles.

We assume a large completely isolated system containing a by comparison extremely small subsystem. This subsystem is allowed to exchange heat with its surroundings and we have

$$E = E_{env} + E_v. \tag{5.1}$$

Here E is the total internal energy of the isolated system, whereas E_v is one particular value which the internal energy of the subsystem may assume. The difference between these internal energies is E_{env}, i.e. the internal energy of the subsystem's environment.

For the moment let us assume that the isolated system, with the subsystem currently removed, contains a gas of particles. If we take a photograph of this gas every once in a while, we observe that the particles move even though E remains constant. If we own a special camera, allowing to record the instantaneous velocity of every gas particle in addition to its position, then each snapshot fully characterizes the gas in the instant the picture is taken. We call this a microstate of the gas. This is a very mechanical point of view, and we know that classical mechanics has its limitations. For instance it is not really possible to determine both the position and the velocity, or rather the momentum, of a gas particle with arbitrary precision according the uncertainty principle of quantum mechanics. We nevertheless assume that the concept of microstates remains valid in the sense that there are many somehow different realizations of our system belonging to the *same* energy E. This is the key premisses of what follows, i.e. every energy value a system assumes can be realized

© Springer Nature Switzerland AG 2022
R. Hentschke, *Thermodynamics*, Undergraduate Lecture Notes in Physics,
https://doi.org/10.1007/978-3-030-93879-6_5

by a vast number of microstates. We call this number $\Omega(E)$. We can apply this microstate-idea also to the above subsystem. In this sense different v-values mean different microstates, i.e. E_v is the internal energy of the subsystem in microstate v.

Having said this we may now continue by studying $\Omega(E_{env}) = \Omega(E - E_v)$, the number of microstates of the environment all possessing the same energy $E - E_v$. Progress requires two additional and important assumptions: (i) all microstates are equally probable; (ii) the probability that our subsystem has the energy E_v, p_v, is proportional to the number of microstates available to the environment under this constraint, i.e.

$$p_v \propto \Omega(E - E_v). \tag{5.2}$$

The first assumption, known as the *postulate of equal a priori probabilities*, sounds quite reasonable, because there is nothing we can think of which favors one particular microstate over another if both have the same energy. The second assumptions corresponds to a *principle of least constraint*, i.e. a subsystem microstate v is more likely than another if the by comparison huge environment suffers a smaller reduction of its available microstates.

A useful expression for p_v can be derived by expanding $\Omega(E - E_v)$ or rather $\ln \Omega(E - E_v)$ in a Taylor series around E, i.e.

$$\ln \Omega(E - E_v) = \ln \Omega(E) - \left. \frac{d \ln \Omega(E)}{dE} \right|_E E_v + \dots. \tag{5.3}$$

Using the definition

$$\beta \equiv \left. \frac{d \ln \Omega(E)}{dE} \right|_E \tag{5.4}$$

and neglecting higher order terms, which we shall justify below, we may write

$$p_v \propto \Omega(E) \exp[-\beta E_v]. \tag{5.5}$$

By introducing another quantity, the so called canonical partition function

$$\boxed{Q_{nVT} = \sum_v \exp[-\beta E_v]}, \tag{5.6}$$

we may use $1 = \sum_v p_v$ to finally express p_v as

$$p_v = \frac{\exp[-\beta E_v]}{Q_{nVT}}. \tag{5.7}$$

In order to understand how this relates to thermodynamics we calculate the average energy of the subsystem

$$\langle E \rangle = \sum_v E_v p_v = \frac{\sum_v E_v p_v}{\sum_v p_v} = \frac{\partial}{\partial(-\beta)} \ln Q_{NVT}. \tag{5.8}$$

For this to be consistent with thermodynamics we must require (cf. Eq. 2.116)

$$\frac{\partial}{\partial(-\beta)} \ln Q_{nVT} = F + TS = F - T\frac{\partial F}{\partial T}\Big|_V = \frac{\partial}{\partial(-1/T)}\left(-\frac{F}{T}\right)\Big|_V. \tag{5.9}$$

Comparing the left side with the last expression on the right we conclude $\beta \propto T^{-1}$ and

$$\boxed{F = -\beta^{-1} \ln Q_{nVT}}. \tag{5.10}$$

The proportionality constant between β and T^{-1} is the gas constant R, i.e.

$$\beta = \frac{1}{RT}, \tag{5.11}$$

if we continue to use moles (n), or Boltzmann's constant k_B, i.e.

$$\beta = \frac{1}{k_B T}, \tag{5.12}$$

if we use the number of particles (N), i.e. atoms, molecules, etc. instead. Eq. (5.10) is an important result. It allows to obtain the free energy, F, from the partition function Q_{nVT}. Q_{nVT} may be computed if the possible energy values, E_v, of our closed subsystem are known.

Example—A Model Magnet Imagine a system consisting of just one magnetic moment variable s. The possible values of s are $s_v = \pm 1 \, (up/down)$. We also assume that $E_v = -J\langle s \rangle s_v$, where $J > 0$ is a coupling constant and $\langle s \rangle$ is the thermal average value of s. Somebody may object that thus far we have assumed macroscopic subsystems, but here the subsystem contains one magnetic moment only. However, what we really do is to assume that there are many s, which do not interact with other s individually but rather with the normalized average magnetization $\langle s \rangle$. This is called a mean field approximation.

The effective one-magnetic moment-partition function simply is

$$Q \underset{=}{\overset{(6.6)}{}} e^{\beta J \langle s \rangle} + e^{-\beta J \langle s \rangle} = 2\cosh(\beta J \langle s \rangle) \tag{5.13}$$

and the average magnetization per moment can be computed via

$$\langle s \rangle = \sum_v s_v p_v = \frac{\partial \ln Q}{\partial(\beta J \langle s \rangle)} = \tanh(\beta J \langle s \rangle). \tag{5.14}$$

This implicit equation for $\langle s \rangle$ has one solution, i.e. $\langle s \rangle = 0$, when $\beta J < 1$. But for $\beta J > 1$ it has two additional solutions, $\langle s \rangle = \pm \langle s \rangle_o \neq 0$ (cf. Fig. 5.1). In this case we must find the stable solution for which the free energy is lowest. The free energy is given by

$$F \underset{=}{^{(6.10)}} -\frac{1}{\beta} \ln \cosh(\beta J \langle s \rangle). \tag{5.15}$$

We see that the solution $\langle s \rangle = 0$ is unstable in comparison to the other two solutions $\langle s \rangle = \pm \langle s \rangle_o$. Because $F(-\langle s \rangle_o) = F(\langle s \rangle_o)$, both solutions are equally stable. If the system is cooled from $T > T_c = J/k_B$ to $T = T_c$ and below, it must "decide" whether to follow the positive (up magnetization) or negative (down magnetization) (cf. the right panel in Fig. 5.1). This decision is made by thermal fluctuations and is called spontaneous symmetry breaking.

It is important to note that Eq. (5.4) together with Eqs. (5.12) and (1.51) yields

$$\boxed{S(E) = k_B \ln \Omega(E)}. \tag{5.16}$$

This relates the entropy, S, of an isolated system with energy E to its number of microstates, $\Omega(E)$. In Sect. 5.4 we had used (5.16) to construct the entropy in macromolecular systems treating the macromolecules as linear paths on a lattice.

Example—Order-to-Disorder Transition in 1D and 2D An example illustrating nicely the significance of Eq. (5.16) is depicted in Figs. 5.2 and 5.3. The upper portion of Fig. 5.2 shows a one-dimensional chain of arrows (or magnetic moments—we recognize the relation to the previous example) all pointing

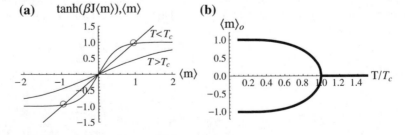

Fig. 5.1 Magnetization and its temperature dependence

Fig. 5.2 Introducing a domain wall into a perfectly ordered chain of arrows

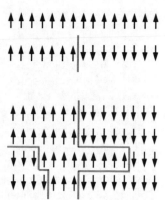

Fig. 5.3 A domain wall in the two-dimensional case

up. This system is fully ordered. The lower portion shows the same row after introduction of a domain wall, which means that all arrows to the left of the domain wall are upside down. We define an internal energy of this system via

$$E = -J \sum_{i=1}^{N-1} s_i s_{i+1} \tag{5.17}$$

(no mean field approximation in this case). Here J is a positive (coupling) constant and $s_i = \pm 1$ ($s_i - 1$ for up-arrows and $s_i = -1$ for down arrows). Notice that this internal energy is invariant under simultaneous inversion of all N arrows. Thus there are two equivalent types of complete orientational ordering—all arrows up and all arrows down. The internal energy difference between the bottom and the top row is

$$\Delta E = E_{\text{bottom}} - E_{\text{top}} = 2J. \tag{5.18}$$

What, however, is the corresponding change in entropy, ΔS? The only distinguishing feature between chains with one domain wall is the position of the domain wall along the chain. In the present case there are $N - 1$ different positions (disregarding left-right symmetry). If we identify the number of different positions with the number of microstates of this system (note that shifting the domain wall position does not alter the chain's energy) we find

$$\Delta S = k_B \ln N \tag{5.19}$$

(we use $N - 1 \to N$ in the limit of $N \to \infty$). Therefore the change of the free energy at constant temperature due to insertion of one domain wall is

$$\Delta F = \Delta E - T\Delta S = 2J - k_B T \ln N. \tag{5.20}$$

In the thermodynamics limit, i.e. $N \to \infty$, we always find $\Delta F \mid_T < 0$ for $T > 0$. However according to (2.130) such change occurs spontaneously. And since this remains true for the insertion of a second, third, and every following domain wall the orientational ordering is completely destroyed, i.e. the disordered state with $\sum_{i=1}^{N} s_i = 0$ is the thermodynamically stable one!

But what happens if we repeat this "experiment" in two-dimensions? Fig. 5.3 shows a two-dimensional lattice of arrows containing two domain walls meandering through the system. We proceed as in the one-dimensional case and compute first the change of the internal energy due to a domain wall consisting of n pairs of arrows, i.e.

$$\Delta E = E_{\text{withdw}} - E_{\text{withoutdw}} = 2Jn. \tag{5.21}$$

The attendant entropy change is

$$\Delta S = k_B \ln p^n. \tag{5.22}$$

Here p is the number of possible orientations of each of the n domain wall segments relative to its predecessor. In our figure this means left turn, right turn, and no turn, i.e. $p = 3$. But this is an overestimate as illustrated in Fig. 5.4. It shows that two domain walls cannot meet. This means that occasionally p is reduced to two orientations or, in rare cases, to just one. Thus the free energy change is

$$\Delta F = (2J - k_B T \ln p)n. \tag{5.23}$$

Clearly, the sign of ΔF does depend not on n but on the term in brackets. It will change at a distinct or critical temperature T_c, i.e.

$$\frac{k_B T_c}{J} = \frac{2}{\ln p} \approx \begin{cases} 1.82 & \text{if } p = 3 \\ 2.89 & \text{if } p = 2 \end{cases} \tag{5.24}$$

For temperatures above T_c the sign of ΔF is negative. Domain walls are created spontaneously by the system destroying any orientational order of the arrows. Below T_c the opposite is true, i.e. domain walls are not stable and

Fig. 5.4 Two domain walls cannot meet

orientational order is the consequence. An exact calculation of this model system, it is called the Ising model, can be done, and the result is that in 2D a finite T_c does exist (Kramers and Wannier 1941; Onsager 1944)—However, it was Rudolf Peierls who first showed that the two-dimensional Ising model has an order-to-disorder transition (Peierls 1936). The exact value, i.e. $k_B T_c / J = 2.269\ldots$, is in fact bracketed by our above estimates for $p = 3$ and $p = 2$!

We conclude that whereas in one dimension no transition is possible the situation is completely different in two (and higher) dimensions. But notice that we could have extended the range of interaction in one dimension to include all arrows. Doing this it is possible to have $\Delta E = \mathcal{O}(N)$ and thus we see that our conclusion does rest on the assumption of a finite interaction range! In fact, the previous example shows this. The mean field approximation implies an infinite interaction range, and because the dimensionality of the system does not enter, it would lead us to conclude that the 1D chain always undergoes a transition at $T_c = J/k_B$.

Remark In the case $J \sim r^{-1-\sigma}$, where r is the distance separating interacting arrows, there exists a critical point in the 1D Ising model for $\sigma < 1$ but it is absent for $\sigma > 1$ (Dyson 1969).

Example—Scaled Particle Theory This is another example in which Eq. (5.16) plays a significant role. Figure 5.5 depicts particles in a gas. Here we assume that the particles are hard spheres. But other compact hard particle shapes are possible too. The figure shows two types of particles—large (grey) ones and one small (black) one. The large particles are spheres with radius R. The small particle is a sphere with radius λR. In the figure the scaling parameter $\lambda \ll 1$. The idea of scaled particle theory (developed in the late 1950s by H. Reiss, H. L. Frisch, and J. L. Lebowitz) is simple: (i) work out the chemical potential of the scaled particle in the two limits $\lambda \ll 1$ and

Fig. 5.5 Hard sphere particles including a scaled particle

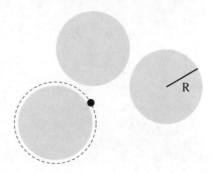

$\lambda \gg 1$; (ii) by interpolation between the two limits derive an approximate chemical potential for $\lambda = 1$; (iii) use this chemical potential to find an approximate equation of state for the gas. As it turns out, this approach yields a pretty good equation of state—even for a dense gas of hard particles.

We start with step (i). The chemical potential of the scaled particle is

$$\mu_{sp} = \mu_{sp,id} + \mu_{sp,ex}. \tag{5.25}$$

The two indices, *id* and *ex*, denote the ideal and excess part of the chemical potential, respectively. Here the ideal part corresponds to the a situation when all ordinary particles are absent. If Ω_{sp} is the number of (micro)states available to the scaled particle, we may, if $\lambda \ll 1$, write

$$\Omega_{sp} = \Omega_{sp,id} \frac{\Omega_{sp}}{\Omega_{sp,id}} \approx \Omega_{sp,id} \frac{V - v_e(\lambda)N}{V} = \Omega_{sp,id}(1 - v_e(\lambda)\rho). \tag{5.26}$$

Notice that $\Omega_{sp}/\Omega_{sp,id}$ is identified with the ratio of the volume available to the center of the scaled particle divided by the total volume of the system, V. The quantity $v_e(\lambda)$ is the volume excluded to the scaled particle's center by the presence of one ordinary particle—indicated by the dashed circle in Fig. 5.5. Thus, in the case of spheres, $v_e(\lambda) = 4\pi R^3(1 + \lambda)^3/3$. If we disregard the overlap between excluded volumes defined in this fashion, then the excluded volume $v_e(\lambda)$ multiplied with the number of ordinary particles, N, is the total available volume ($\rho = N/V$). Notice also that the approximation becomes exact in the limit $\lambda \to 0$. Using Eq. (5.16) we may write

$$\mu_{sp} = \mu_{sp,id} - RT \ln[1 - v_e(\lambda)\rho] \qquad (\lambda \ll 1) \tag{5.27}$$

for very small λ. Due to the hard particle assumption there is no enthalpic contribution to the free enthalpy of the scaled particle. Now we consider the opposite limit, i.e. $\lambda \gg 1$. This means that the scaled particle is inflated like a ballon against the constant pressure, P, exerted by the ordinary particles. Insertion of the scaled particle into the system therefore requires the (reversible) work $Pv_{sp}(\lambda)$, where $v_{sp}(\lambda) = 4\pi R^3 \lambda^3/3$. Thus in this limit

$$\mu_{sp} = \mu_{sp,id} + Pv_{sp}(\lambda) \qquad (\lambda \gg 1). \tag{5.28}$$

Step (ii) is the interpolation between the two limits via

$$\mu_{sp,ex}(\lambda) = c_o + c_1\lambda + c_2\lambda^2 + Pv_{sp}(\lambda). \tag{5.29}$$

The coefficients c_i are obtained by expanding $v_e(\lambda)$ at $\lambda = 0$ to second order in λ. We find

$$c_o = -\ln[1 - v] \qquad c_1 = \frac{3v}{1 - v} \qquad c_2 = \frac{1}{2}\left(\frac{6v}{1 - v} + \frac{9v^2}{(1 - v)^2}\right), \quad (5.30)$$

where v is the so called volume fraction $v = b\rho$, and b is the ordinary particle volume.

The final step, step (iii), consists in setting $\lambda = 1$. The scaled particle now is an ordinary particle too and its chemical potential, μ, also is that of an ordinary particle. The ideal part of the chemical potential is given by Eq. (5.66), an equation yet to be derived, plus the excess part given by Eq. (5.29), where $\lambda = 1$. We obtain the pressure in our hard sphere gas via integration of the Gibbs-Duhem equation, i.e.

$$\frac{\partial (P/RT)}{\partial \rho}\bigg|_T = \rho \frac{\partial (\mu/RT)}{\partial \rho}\bigg|_T. \qquad (5.31)$$

Straightforward integration yields the desired equation of state:

$$\frac{P}{RT\rho} = \frac{1 + v + v^2}{(1 - v)^3}. \qquad (5.32)$$

How can we test this result? Surely, small molecules are not hard spheres. However, in Fig. 3.10 we had discussed osmotic pressure data obtained from hemoglobin in aqueous solution. Hemoglobin is rather large and roughly spherical. In addition, we had argued on page 205 that the osmotic pressure can be approximated by equations like Eq. (5.32), i.e. the right side of this equation multiplies the van't Hoff equation, and the result should yield an improved osmotic pressure. This is indeed the case. The solid line in Fig. 3.10 is obtained in this fashion by adjusting the hemoglobin volume $b = b_{Hb}$. A good fit to the experimental data requires a hemoglobin diameter of ≈ 5.5 nm—in good accord with its linear dimension obtained via more detailed considerations.

Remark Scaled particle theory is a clever way to obtain an approximate equation of state for a non-ideal gas of hard bodies, for which we can work out the excluded volume, i.e. the equivalent of the dashed line in Fig. 5.5. But because it is an excluded volume theory, i.e. there is no attractive interaction as in the van der Waals theory, it cannot describe a gas-liquid phase transition. It is limited to situations when the phenomenon of interest is governed by excluded volume interaction. This is not the case for gases of small molecules. Perhaps the best example are lyotropic liquid crystalline systems (e.g., Odijk 1986). These are solutions containing large molecules or molecular aggregates. The excluded volume interaction here can lead to the spontaneous formation of anisotropic phases. In the simplest case the orientation of rod-like large molecules or large molecular aggregates, which at low solute concentration is isotropic, spontaneously becomes nematic, i.e. the "rods" on average align along a certain direction in space (called the director), when the concentration is increased (e.g., Herzfeld 1996).

Example—Rubber Elasticity and Thermal Contraction According to experience most materials expand when their temperature increases, i.e. their thermal expansion coefficient, α_P, (cf. Eq. (2.5)), is positive. However, take a rubber band fixed at one end and stretched by a weight attached to its other end. Upon heating of the rubber band, using a heat gun or a hair dryer capable of producing sufficient heat, a significant contraction is observed. Why does this happen? Once again Eq. (5.16) helps to find the answer. First however we must tie the entropy to the thermal contraction just described.

Equation (1.4) describes the work done by the elastic forces inside an elastic body. The attendant free energy is $F = \int_V dVf$, where the integration is over the volume of the rubber band, and the differential free energy density is

$$df = -s\,dT + \sigma_{\alpha\beta}\,du_{\alpha\beta}. \tag{5.33}$$

Here $-s = \partial f/\partial T|_{u_{\alpha\beta}}$ and $\sigma_{\alpha\beta} = \partial f/\partial u_{\alpha\beta}|_T$. $\sigma_{\alpha\beta}$ and $u_{\alpha\beta}$ are the components of the stress and the strain tensors, respectively. Note that we use the summation convention. It is useful to introduce the free enthalpy density

$$g = f - \sigma_{\alpha\beta}u_{\alpha\beta}, \tag{5.34}$$

i.e.

$$dg = df - d(\sigma_{\alpha\beta}u_{\alpha\beta}) = -s\,dT + \sigma_{\alpha\beta}\,du_{\alpha\beta} - d(\sigma_{\alpha\beta}u_{\alpha\beta}) = -s\,dT - u_{\alpha\beta}\,d\sigma_{\alpha\beta}.$$

Taking the derivative of $s = -\partial g/\partial T|_{\sigma_{\alpha\beta}}$ with respect to σ_{zz}, z being the direction parallel to the rubber band, we obtain

$$\left.\frac{\partial s}{\partial \sigma_{zz}}\right|_T = -\frac{\partial}{\partial \sigma_{zz}}\left.\frac{\partial g}{\partial T}\right|_{\sigma_{\alpha\beta}}\bigg|_T = \left.\frac{\partial u_{zz}}{\partial T}\right|_{\sigma_{\alpha\beta}} = \alpha_{\sigma,1D}. \tag{5.35}$$

Note that $u_{zz} = \delta L/L_o$, where δL is the elongation of the rubber band and L_o is its unstrained length, i.e. $\partial u_{zz}/\partial T|_{\sigma_{\alpha\beta}} = L^{-1}\partial L/\partial T|_{\sigma_{\alpha\beta}} = \alpha_{\sigma,1D}$ is the one-dimensional analog of Eq. (2.5).

At this point we need an expression for the entropy, S, of the rubber band. Figure 5.6 shows a cartoon of a linear polymer molecule of which rubber is made of. As before in the context of phase equilibria in macromolecular

Fig. 5.6 A polymer chain on a lattice

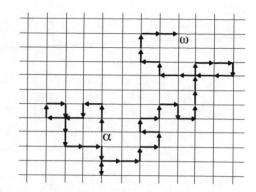

systems we model the polymer as a random path on a cubic lattice. We ask the question: What is the probability $p(R)$ that the two ends, labeled α and ω, do have the separation R? The answer is $p(R) = \Omega(R) / \sum_R \Omega(R)$. Here $\Omega(R)$ is the number of different paths of length n originating from the same lattice point and ending a distance R from the origin. The denominator consequently is the sum over all possible paths of length n originating from the same lattice point. Using Eq. (5.16) we have

$$S(R) - S = k_B \ln p(R), \tag{5.36}$$

where $S(R) = k_B \ln \Omega(R)$ and $S = k_B \ln \sum_R \Omega(R)$. The end-to-end vector $\vec{R}$ is given by

$$\vec{R} = \sum_{i=1}^n (x_i, y_i, z_i) = \left(\sum_{i=1}^n x_i, \sum_{i=1}^n y_i, \sum_{i=1}^n z_i \right), \tag{5.37}$$

where x_i, y_i, and z_i are random variables. Each of these may assume the values $\{-a, 0, 0, 0, 0, a\}$ with equal likelihood. Here a is the lattice spacing and the six values correspond to the six possible orientations of the step-arrows (in Fig. 5.6) along the main axes of the cubic lattice. We obtain $p(R)$ via an important mathematical theorem—the central limit theorem. This theorem states that if s_i are random variables with the average μ_s and the mean square fluctuation σ_s^2 then the new random variable,

$$S_n = \frac{\sum_{i=1}^n s_i - n\mu_s}{\sigma_s \sqrt{n}}, \tag{5.38}$$

possesses the probability density,

$$f(S_n) = \frac{1}{\sqrt{2\pi}} \exp[-S_n^2/2] \,, \tag{5.39}$$

in the limit of infinite n. However, this remains a very good approximation even if n is not very large (The reader may confirm this by generating S_5 from random numbers $s_i \in (0, 1)$ (i.e. $\mu_s = 1/2$ and $\sigma_s = 1/\sqrt{12}$). Construction of a distribution histogram based on $10^4 S_5$-values generated in this fashion closely approximates a Gaussian distribution with zero mean and standard deviation equal to unity.). Based on the central limit theorem we immediately conclude

$$p(R) \approx \frac{f(X_n)f(Y_n)f(Z_n)}{(na^2/3)^{3/2}} = \left(\frac{3}{2\pi na^2}\right)^{3/2} \exp\left[-\frac{3}{2}\frac{R^2}{na^2}\right], \tag{5.40}$$

where $\mu_x = \mu_y = \mu_z = 0$, $\sigma_x^2 = \sigma_y^2 = \sigma_z^2 = a^2/3$, and $4\pi \int_0^\infty dR R^2 p(R) = 1$. Using this expression in Eq. (5.36) we obtain

$$S(R) - S(0) = -\frac{3k_B R^2}{2na^2}. \tag{5.41}$$

At this point we know how the conformation entropy of a polymer chain, containing n links, changes when its end-to-end distance R is changed. Even though our model is a rough coarse-grained model of a real polymer, it is still a model of a single polymer chain and not yet a continuum model of the rubber band.

Real rubber is a complex material. It consists largely of linear polymer chains, but the chains are cross-linked. These cross-links can be chemical bonds (e.g. sulfur bridges formed during a process called vulcanization) between different polymer chains (or even within the same chain). They may also be physical entanglements. We may view the points labeled α and ω in Fig. 5.6 as the positions of two such cross-links. Thus, when we stretch rubber, we really stretch a complex flexible network called elastomer. We have also ignored that the polymer chains are real molecules interacting via specific microscopic interactions. Even though rubber deforms easily, its compressibility is that of a liquid, i.e. its volume hardly changes under deformation.

A cartoon of a rubber volume element is depicted in Fig. 5.7a. Overall the rubber band is a network of ν cross-linked chain segments. Every segment contains n links, whose individual $S(R_i)$ ($i = 1, \ldots, \nu$) are given by Eq. (5.41). If a macroscopic volume element inside the rubber band, possessing the edge lengths L_x, L_y, and L_z, is deformed, its new edge lengths are $L_x' = \lambda_x L_x$, $L_y' = \lambda_y L_y$, and $L_z' = \lambda_z L_z$. We assume that the segment end-to-end vectors

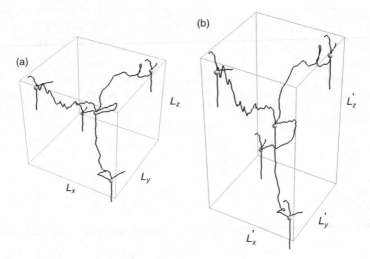

Fig. 5.7 A rubber volume element containing cross-linked chain segments in the relaxed (**a**) and in the deformed (**b**) state. Cross-links are indicated by the dots.

contained in this volume element will change their components analogously. Hence

$$\frac{L_\alpha}{R_{i,\alpha}} = \frac{L'_\alpha}{R'_{i,\alpha}}, \tag{5.42}$$

which is an affine deformation (cf. Fig. 5.7b). Rubber has a very small compressibility and consequently we also assume that $V' = L'_x L'_y L'_z = L_x L_y L_z = V$. If we stretch the rubber band by a factor $\lambda_z \equiv \lambda$ in z-direction, i.e. $L'_z = \lambda L_z$, volume conservation, i.e. $\lambda_x \lambda_y \lambda_z = 1$, implies $\lambda_x = \lambda_y = \lambda^{-1/2}$.

With this we are able to calculate the total elastic entropy change ΔS_{el} during stretching of the rubber band in terms of the entropy changes experienced by the individual network segments. According to Eq. (5.41) we have

$$\Delta S_{el} = \sum_{i=1}^{v} \left(S(R'_i) - S(R_i) \right) = -\frac{3k_B}{2na^2} \sum_{i=1}^{v} \left(R'^2_i - R^2_i \right). \tag{5.43}$$

Replacing R'^2_i by $R'^2_i = \lambda^{-1} R^2_{x,i} + \lambda^{-1} R^2_{y,i} + \lambda^2 R^2_{z,i}$ yields

$$\Delta S_{el} = -\frac{3k_B}{2na^2} \sum_{i=1}^{v} \left((\frac{1}{\lambda} - 1) R^2_{i,x} + (\frac{1}{\lambda} - 1) R^2_{i,y} + (\lambda^2 - 1) R^2_{i,z} \right). \tag{5.44}$$

Because the number of segments, ν, is a very large number, we can write $\frac{1}{\nu}\sum_{i=1}^{\nu} R_{i,\alpha}^2 = \langle R_\alpha^2 \rangle$. Here $\langle R_\alpha^2 \rangle$ is the ensemble average of the square of the α-component of the end-to-end separation of the segments. Since $\langle R_\alpha^2 \rangle$ refers to the relaxed state, all components behave equally and thus $\langle R_\alpha^2 \rangle = \langle R^2 \rangle/3 = na^2/3$. The last equality follows directly from $p(R)$. Hence

$$\Delta S_{el} = -\frac{k_B \nu}{2}\left(\frac{2}{\lambda} + \lambda^2 - 3\right). \tag{5.45}$$

It is important to note that ΔS_{el} does not comprise the entire entropy of the polymer chains forming the network. However, in the following we assume that it is the most important part of this entropy and neglect the rest. In particular we compute σ_{zz} via

$$\sigma_{zz} = \left.\frac{\partial f}{\partial u_{zz}}\right|_T \approx -\frac{1}{TV}\frac{\partial \Delta S_{el}}{\partial(\lambda - 1)} = \frac{k_B T \nu}{V}\left(\lambda - \frac{1}{\lambda^2}\right). \tag{5.46}$$

We can now compute $\alpha_{\sigma,1D}$ with the help of Eq. (5.35), i.e.

$$\alpha_{\sigma,1D} = \left.\frac{\partial s}{\partial \sigma_{zz}}\right|_T = \left.\frac{\partial s}{\partial \lambda}\right|_T \left(\left.\frac{\partial \sigma_{zz}}{\partial \lambda}\right|_T\right)^{-1}. \tag{5.47}$$

Again $s \approx \Delta S_{el}/V$ and we find for $\alpha_{\sigma,1D}$

$$\alpha_{\sigma,1D} = -\frac{1}{T}\frac{\lambda(\lambda^3 - 1)}{\lambda^3 + 2}. \tag{5.48}$$

Let us discuss this. Note first the $1/T$-dependence of $\alpha_{\sigma,1D}$. This is the same temperature dependence we had obtained for the thermal expansion coefficient of the ideal gas, which is not too surprising because here we also neglect all direct interactions. The first important step along the way to Eq. (5.48) is Eq. (5.41). It tells us that the entropy of a polymer chain is reduced when it is stretched. This is because increasing the end-to-end distance reduces the number of possible paths the chain can assume. In principle this the reason why $\alpha_{\sigma,1D}$ is negative, i.e. increasing the temperature causes contraction of the rubber band. However, if one tries to carry out the experiment mentioned in the beginning, one may be surprised to find that the rubber band behaves conventionally and expands with increasing temperature. Just like any simple liquid, the thermal expansion coefficient of relaxed rubber is positive (due to anharmonicity of the microscopic interactions between the monomers in the polymer chain). But notice that $\alpha_{\sigma,1D}$ according to Eq. (5.48) increases if λ is increased (e.g. for $T = 350$ K and 1% strain, i.e. $\lambda = 1.01$, we find $\alpha_{\sigma,1D} \approx -3 \cdot 10^{-5}$ K^{-1}; at the same T but for 30% strain, i.e. $\lambda = 1.3$, $\alpha_{\sigma,1D} \approx -10^{-3}$

K^{-1}; typical for most liquids is $\alpha_{P,3D} = 3\,\alpha_{P,1D} \approx 10^{-3}$ K^{-1}). With increasing strain one observes what is called thermoelastic inversion (cf. Strobl 1997). This means that the ordinary thermal expansion is overtaken by a contraction in response to the reduction of the chains conformation entropy as we have just discussed.

Remark 1 Expanding the right hand side of Eq. (5.46) to first order in the strain, i.e. in $\lambda - 1$, yields

$$\sigma_{zz} \approx \frac{3k_B T v}{V}(\lambda - 1). \tag{5.49}$$

Essentially this is Hook's law and the factor multiplying $\lambda - 1$ is the elastic modulus of the rubber. Note that the latter is proportional to temperature and to the density of segments or, equivalently, to the cross-link density. The more cross-links there are the stiffer the rubber becomes. It is worth noting that the right hand side of Eq. (5.46) is a good description of reality at small λ but fails when λ becomes large, mainly because of the finite extensibility of a polymer chain. The meaning of "small" versus "large" and an in depth discussion of other points beyond the scope of this (idealized) example can be found in Rubinstein and Colby (2003).

Remark 2 The formulas (5.43) to (5.45) for ΔS_{el} are not correct under conditions when the volume is not constant. An example is the swelling of a polymer network by a solvent. The swelling caused by the establishment of chemical equilibrium of solvent inside and outside the polymer network stretches the chain segments uniformly and reduces their conformational entropy—which physically is the same effect as before. In this case however $V' \neq V$ and $\lambda_x = \lambda_y = \lambda_z \equiv \lambda$. In general (5.45) is replaced by

$$\Delta S_{el} = -\frac{k_B v}{2}\left(\lambda_x^2 + \lambda_y^2 + \lambda_z^2 - 3\right) + \frac{k_B v}{2}\ln(\lambda_x \lambda_y \lambda_z). \tag{5.50}$$

For $\lambda_x = \lambda_y = \lambda^{-1/2}$, and $\lambda_z = \lambda$ the ln-term vanishes and we obtain (5.45). But if $V' \neq V$ the ln-term does not vanish and must be included. This term has two sources. The first one is the normalization of (5.40). If we want to compare entropies in the relaxed and the deformed state, we must do this in the same coordinate system. This means that there is a Jacobian, connecting the unprimed with the primed system, giving rise to a term $k_B v \ln(\lambda_x \lambda_y \lambda_z)$. Source number two yields an extra term $-k_B v \ln(\lambda_x \lambda_y \lambda_z)/2$, which is an extra entropy change $k_B \ln[(\delta V/V')^{v/2}/(\delta V/V)^{v/2}]$ (and again an application of

Eq. (5.16)). Here δV is the volume occupied by a cross-link, i.e. $\delta V/V$ and $\delta V/V'$ are the respective probabilities to find the cross-link in a particular δV in the undeformed and in the deformed system. Since the system contains v chain segments, this implies that it contains $v/2$ cross-links (look again at Fig. 5.7 and think about this!). Assuming independent cross-links, the total probability for finding the cross-links in their δV-cells is $(\delta V/V)^{v/2}$ and $(\delta V/V')^{v/2}$, respectively. This term, originally due to P. Flory, of course is an approximation ignoring all correlations.

Remark 3 We can use Eq. (5.46) to measure the density of chain segments or cross-links in a piece of rubber. First we adjusting λ to a value which is sufficiently large so that the ordinary thermal expansion is small compared to the expected contraction. For this λ-value (and perhaps additional ones) we measure σ_{zz} in a range of temperatures. The slope of a line through these data points, divided by $k_B T(\lambda - \lambda^{-2})$, yields v/V. However, there is another method for measuring v/V, which is used frequently in industry laboratories. Since we have all the necessary ingredients available, it is worth to briefly talk about this method as well.

Equation (4.153) describes the mixing free enthalpy of a polymer-solvent system on a lattice, where ϕ is the volume fraction polymer. If we express the solvent volume fraction $1 - \phi$ via $1 - \phi = N_s/N$, where N_s is the number of solvent cells and N is the total number of cells, the result is

$$\frac{\Delta G}{k_B T} = v \ln \phi + N_s \ln(1 - \phi) + \chi \phi N_s. \tag{5.51}$$

The quantity v is the number of polymer chains.

In a simple approximate theory of polymer network swelling due to Flory and Rehner (1943) the translational entropy term $v \ln \phi$ is replaced by the elastic entropy contribution $\frac{3v}{2}(\lambda^2 - 1\hat{A} - \ln \lambda)$ when the network undergoes uniform swelling, i.e. $\lambda_x = \lambda_y = \lambda_z \equiv \lambda$. Minimizing this new ΔG with respect to N_s, i.e. $0 = \partial(\Delta G/(k_B T))/\partial N_s$, which means equating the solvent chemical potentials inside and outside the network, yields the Flory-Rehner equation in its standard form:

$$\rho_P \frac{v_s}{m_P} = -\frac{\ln(1-\phi) + \phi + \chi\phi^2}{\phi^{1/3} - \phi/2}. \tag{5.52}$$

The quantity ρ_P is the mass density of the (dry) polymer, v_s is the molecular volume of the solvent, and m_P is the (average) mass of a network segment. Note that $\lambda^3 = V_{swollen}/V_P = 1/\phi = (V_P + N_s v_s)/V_P$, where $V_{swollen}$ is the equilibrium volume of the swollen network (or gel) and V_P is the dry polymer volume. Note also that $\partial\phi/\partial N_s = -\phi^2 v_s/V_P$ (caveat: It is incorrect to use the above formula $\phi = 1 - N_s/N$, assuming $N = const$, for the derivative of ϕ with respect to N_s. This means that the volume is kept constant—which is not the case.), $\partial\lambda/\partial N_s = v_s/(3V_P\lambda^2)$, as well as $\rho_P \frac{v_s}{m_P} = \frac{v_s}{V_P} v$. In the original paper, cf. Eq. (11) in Flory and Rehner (1943), the $-\phi/2$-term in the denominator of Eq. (5.52), which results from the $-\ln\lambda$-term in the elastic entropy, is not included. This persisted for some time (e.g. Eq. (5.9) in Treloar (1973), but finally "converged" to the above form. However, the significance of the $-\phi/2$-term is somewhat questionable. The factor 1/2 resulted from a rather approximate entropy contribution of the network nodes and, in addition, is based on the assumption of a regular 4-fold coordinated network. In addition, in many cases of practical importance $\phi^{1/3}$ dominates over $\phi/2$ and the latter can be neglected. Nevertheless, Eq. (5.52) ties the quantity v to the swollen volume of a polymer network. Therefore it provides another means for the determination of the cross-link density.

5.1.1 Entropy and Information

This is a good place to briefly talk about entropy and information. Figure 5.8 shows a chessboard with a single pawn on $c2$. Imagine somebody who wants to find the pawn's position without looking at the board—just by asking another person, who can look at the board, questions requiring "yes" or "no" as answer. The questioner might proceed as follows: Q1—is the pawn somewhere on files A through D? A1— yes; Q2—is the pawn somewhere on rows 1 through 4? A2—yes; Q3—is the pawn on files A or B? A3—no; ...The numbered dashed lines on the right board illustrate how via bisection the location of the pawn is found after six questions. This is because there are 64 squares on the board and $64 = 2^6$. If we identify the number of possible squares with the quantity Ω in Eq. (5.16), then we may define an entropy for the pawn/chessboard system via

$$S = \log_2 64 = 6. \tag{5.53}$$

This entropy is the number of yes/no-questions we need to ask in order to acquire total knowledge about the system. Or we may say: S is a measure for our lack of information—the more questions we must ask the bigger is our information deficit.

The entropy in Eq. (5.19) of our first example had the same quality. The number N denotes the possible positions of the domain wall—quite analogous to the 64 squares on the above board. In the second example the quantity Ω_{sp} is a measure for the number of possible positions of the scaled particle in the system. Again the analogy is obvious. Finally in the third example $\Omega(R)$ is the number of different possible paths of total length R. Thus, in all examples Ω is the number of qualitatively identical alternatives. The more of these alternatives there are, the greater the associated entropy becomes, i.e. the greater is the lack of information regarding one particular alternative. Often it is said that the increase of entropy signals increasing "disorder". What is meant by disorder is the lack of information.

5.1.2   E and the Hamilton Operator $\mathcal{H}$

At this point we return to our main subject and talk about the calculation of the E_v in quantum mechanics. It appears reasonable to identify the E_v with the eigenvalues of the subsystem Hamilton operator H fulfilling the stationary Schrödinger's equation

$$H|v> \; = E_v|v>. \tag{5.54}$$

Here $|v>$ is the appropriate eigenket. Therefore Eq. (5.6) may be expressed via the $|v>$:

$$
\begin{aligned}
Q_{NVT} &= \sum_v \exp[-\beta E_v] \\
&= \sum_v <v|\exp[-\beta H]|v> \; = Tr(\exp[-\beta H]).
\end{aligned}
\tag{5.55}
$$

Tr is the trace of the quantum mechanical operator $\exp[-\beta H]$. In particular

$$\langle E \rangle = \frac{-\frac{\partial}{\partial \beta} \sum_v <v|\exp[-\beta H]|v>}{\sum_v <v|\exp[-\beta H]|v>} = \frac{Tr(He^{-\beta H})}{Tr(e^{-\beta H})} \equiv \; <H>, \tag{5.56}$$

where $<H>$ is the quantum mechanical expectation value of H. Even though the subsystem is in thermal contact with its environment, the calculation of the partition function requires knowledge of the quantum states of the subsystem only.

5.1.3 The Ideal Gas Revisited

The quantum mechanical result for E_v in the case of a particle confined to a one-dimensional box with volume (or length) L is[1]

$$E_v = \frac{\hbar^2 k_v^2}{2m} \tag{5.57}$$

Here $\hbar = h/(2\pi)$ is Planck's constant divided by 2π and m is the mass of the particle. We also have $v = (L/\pi)k_v$ ($v = 1, 2, \ldots$). Because L/π is a large number, we replace the sum over v in the partition function by an integration over k as follows

$$\sum_{v=1}^{\infty} = \frac{L}{\pi} \int_0^{\infty} dk. \tag{5.58}$$

The partition function becomes

$$\sum_{v=1}^{\infty} \exp[-\beta E_v] = \frac{L}{\pi} \int_0^{\infty} dk \exp\left[-\beta \frac{\hbar^2 k^2}{2m}\right] = \frac{L}{\Lambda_T}, \tag{5.59}$$

where

$$\Lambda_T = \sqrt{\frac{2\pi \hbar^2 \beta}{m}} \tag{5.60}$$

is called the thermal wavelength. Inserting the result (5.59) into the free energy Eq. (5.10) we obtain for the pressure, $P = -\partial F/\partial L|_T$,

$$PL = \beta^{-1}. \tag{5.61}$$

This is exactly what we expect for a single particle in a one-dimensional box.

But what about many particles in a three-dimensional volume? We may extend Eq. (5.59) to the three dimensions via

---

[1] Thus far E_v corresponded to the energy of a system. And systems do contain large number of particles. Now there is only one! We assume that there is so little interaction that each particle in a large system may be studied individually. But we also require that there is just sufficient interaction between this particle an its surroundings for it to reach thermal equilibrium. The idea is that one can collect instantaneous but uncorrelated (!) copies of this one particle, which, after one has obtained very many copies, are combined into one system and that this system is a system at equilibrium in the thermodynamic sense.

$$Q_{NVT} = \sum_{v_x,v_y,v_z=1}^{\infty} \exp[-\beta E_{v_x,v_y,v_z}]$$

$$= \frac{1}{N!} \left(\frac{L}{\pi} \int_0^{\infty} dk_x \exp\left[-\beta \frac{\hbar^2 k_x^2}{2m}\right] \right)^{3N}$$

$$= \frac{1}{N!} \left(\frac{V}{(2\pi)^3} \int d^3k \exp\left[-\beta \frac{\hbar^2 k^2}{2m}\right] \right)^{N} \qquad (5.62)$$

$$= \frac{1}{N!} \left(\frac{V}{2\pi^2} \int_0^{\infty} dk k^2 \exp\left[-\beta \frac{\hbar^2 k^2}{2m}\right] \right)^{N}$$

$$= \frac{1}{N!} \left(\frac{V}{\Lambda_T^3} \right)^{N}.$$

$3N$ is justified easily, because for a particle in a three-dimensional box we have from quantum mechanics $k^2 = k_x^2 + k_x^2 + k_x^2$ and for each k-component i we have $v_i = (L/\pi)k_{v_i}$ (note: $V = L^3$) with $v_i = 1, 2, \ldots$. The factor $N!^{-1}$ results from the proper normalization of the N-particle state $|v>$ to be used in Eq. (5.55) instead of the single particle state. However, there is alternative classical motivation for this factor in the context of the so called Gibbs paradox. But let us first proceed with the three dimensional ideal gas. Inserting the partition function (5.62) into the free energy Eq. (5.10) immediately yields for the pressure, $P = -\partial F/\partial V|_T$,

$$PV = Nk_B T, \qquad (5.63)$$

i.e. the ideal gas law. Using Eq. (5.8) we obtain for the internal energy of the ideal gas

$$\langle E \rangle = \frac{3}{2} N k_B T \qquad (5.64)$$

and for the heat capacity at constant volume

$$C_V = \frac{3}{2} N k_B \qquad (5.65)$$

(cf. the footnote on page 56). Another quantity of interest is the chemical potential, which here follows most conveniently via $N\mu = F + PV$, i.e.

$$\mu = k_B T \ln \rho \Lambda_T^3, \qquad (5.66)$$

where $\rho = N/V$ is the number density and we have used the Stirling approximation (4.135) (here: $\ln N! \approx N \ln N - N$). Via $F = N\mu - PV = E - TS$ and again using the Stirling approximation we quickly obtain the entropy

$$S = Nk_B \ln \left[\frac{e^{5/2}}{\rho \Lambda_T^3} \right]. \tag{5.67}$$

This equation is the Sackur-Tetrode equation (2.32).

5.1.4 Gibbs Paradox

Let us discuss the following experiment. We consider two identical boxes. Their respective volumes are $V = L^3$ and they share one common wall which is a movable partition. Assuming we can move the partition in and out without doing work, i.e. the partition slides easily back and forth, we find that the entropy of the combined system is equal to the sum of the entropies of the individual systems. Thus we have

$$\Delta S = S(2N, 2V) - 2S(N, V) = 0. \tag{5.68}$$

Computing the entropy via $S = -\partial F / \partial T|_V$ we find

$$\Delta S/k_B = -\ln(2N)! + 2N \ln 2 + 2 \ln N! \tag{5.69}$$

For small N we easily find that $\Delta S \neq 0$. So what is wrong? First we note that in a macroscopic system N is large, i.e. $N \sim 10^{23}$.[2] Computing ΔS for large N requires that we use the Stirling approximation (4.135). The result is

$$\Delta S/k_B = \frac{1}{2} \ln(\pi N). \tag{5.70}$$

Again we obtain $\Delta S \neq 0$[3] but the point to notice is that the ratio $\Delta S/S$ tends to zero as N grows, i.e.

$$\frac{\Delta S}{S} \sim \frac{\ln N}{N} \xrightarrow{N \to \infty} 0. \tag{5.71}$$

---

[2] The assumption of large N already entered our formalism via the truncated expansion (5.3).

[3] We use the Stirling approximation including $\sqrt{2\pi N}$. Otherwise the result is $\Delta S = 0$.

In this limit we therefore obtain the desired result. Without the extra factor $N!^{-1}$, which we introduced into the partition function, the result would have been different even for large N, i.e. $\Delta S/S \sim 1/\left(\ln\left[V/\Lambda_T^3\right]\right)$. The convergence is so slow ($V \sim N$) that the missing factor is noticeable on the macroscopic scale. This is called the Gibbs paradox. Therefore we could have guessed this factor on purely classical grounds. Notice that $N!$ is the number of indistinguishable permutations in the case of N identical objects. The factor $N!^{-1}$ thus accounts for the fact that our particles can exchange places with each other without loss of information.

5.1.5   Ideal Gas Mixture

From the preceeding discussion we conclude that the partition function of an ideal gas mixture is

$$Q_{NVT} = \prod_j Q_{NVT,j} = \prod_j \frac{1}{N_j!}\left(\frac{V}{\Lambda_{T,j}}\right)^{N_j}. \tag{5.72}$$

where $N = \sum_j N_j$ is the total particle number. Notice that the thermal wavelength depends on a particle's mass and thus on j. The partial pressure of component i is

$$P_i = -\left.\frac{\partial F_i}{\partial V}\right|_{T,N_j} \quad \text{with} \quad F_i = -k_B T \ln Q_i \tag{5.73}$$

i.e. we recover Dalton's law because obviously $P = \sum_i P_i$. Another short calculation yields the chemical potential of an individual component

$$\mu_i = \left.\frac{\partial F_i}{\partial N_i}\right|_{T,N_{j\neq i}} = k_B T \ln \rho_i \Lambda_{T,i}^3 \tag{5.74}$$

where $\rho_i = N_i/V$. This chemical potential we had assumed in the context of the Saha equation on page 130.

5.1.6   Energy Fluctuations

We want to compute the mean square energy fluctuation based on Eq. (5.7). Thus we write

$$\langle(\delta E)^2\rangle = \langle(E - \langle E\rangle)^2\rangle = \langle E^2\rangle - \langle E\rangle^2$$

$$= \sum_v p_v E_v^2 - \left(\sum_v p_v E_v\right)^2$$

$$= \frac{1}{Q_{NVT}} \frac{\partial^2 Q_{NVT}}{\partial \beta^2}\bigg|_{N,V} - \frac{1}{Q_{NVT}^2} \left(\frac{\partial Q_{NVT}}{\partial \beta}\bigg|_{N,V}\right)^2$$

$$= \frac{\partial^2 \ln Q_{NVT}}{\partial \beta^2}\bigg|_{N,V},$$

i.e.

$$\langle(\delta E)^2\rangle = -\frac{\partial \langle E\rangle}{\partial \beta}\bigg|_{N,V}, \qquad (5.75)$$

and therefore

$$\langle \delta E^2\rangle = k_B T^2 C_V. \qquad (5.76)$$

Note that in large systems such fluctuations are relatively small. We see this if we study the ratio $(\langle \delta E^2\rangle)^{1/2}/\langle E\rangle$. Using Eq. (5.76) together with $\langle E\rangle \propto N k_B T$ we find

$$\frac{\sqrt{\langle \delta E^2\rangle}}{\langle E\rangle} \propto \frac{1}{\sqrt{N}}. \qquad (5.77)$$

For macroscopic systems with $N \approx N_A$ the relative energy fluctuations are vanishingly small.

5.1.7   The Likelihood of Energy Fluctuations

At the beginning of this chapter we considered an isolated system. When this system has the energy E then there are $\Omega(E)$ different microstates with this energy. If the system is not isolated but coupled to an external heat bath with temperature T then it becomes a subsystem within this heat bath. The probability for the system to have the energy E is then determined by two factors, $\Omega(E)$ and, as we have just seen, $\exp[-\beta E]$. Thus we have

$$p(E) \propto \Omega(E)\exp[-\beta E]. \qquad (5.78)$$

Note that $p(E)$ is different from $p(E_v)$ even if $E = E_v$. This is because $p(E_v)$ is the probability of the subsystem being in microstate v, whereas $p(E)$ is the probability of measuring the internal E in the subsystem.

Without much knowledge about $\Omega(E)$ we still may infer an important piece of information regarding the general shape of $p(E)$. Again we start by expanding $\ln[\Omega(E)\exp[-\beta E]]$ around $\langle E \rangle$. The result is

$$\ln[\Omega(E)e^{-\beta E}] = \ln\Omega(\langle E \rangle) + \left(\frac{d}{dE}\ln\Omega(E)\right)_{E=\langle E \rangle}\delta E$$

$$+ \frac{1}{2}\frac{d}{dE}\left(\frac{d\ln[\Omega(E)]}{dE}\right)_{E=\langle E \rangle}\delta E^2 - \beta\langle E \rangle - \beta\delta E + \ldots$$

Using Eq. (5.16) together with Eq. (1.52) yields

$$\ln[\Omega(E)e^{-\beta E}] = \ln\Omega(\langle E \rangle) - \beta\langle E \rangle - \frac{1}{2}\frac{\delta E^2}{k_B T^2 C_V} + \ldots \tag{5.79}$$

and thus

$$p(E) = p(\langle E \rangle)\exp\left[-\frac{1}{2}\frac{\delta E^2}{k_B T^2 C_V}\right]. \tag{5.80}$$

Again we find $\langle \delta E^2 \rangle = k_B T^2 C_V$—as we should. Just how unlikely deviations from the average energy are, meaning that $p(E)$ is sharply peaked around $\langle E \rangle$, becomes clear if we put in some numbers. We consider 10^{-3} moles of a gas. With $\delta E = 10^{-6}\langle E \rangle$ and $\langle E \rangle \approx Nk_B T$ as well as $C_V \approx k_B N$ we find

$$\frac{p(E)}{p(\langle E \rangle)} \approx \exp[-10^{-12}N] = \exp[-10^{-12}0.001N_A] \approx \exp[-10^8].$$

5.1.8 Harmonic Oscillators and Simple Rotors

There are two simple models, we may envision them as two different types of "particles", which we should discuss, because they frequently enter into the description of more complex systems. Our discussion will be analogous to the treatment of the ideal gas in Sect. 5.1.3.

The first model is the one-dimensional harmonic quantum oscillator, which, as we already know from introductory quantum theory, has the energy eigenvalues

$$E_v = \hbar\omega\left(v + \frac{1}{2}\right),\tag{5.81}$$

where ω is the oscillator's frequency and $v = 0, 1, 2, \ldots$. It is not difficult to obtain the partition function

$$Q^{1D-osc} = \sum_{v=0}^{\infty} \exp[-\beta E_v] = \frac{e^{-\beta\hbar\omega/2}}{1 - e^{-\beta\hbar\omega}} = \frac{1}{2\sinh\frac{\beta\hbar\omega}{2}},\tag{5.82}$$

where we use the geometric series $\sum_{v=0}^{\infty} q^v = (1 - q)^{-1}$ $(q < 1)$. Straightforward differentiation first yields the average internal energy,

$$\langle E \rangle = \frac{\partial}{\partial(-\beta)}\ln Q^{1D-osc} = \frac{\hbar\omega}{2}\coth\frac{\beta\hbar\omega}{2},\tag{5.83}$$

and subsequently the heat capacity of the oscillator

$$\frac{1}{k_B}C_V = \frac{1}{k_B}\frac{\partial}{\partial T}\langle E \rangle = \left(\frac{T_{vib}}{2T}\ \mathrm{csch}\ \frac{T_{vib}}{2T}\right)^2,\tag{5.84}$$

where

$$T_{vib} = \frac{\hbar\omega}{k_B}\tag{5.85}$$

is a characteristic temperature. To better understand the meaning of T_{vib} we should try to work out the classical partition function for the oscillator.

Looking back at Eq. (5.62) we notice that the argument of the exponential function is the kinetic energy. Taking this one step further we replace the kinetic energy with the Hamilton function $\mathcal{H}$. If in addition we express momentum via $p = \hbar k$ and the box size via $\int dx$ we may write for the 1D harmonic oscillator

$$Q_{cl}^{1D-osc} = \int_{-\infty}^{\infty}\frac{dp}{2\pi\hbar}\int_{-\infty}^{\infty}dx\exp[-\beta(\mathcal{H}(p,x))],\tag{5.86}$$

where

$$\mathcal{H}(p,x) = \frac{p^2}{2m} + \frac{1}{2}m\omega^2 x^2.\tag{5.87}$$

Here m is the oscillator's mass, p its momentum, and x its displacement from equilibrium. An easy integration yields

$$Q_{cl}^{1D-Osc} = \frac{1}{\beta\hbar\omega}. \tag{5.88}$$

Example—van der Waals Equation   Before we proceed with the discussion of Eq. (5.88) we want to briefly consider a different potential. The classical particle is confined to a one-dimensional box of length L_1, inside of which the potential energy is $\mathcal{U} = -a\rho$. Here $a\rho >$ is a constant. Notice that the total extend of the box is L_1 but we subtract a length b, because the particle itself posses this size and therefore its center can only access a smaller "volume" $L_1 - b$. The partition function therefore is

$$Q(1) = \frac{1}{2\pi\hbar} \int_{-\infty}^{\infty} dp \int_{-(L_1-b)/2}^{(L_1-b)/2} dx e^{-\beta\mathcal{H}} = \frac{L_1 - b}{\Lambda_T} e^{\beta a\rho}. \tag{5.89}$$

Let us also assume that we have N such particles in independent boxes. Their partition function is

$$Q(N) = Q(1)^N = \left(\frac{1}{N}\frac{L - Nb}{\Lambda_T}\right)^N e^{N\beta a\rho}, \tag{5.90}$$

where $L = NL_1$. At this point we set the quantity ρ equal to $1/L_1 = N/L$. If we calculate the pressure analogous to Eq. (5.61) the result is

$$P = \frac{Nk_BT}{L - Nb} - a\left(\frac{N}{L}\right)^2. \tag{5.91}$$

This generalization of Eq. (5.61) is a one-dimensional version of the van der Waals Eq. (4.1). The necessary ingredients are (i) each particle reduces the available "volume" by b; (ii) each particle has a negative potential energy contributed by all other particles according to their mean density ρ (cf. footnote 25 in Chap. 4). There is no factor $N!^{-1}$ in $Q(N)$. This is because every particle is in its own cell. In principle the cells are distinguishable (even though this is not an essential ingredient). This type of approach is known as cell theory (Hirschfelder 1954).

Now we continue with Eq. (5.88). The classical thermal energy of the oscillator therefore is $\langle E \rangle = k_BT$ and the attendant heat capacity $C_V = k_B$. Figure 5.9 shows the comparison of this value to the quantum result plotted versus the reduced temperature T/T_{vib}. Above T_{vib} the oscillator is well described by the classical result, whereas below T_{vib} the quantum behavior dominates. Had we included only

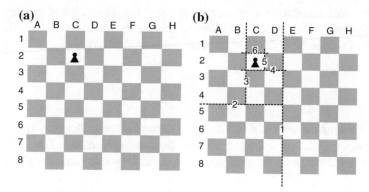

Fig. 5.8 A single pawn on a chessboard

the ground state in the quantum partition function the result would have been $C_V = 0$. This means that (not too far) below T_{vib} the oscillator is "frozen".

The oscillator model is useful in the context of molecules. Classically the atoms in molecules vibrate according to collective or normal modes. Normal mode analysis shows that the Hamilton function of a molecule containing N atoms may be transformed into

$$\mathcal{H} = \mathcal{H}_o + \sum_{i=1}^{N_f} \left(A_i P_i^2 + B_i X_i^2 \right), \tag{5.92}$$

where X_i are normal mode coordinates and P_i are the conjugate momenta. A_i and B_i are coefficients. The term $\mathcal{H}_o$ describes the molecule at (vibrational) rest. Equation (5.92) is a simplification which holds only for sufficiently small amplitudes, i. e. for small deformations relative to the equilibrium molecular "shape". The number of vibrational modes, N_f, generally is equal to $3N - 6$. The -6 is due to the three translational and three rotational degrees of freedom which must be subtracted. In the case of linear molecules there is one less rotation and $N_f = 3N - 5$. In this sense molecules may be considered as collection of N_f independent one-dimensional oscillators with normal mode frequencies ω_i. The vibrational partition function of a molecule therefore is (approximately) given by

$$Q^{vib} = \prod_{i=1}^{N_f} Q^{1D-Osc}(\omega_i). \tag{5.93}$$

We may let the number of atoms become arbitrarily large and conclude that Eq. (5.93) remains valid for solids as well. The difference is that the normal mode frequencies of small molecules usually are quite high so that $T_{vib} \sim 10^3$ K. At room temperature this means that small molecules are frozen in their vibrational ground states. In solids this is not the case. The example at the end of this section illustrates this distinction.

We just mentioned molecular rotation. Is there a rotational partition function? If yes—how does it look like? From classical mechanics we know that the Hamilton function of a rigid body freely rotating in space is given by

$$\mathcal{H} = \sum_{i=1}^{3} \frac{L_i^2}{2\mathcal{I}_i}. \tag{5.94}$$

Here i denote the axes of a coordinate system attached to the body in which the moment of inertia tensor is diagonal. These diagonal elements are $\mathcal{I}_i$ and L_i^2 are the attendant angular momentum components squared. Quantum mechanically we have

$$\underline{\mathcal{H}} = \sum_{i=1}^{3} \frac{\underline{L_i^2}}{2\mathcal{I}_i}, \tag{5.95}$$

where the underlined quantities are operators. Table 5.1 summarizes the simple cases:

The rotational partition function therefore is

$$Q^{rot} = \sum_{l=0}^{\infty} g_l \exp[-\beta E_{l,m}]. \tag{5.96}$$

Analogous to the above characteristic temperature of vibration it is now sensible to define a characteristic temperature of rotation via

$$T^{rot} = \frac{\hbar^2}{2k_B \mathcal{I}}. \tag{5.97}$$

A rough estimate using the atomic mass unit (Appendix A) times 1 Å^2 for the moment of inertia reveals that $T^{rot} \sim 10$ K. This is small compared to room temperature—and in most cases it means that we can use the classical rotation partition function. Momentarily however we proceed working out Eq. (5.96). The simplest approach is the straightforward summation over a limited number of l-values. The resulting heat capacity, $C_V/k_B = T\partial^2/\partial T^2 T \ln Q^{rot}$, is shown in Figs. 5.10 and 5.11. Figure 5.10 shows both the heat capacities for the linear and the spherical rotor. The

Table 5.1 Energy eigenvalues and attendant degeneracies for different rotors

Rotor type	Description	Energy eigenvalues $E_{l,m}/(\frac{\hbar^2}{2\mathcal{I}})$	Degeneracy g_l
Linear	$\mathcal{I} \equiv \mathcal{I}_1 = \mathcal{I}_2; \mathcal{I}_3 = 0$	$l(l+1)$	$2l+1$
Spherical	$\mathcal{I} \equiv \mathcal{I}_1 = \mathcal{I}_2 = \mathcal{I}_3$	$l(l+1)$	$(2l+1)^2$
Symmetric	$\mathcal{I} \equiv \mathcal{I}_1 = \mathcal{I}_2; \mathcal{I}_3 \neq 0$	$l(l+1) + (\frac{\mathcal{I}}{\mathcal{I}_3} - 1)m^2$	$\kappa(2l+1)$
		$l = 0, 1, 2, \ldots$	$\kappa = 1(m = 0)$
		$m = -l, -l+1, \ldots, l-1, l$	$\kappa = 2(m \neq 0)$

Fig. 5.9 Heat capacity of the 1D harmonic oscillator versus reduced temperature

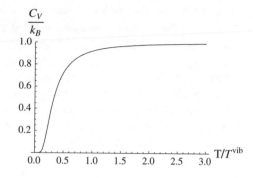

Fig. 5.10 Heat capacities of the linear and the spherical rotor versus reduced temperature

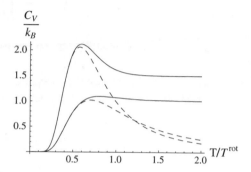

former approaches 1 and later 3/2 at high temperatures. Note that the dashed lines shows corresponding results if only the terms $l = 0, 1$ are taken into account, whereas the solid lines are the numerically exact results. Figure 5.11 compares the heat capacity of a prolate symmetric rotor ($\mathcal{I}/\mathcal{I}_3 - 1 = 2$; broad maximum) to the one of an oblate symmetric rotor ($\mathcal{I}/\mathcal{I}_3 - 1 = -1/2$; narrow maximum).

We often encounter experimental situations when T^{rot} is low compared to the relevant T. It therefore is useful to also work out the classical partition function, Q_{cl}^{rot}. We find Q_{cl}^{rot} via generalization of Eq. (5.86), i.e.

$$Q_{cl} = \frac{1}{\sigma} \int_{V_q} \int_{-\infty}^{\infty} \frac{d^n q \, d^n p_q}{(2\pi\hbar)^n} \exp[-\beta \mathcal{H}(\{p\}, \{q\})]. \qquad (5.98)$$

Here $\{q\}$ is a set of coordinates and $\{p_q = \partial \mathcal{L}(\{\dot{q}\}, \{q\})/\partial \dot{q}\}$ are their conjugate momenta. $\mathcal{L}$ is the Langrangian of the system. In the case of a system of N point particles the factor $1/\sigma$ is equal to $1/N!$ as we have seen. If we consider one molecule only, then σ is a symmetry number, i.e. the number of rotations mapping the molecule onto itself. In the case of the water molecules in the next example $\sigma = 2$. This accounts for the 2-fold rotational symmetry with respect to the symmetry axis of the molecule. If we describe the rotation of a small molecule like water the proper set of coordinates, $\{q\}$, are the Euler angles, i.e. $0 \leq \varphi \leq 2\pi$, $0 \leq \theta \leq \pi$, and $0 \leq \psi \leq 2\pi$. The difficult part is to work out the equations between the conjugate momenta and the angular velocities ω_i, where the index refers to the same axes as the index i in Eq. (5.94), because the Langrangian is $\mathcal{L} = \sum_{i=1}^{3} \frac{1}{2} \mathcal{I}_i \omega_i^2$. These equations usually are discussed in lectures on classical mechanics of rigid bodies. They are: $p_\varphi = \mathcal{I}_1 \omega_1 \sin\theta \sin\psi + \mathcal{I}_2 \omega_2 \sin\theta \cos\psi + \mathcal{I}_3 \omega_3 \cos\theta$, $p_\theta = \mathcal{I}_1 \omega_1 \cos\psi - \mathcal{I}_2 \omega_2 \sin\psi$, and $p_\psi = \mathcal{I}_3 \omega_3$. Rather than calculating the Hamiltonian and integrating over the momenta it is easier to change to the angular velocities via $dp_\varphi dp_\theta dp_\psi = | J | d\omega_1 d\omega_1 d\omega_1$ (e.g. Pauli 1972). The determinant J is given by $J = \mathcal{I}_1 \mathcal{I}_2 \mathcal{I}_3 \sin\theta$. Because we now can use the Lagrangian instead of the Hamiltonian in the exponential in (5.98), the integrations become independent and we obtain

$$Q_{cl}^{rot} = \frac{\sqrt{\pi}}{\sigma} \prod_{i=1}^{3} \sqrt{\frac{2\mathcal{I}_i}{\beta\hbar^2}} \tag{5.99}$$

as final result for the classical rotation partition function. Via $\langle E \rangle = \partial/\partial(-\beta) \ln Q$ we find $\langle E \rangle = \frac{3}{2} k_B T$ and thus $C_V = \frac{3}{2} k_B$. Note that this agrees with the high temperature limit of the spherical and symmetric rotors in Figs. 5.10 and 5.11. The classical partition function for the linear rotor is[4]

$$Q_{cl}^{rot} = \frac{1}{\sigma} \prod_{i=1}^{2} \sqrt{\frac{2\mathcal{I}_i}{\beta\hbar^2}} \tag{5.100}$$

Of course we have $\mathcal{I}_1 = \mathcal{I}_2$. Now we obtain $\langle E \rangle = k_B T$ and thus $C_V = k_B$ - again in agreement with the quantum result at high temperatures.[5] The following is a nice example combining all of the above in one problem.

Example—Vapor Pressure of Ice Van der Waals' theory allows to approximate the vapor-liquid phase coexistence of a pure substance in the T-P-plane. Here we want to approximately determine the coexistence line between vapor and solid (sublimation line)—for water. We model water as a rigid molecule, because its three vibrational modes correspond to characteristic temperatures, $T_{vib,i} = \hbar\omega_i/k_B$, of roughly 5400, 5300, and 2300 K. The temperatures of interest in this example are between 160–270 K. Thus molecular vibrations may be neglected (in the sense discussed above). We proceed as follows: In part (a) the water chemical potential in the (ideal) gas phase is estimated. In part (b) the same is done for frozen water. Finally in part (c) we use the equality of the chemical potentials to relate the gas pressure to the temperature at coexistence.

---

[4] In this case the angle ψ does not enter and the above equations relating the momenta to the angular velocities reduce to $p_\varphi = \mathcal{I}_2 \omega_2 \sin\theta$ and $p_\theta = \mathcal{I}_1 \omega_1$.

[5] Of course, all this is expected because of the equipartition theorem of statistical mechanics stating that every term in the sum in Eq. (5.92) contributes $k_B/2$ to the heat capacity.

(a) The gas phase chemical potential is $\mu_{H_2O}^{vapor} = \mu_{H_2O}^{trans} + \mu_{H_2O}^{rot}$. $\mu_{H_2O}^{trans}$ is given by Eq. (5.66), which may be rewritten as

$$\beta\mu_{H_2O}^{trans} = \frac{5}{2}\ln\frac{T_o^{trans}}{T} + \ln\frac{P}{P_o}. \tag{5.101}$$

Here $P_o = 1bar$ is an arbitrary reference pressure, $(T_o^{trans})^{5/2} = \left(\frac{2\pi\hbar^2}{m_{H_2O}k_B}\right)^{3/2}\frac{P_o}{k_B}$, i.e. $T_o^{trans} \approx 0.76$ K, and m_{H_2O} is the molecular weight of water. $\mu_{H_2O}^{rot}$ is calculated via the classical molecular partition function Q_{cl}^{rot} given by Eq. (5.99). It is convenient to express Q_{cl}^{rot} in terms of the characteristic temperatures, i.e. $Q_{cl,H_2O}^{rot} = \prod_{i=1}^{3}\left(T/T_i^{rot}\right)^{1/2}$, where $T_i^{rot} = (4/\pi)^{1/3}\hbar^2/(2k_B\mathcal{I}_i)$. Here $\mathcal{I}_i$ are the moments of inertia with respect to the principal axes of rotation obtained via diagonalization of the moment of inertia tensor. The components of the latter are $\mathcal{I}_{\alpha\beta} = \sum_{k=1}^{3}m_k(r_k^2\delta_{\alpha\beta} - x_{\alpha,k}x_{\beta,k})$. Using an OH bond length of 1 Å and a HOH angle of 109.5^o one obtains $T_1 \approx 44$ K, $T_2 \approx 20$ K, and $T_3 \approx 14$ K. These temperatures are low compared to the above temperature range of interest justifying the use of the classical partition function. Our final result is

$$\beta\mu_{H_2O}^{rot} = \frac{1}{2}\sum_{i=1}^{3}\ln\frac{T_i^{rot}}{T}. \tag{5.102}$$

(b) The chemical potential of water molecules in ice is estimated via $\mu_{H_2O}^{ice} = \mu_o + \mu_{ice}^{vib}$. Here $\beta\mu_{ice}^{vib} = -\partial\ln Q^{vib}/\partial N\mid_{T,V}$ and $Q^{vib} = \prod_{j=1}^{3N}Q_j^{1D-osc}$. The summation is over the $3N - 6 \approx 3N$ normal modes of the crystal, where in the present case N is the number of rigid water molecules. Q_j^{1D-osc} is the quantum mechanical partition function of a one-dimensional harmonic oscillator with frequency ω_j given in Eq. (5.82). In contrast to the normal modes of the individual water molecule these frequencies are low. Because N is large, i.e. there are many normal modes with wave vectors $\vec{k}$, we write

$$3N = \sum_{j=1}^{3N} = 3\frac{V}{(2\pi)^3}4\pi\int_0^{k_D}dkk^2. \tag{5.103}$$

This is quite analogous to the above conversion of the summation into an integration for a particle in a 3D box. Two things a different nevertheless. The integration is cut off at k_D, because of the finite number of modes. And there is an extra factor 3 accounting for the three possible types of vibrational polarization—$2\times$ transversal and $1 \times$ longitudinal. A simple relation tying the k-values to oscillator frequencies $\omega_j \rightarrow \omega(k)$ is $\omega = v_sk$, where v_s is the average velocity of sound in the crystal, i.e. $3/v_s^3 = 1/v_{s,t1}^3 + 1/v_{s,t2}^3 + 1/v_{s,l}^3$ (here: $v_s \approx 3300$ m/s). The details of this approximation may be found in textbooks on solid state physics in the context of the Debye model of the low temperature heat capacity in insulators. Putting everything together we find

$$-\ln Q^{vib} = \frac{3V}{2\pi^2}\left(\frac{2}{\beta\hbar v_s}\right)^3 \int_0^{x_D} dx x^2 \ln[2\sinh x], \qquad (5.104)$$

where $x_D = (\beta\hbar v_s/2)(6\pi^2 N/V)^{1/3}$. Taking the derivative with respect to N at constant T, V yields

$$\beta\mu_{ice}^{vib} = 3\ln\left[2\sinh\frac{T_{ice}^{vib}}{T}\right] \qquad (5.105)$$

with $T_{ice}^{vib} = (\hbar v_s/(2k_B))(6\pi^2 N/V)^{1/3} \approx 158K$. Note that $N/V \approx N_A/18 \text{cm}^3$ is the number density of water in ice.

(c) Chemical equilibrium, i.e. $\mu_{H_2O}^{vapor} = \mu_{H_2O}^{ice}$, yields the following relation between vapor pressure, $P(T)$, and temperature, T,

$$\frac{P}{P_o} = 8\left(\frac{T}{T_{1bar}^{trans}}\right)^{5/2} \prod_{i=1}^3 \left(\frac{T}{T_i^{rot}}\right)^{1/2} \sinh^3\left(\frac{T_{ice}^{vib}}{T}\right) e^{-\epsilon/T}, \qquad (5.106)$$

where $\beta\mu_o = -\epsilon/T$. Figure 5.12 shows a comparison of this formula to experimental vapor pressure data (crosses) from HCP. Here $\epsilon = 6500$ K, which corresponds to about 54 kJ/mol. This is a meaningful number, because each water molecule participates in 4 hydrogen bonds stabilizing the tetrahedral crystal (ice I). The cohesive energy per water molecule therefore corresponds to two hydrogen bonds. Our value of 27 kJ/mol is in quite reasonable agreement with energies for HO..H hydrogen bonds obtained by other methods.

Fig. 5.11 Heat capacities of the prolate and the oblate symmetric rotor versus reduced temperature

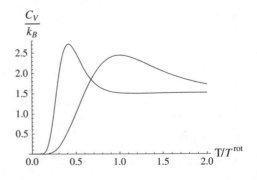

Remark 1—low temperature heat capacity of insulators The integral in Eq. (5.104) may be rewritten as

$$\int_0^{x_D} dx x^2 (\ln[2\sinh x] - x + x) = \frac{x_D^4}{4} + \underbrace{\int_0^{x_D} dx x^2 (\ln[2\sinh x] - x)}_{\approx -0.270581\,(x_D \to \infty)}$$

allowing to work out the limit of large x_D or low temperatures. Thus the vibrational free energy in this limit is given by, $F^{vib} \approx c_o + c_1 T^4$, where c_o and c_1 are independent of temperature. Consequently the contribution of F^{vib} to the heat capacity, $C_V = -T \partial^2 F / \partial T^2 \,|_{V,N}$, of the crystal is $\propto T^3$ as $T \to 0$. This is a famous result describing correctly the temperature dependence of the heat capacity of insulators at low temperatures due to quantized vibrational crystal excitations (Debye's T^3-law).

Remark 2—black body radiation Equation (5.104) may be applied to yet another important problem—the one that initiated quantum theory—black body radiation. In classical electrodynamics one can show that the energy of an electromagnetic field may be written as

$$E = \sum_{\vec{k},\alpha} \frac{1}{2} \left(p_{\vec{k},\alpha}^2 + \omega^2 q_{\vec{k},\alpha}^2 \right). \tag{5.107}$$

Here the "momenta", $p_{\vec{k},\alpha}$, and "coordinates", $q_{\vec{k},\alpha}$, are suitable combinations of Fourier coefficients in a Fourier decomposition of the vector potential (e.g., Hentschke 2009). In this fashion the electromagnetic field energy of a certain volume is a collection of $\sum_{\vec{k},\alpha}$ independent one-dimensional harmonic oscillators. Here $\vec{k}$ denotes the possible modes and α denotes their polarizations. This sum is infinite and since every term contributes on average $\frac{1}{2} k_B T$ to the energy (equipartition theorem) the result is infinite! A solution to the divergency problem was suggested by Max Planck in 1900. His solution amounts to treating the oscillators as quantum oscillators—just as in the case of Eq. (5.104). Nevertheless some modifications are necessary: (i) 3 V must be replaced by 2 V, because there are two polarization directions only; (ii) v_s is replaced by the speed of light c; (iii) x_D is replaced by ∞; (iv) the zero point energy must be subtracted, because it is not part of the radiation field, i.e. $(2\sinh x)^{-1}$ is replaced by $(2\sinh x)^{-1} \exp[x]$. The resulting black body radiation version of Eq. (5.104) becomes

$$-\ln Q^{vib} = \frac{2V}{2\pi^2} \left(\frac{2}{\beta \hbar c} \right)^3 \int_0^\infty dx x^2 \ln[1 - e^{-2x}]. \tag{5.108}$$

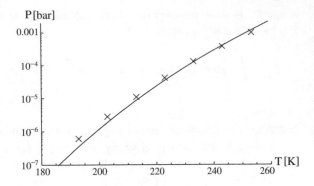

Fig. 5.12 Theoretical and experimental vapor pressure of ice

Via partial integration we rewrite the integral into its more common form

$$\int_0^\infty dx x^2 \ln[1 - e^{-2x}] = -\frac{2}{3} \int_0^\infty dx \frac{x^3}{e^{2x} - 1} = -\frac{2}{45} \left(\frac{\pi}{2}\right)^4.$$

The thermal energy density of the black body radiation is now finite, i.e.

$$\frac{\langle E \rangle}{V} = \frac{1}{V} \frac{\partial}{\partial(-\beta)} \ln Q^{vib} = \frac{\pi^2}{15} \frac{(k_B T)^4}{\hbar^3 c^3}. \tag{5.109}$$

Notice that we have derived the temperature dependence before based on purely thermodynamic considerations (cf. Eq. 2.39). But this time we also obtain the coefficient.

A spectacular experiment was performed in 1989 measuring the cosmic background radiation spectrum to high precision (Mather et al. 1990). The expected frequency dependence, intensity per frequency interval $d\omega$, should follow $[\omega^3/(\exp[\beta\hbar\omega] - 1)]$ or, if converted to wavelength, i.e. intensity per wavelength interval $d\lambda$, $[\lambda^{-5}/(\exp[\beta hc/\lambda] - 1)]$. The cosmic background radiation spectrum is found to be in complete agreement with Planck's prediction. This is shown in Fig. 5.13 comparing the measured data to the theory at a background radiation temperature of 2.725 K.

5.2 Generalized Ensembles

Once again we return to the isolated system divided into subsystems which we had introduced at the beginning of previous section. In addition to energy, E, we now allow the exchange of another extensive quantity, X, between the subsystems, i.e. E and X both fluctuate around their equilibrium values. X may be any of the variable quantities $(d \ldots)$ on the right side in Eq. (1.51). As in the case of the canonical ensemble we write

$$p_v \propto \Omega(E - E_v, X - X_v).$$ (5.110)

The probability p_v again is proportional to the number of environmental microstates compatible with the values of E_v and X_v. We expand (cf. Eq. 5.2) around E and X, i.e.

$$p_v \propto \exp[\ln \Omega(E - E_v, X - X_v)]$$
$$= \exp \left[\ln \Omega(E, X) - E_v \underbrace{\frac{\partial \ln \Omega(E, X)}{\partial E} \bigg|_X}_{= \beta} - X_v \underbrace{\frac{\partial \ln \Omega(E, X)}{\partial X} \bigg|_E}_{= \xi} + \cdots \right]$$

Remark Higher oder terms may be neglected[6]) and thus

$$p_v = \frac{\exp[-\beta E_v - \xi X_v]}{\Xi}$$ (5.111)

with

$$\Xi = \sum_v \exp[-\beta E_v - \xi X_v].$$ (5.112)

The thermodynamic quantities E and X are the averages

$$\langle E \rangle = \sum_v p_v E_v = \frac{\partial \ln \Xi}{\partial(-\beta)} \bigg|_{\xi, Y}$$ (5.113)

and

---

[6] Consider for instance:

$$\frac{1}{2} E_v^2 \left(\frac{d^2}{dE^2} \ln \Omega(E) \right)_E = \frac{1}{2} E_v^2 \left(\frac{dE}{d\beta} \right)_E^{-1} = -\frac{1}{2} E_v^2 \left(k_B T^2 C_V^{Syst} \right)^{-1}.$$

We have however

$$E_v^2 \left(k_B T^2 C_V^{Syst} \right)^{-1} \propto \beta E_v N / N^{Syst}.$$

Therefore this term is negligible compared to the leading one .

$$\langle X \rangle = \sum_\nu p_\nu X_\nu = \left. \frac{\partial \ln \Xi}{\partial (-\xi)} \right|_{\beta, Y}. \tag{5.114}$$

Here Y represents all non-fluctuating (extensive) variables. Thus we also have

$$d \ln \Xi = -\langle E \rangle d\beta - \langle X \rangle d\xi. \tag{5.115}$$

In order to tie all this to thermodynamics we momentarily consider the quantity

$$\frac{\varphi}{k_B} = -\sum_\nu p_\nu \ln p_\nu$$

$$\overset{(5.104)}{=} -\sum_\nu p_\nu [-\ln \Xi - \beta E_\nu - \xi X_\nu] \tag{5.116}$$

$$= \ln \Xi + \beta \langle E \rangle + \xi \langle X \rangle.$$

The differential $d(\varphi/k_B)$ is

$$d(\varphi/k_B) = -\langle E \rangle d\beta - \langle X \rangle d\xi + \beta d\langle E \rangle + \langle E \rangle d\beta + \xi d\langle X \rangle + \langle X \rangle d\xi, \tag{5.117}$$

and therefore

$$d\varphi = k_B \beta d\langle E \rangle + k_B \xi d\langle X \rangle. \tag{5.118}$$

The comparison with Eq. (1.51) now suggests that φ is the entropy, i.e.

$$S = -k_B \sum_\nu p_\nu \ln p_\nu. \tag{5.119}$$

Equation (5.119) is called the Gibbs entropy equation.

Question: Is this equation consistent with our previous expression of the entropy in terms of the number of microstates (cf. Eq. 5.16)? The answer is yes. In the previous case of an isolated system the index ν runs over the individual microstates and $p_\nu = 1/\Omega(E)$. Inserting this into Eq. (5.119) yields $S = k_B \sum_\nu p_\nu \ln \Omega(E) = k_B \ln \Omega(E) \sum_\nu p_\nu = k_B \ln \Omega(E)$.

5.2.1 Fluctuation of X

As in the case of E in Sect. 5.1.6 we are interested in the mean square fluctuation of X. However this time we derive a general and quite useful formula relating $\langle (\delta X)^2 \rangle$ to the mean of X. We write

$$\langle (\delta X)^2 \rangle = \langle (X - \langle X \rangle)^2 \rangle = \langle X^2 \rangle - \langle X \rangle^2$$
$$= \sum_v X_v^2 p_v - \sum_{v,v'} X_v X_{v'} p_v p_{v'}$$
$$= \frac{\partial^2}{\partial(-\xi)^2} \ln \Xi \bigg|_{\beta,Y} = \frac{\partial}{\partial(-\xi)} \langle X \rangle \bigg|_{\beta,Y}.$$

And thus

$$\frac{\partial}{\partial(-\xi)} \langle X \rangle \bigg|_{\beta,Y} = \langle (\delta X)^2 \rangle. \tag{5.120}$$

Example—Dielectric Constant and Polarization Fluctuation We apply this formula to polarization fluctuations in an isotropic dielectric medium. Based on Eq. (1.21) we write

$$\delta S = \ldots - \frac{1}{T} \int_V dV \vec{E} \cdot \delta \vec{P} = \ldots - \frac{1}{T} \vec{E} \cdot \delta \vec{p}. \tag{5.121}$$

Here $\vec{p} = \int_V dV \vec{P}$ is the dipole moment of the material inside the volume V and $\vec{P}$ is the attendant polarization. We also assume that the average (macroscopic) electric field, $\vec{E}$, in the dielectric is constant throughout V. Setting $X = p_\alpha$ we find $\xi = k_B^{-1} \partial S / \partial p_\alpha = -\beta E_\alpha$. Here the index α denotes vector components. Now we use Eq. (5.120), i.e.

$$\frac{\partial}{\partial(-\xi)} \langle X \rangle = \frac{\partial p_\alpha}{\partial(-\beta E_\alpha)} = \langle \delta p_\alpha^2 \rangle. \tag{5.122}$$

At this point we make use of the equation, $\vec{P} = \frac{1}{4\pi}(\varepsilon_r - 1)\vec{E}$, where ε_r is the dielectric constant of the medium to obtain

$$\varepsilon_r = 1 + \frac{4\pi}{3} \frac{\beta}{V} \langle \delta p^2 \rangle. \tag{5.123}$$

Note that $\langle \delta p^2 \rangle = 3 \langle \delta p_\alpha^2 \rangle$ in an isotropic system. Equation (5.123) relates the dielectric constant to the equilibrium fluctuations of the dipole moment taken over the volume V.

Notice the following useful extension of the above. Consider $\langle E \rangle = \sum_v E_v e^{-\beta E_v - \xi X_v} / \sum_v e^{-\beta E_v - \xi X_v}$. Partial differentiation yields

$$\frac{\partial}{\partial(-\xi)} \langle E \rangle \bigg|_{\beta,Y} = \langle EX \rangle - \langle E \rangle \langle X \rangle. \tag{5.124}$$

Using Eqs. (5.120) and (A.2) we obtain

$$\frac{\partial}{\partial \langle X \rangle} \langle E \rangle \bigg|_{\beta,Y} = \frac{\langle EX \rangle - \langle E \rangle \langle X \rangle}{\langle (\delta X)^2 \rangle}. \tag{5.125}$$

An example application of this equation is discussed in the next section.

5.3 Grand-Canonical Ensemble

We consider the special choice $X = N$. In this case we speak of the grand-canonical ensemble. With $\xi \overset{(1.51)}{=} - \beta \mu$ follows

$$p_v = \frac{\exp[-\beta(E_v - \mu N_v)]}{Q_{\mu VT}} \tag{5.126}$$

and

$$Q_{\mu VT} = \sum_v \exp[-\beta(E_v - \mu N_v)]. \tag{5.127}$$

This is the grand-canonical partition function.

5.3.1 *Pressure*

Insertion of Eq. (5.126) into the Gibbs entropy equation yields

$$TS = -\beta^{-1} \sum_v p_v \left(- \ln Q_{\mu VT} - \beta E_v + \beta \mu N_v \right)$$

$$= \beta^{-1} \ln Q_{\mu VT} + \langle E \rangle - \underbrace{\mu \langle N \rangle}_{=G}.$$

With $G = H - TS$ or $TS = E + PV - G$ follows

$$PV = \beta^{-1} \ln Q_{\mu VT}. \tag{5.128}$$

5.3.2   Fluctuating Particle Number and Energy

In the case of the particle number Eq. (5.120) yields

$$\langle (\delta N)^2 \rangle = \left. \frac{\partial \langle N \rangle}{\partial (\beta \mu)} \right|_{\beta, V} . \tag{5.129}$$

This equation may be transformed using various thermodynamic equations. First we apply the Gibbs-Duhem Eq. (2.168), i.e. $d\mu = (V/\langle N \rangle) dP|_T$, to obtain

$$\left. \frac{\partial (\beta \mu)}{\partial \langle N \rangle} \right|_{\beta, V} = \beta \frac{V}{\langle N \rangle} \left. \frac{\partial P}{\partial \langle N \rangle} \right|_{\beta, V} . \tag{5.130}$$

Here $\langle N \rangle$ is identical to the thermodynamic particle number N. Using Eqs. (A.2) and (2.6) we find

$$\beta \frac{V}{\langle N \rangle} \left. \frac{\partial P}{\partial \langle N \rangle} \right|_{\beta, V} \overset{(A.2)}{=} -\beta \frac{V}{\langle N \rangle} \left. \frac{\partial P}{\partial V} \right|_{\beta, \langle N \rangle} \left. \frac{\partial V}{\partial \langle N \rangle} \right|_{\beta, P}$$

$$= -\beta \frac{V}{\langle N \rangle} \left. \frac{\partial P}{\partial V} \right|_{\beta, \langle N \rangle} \frac{V}{\langle N \rangle} \overset{(2.6)}{=} \frac{1}{\langle N \rangle} \frac{\kappa_T^{ideal}}{\kappa_T} ,$$

where $\kappa_T^{ideal} = \beta V / \langle N \rangle$ (cf. Eq. 2.9). Combination of this result with Eqs. (5.129) and (5.130) yields

$$\frac{\sqrt{\langle (\delta N)^2 \rangle}}{\langle N \rangle} = \sqrt{\frac{\kappa_T}{\kappa_T^{ideal}}} \frac{1}{\sqrt{\langle N \rangle}} . \tag{5.131}$$

In a normal situation, i.e. $\sqrt{\kappa_T / \kappa_T^{ideal}}$ is finite, we see immediately that the right side vanishes as $\langle N \rangle$ approaches infinity. In the case of $(\langle \delta E^2 \rangle)^{1/2} / \langle E \rangle$ Eq. (5.75) remains valid. As before in the canonical ensemble we find again that in the thermodynamic limit the micro-canonical ensemble, i.e. E and N are constants, is approached.

Example—Isosteric Heat of Adsorption for Methane on Graphite   One typical application of the (classical) grand-canonical ensemble is equilibrium adsorption. Reconsider our discussion of the isosteric heat of adsorption, q_{st}, on page 106. Using Eq. (5.125) together with $X = N$ we may derive a formula for q_{st} useful for concrete calculations.

In order to combine the definition of q_{st} in Eq. (3.62) with Eq. (5.125) we use Eq. (A1) to rewrite the second term in (3.62), i.e.

$$
\frac{\partial \mu}{\partial T}\Big|_P = \frac{\partial \mu}{\partial T}\Big|_V + \underbrace{\frac{\partial \mu}{\partial V}\Big|_T}_{=-\frac{\partial P}{\partial N}\Big|_{T,V}} \underbrace{\frac{\partial V}{\partial T}\Big|_P}_{=V\alpha_P}.
\tag{5.132}
$$

Using $F = E - TS$ and $\mu = \partial F / \partial N\,|_{T,V}$ we have

$$
\mu = \frac{\partial E}{\partial N}\Big|_{T,V} + T\frac{\partial \mu}{\partial T}\Big|_{V,N},
\tag{5.133}
$$

and q_{st} becomes

$$
q_{st} = -\frac{\partial E_s}{\partial N_s}\Big|_{T,V_s} + \frac{\partial E_b}{\partial N_b}\Big|_{T,V_b} + TV_b\frac{\partial P_b}{\partial N_b}\Big|_{T,V_b}\alpha_P.
\tag{5.134}
$$

At this point we may employ Eq. (5.125) with $X = N$ to obtain

$$
q_{st} = -\frac{\langle U_s N_s\rangle - \langle U_s\rangle\langle N_s\rangle}{\langle(\delta N_s)^2\rangle} + \frac{\langle U_b N_b\rangle - \langle U_b\rangle\langle N_b\rangle}{\langle(\delta N_b)^2\rangle} + TV_b\frac{\partial P_b}{\partial N_b}\Big|_{T,V_b}\alpha_P.
\tag{5.135}
$$

Here we have used $E_{kinetic} \propto N$ so that only the potential energies of the respective systems, U_s and U_b, remain, whereas the kinetic energies, $E_{kinetic}$, drop out.

If the bulk gas is ideal, i.e. $U_b = 0$, this equation simplifies to

$$
q_{st} = -\frac{\langle U_s N_s\rangle - \langle U_s\rangle\langle N_s\rangle}{\langle(\delta N_s)^2\rangle} + RT.
\tag{5.136}
$$

Notice that an ideal bulk gas does not imply ideality in the interface. Equation (5.136) as well as the previous one are useful for calculating q_{st} from computer simulations. This is because the necessary averages are fairly easy to calculate.

Figure 5.14 shows an example (A computer program which may be modified to generate the necessary data is included in the appendix. The theoretical background needed to understand the program is discussed in Chap. 6). The box contains molecules (black dots) interacting with a solid surface (the bottom face of the box). We may imagine that this picture shows a snapshot taken of a gas interacting with an adsorbing surface—notice that the number density is highest near the bottom! If we define the long axis of the box as being the z-direction, we may sort the particles into a histogram

Fig. 5.13 Cosmic background radiation intensity and Planck's prediction versus wavelength

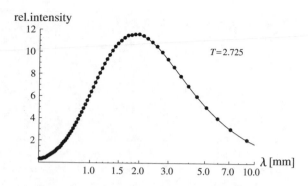

according to their height above the surface. Subsequently we average over histograms for many such snapshots and obtain Fig. 3.11.

The box does not show a real gas but rather the simulation of a gas. The molecules interact pairwise via a so-called Lennard–Jones (LJ) potential, i.e.,

$$u_{LJ,ij} = 4\epsilon \left[\left(\frac{\sigma}{r_{ij}} \right)^{12} - \left(\frac{\sigma}{r_{ij}} \right)^{6} \right]. \tag{5.137}$$

Here r_{ij} is the distance between molecules i and j. For $r_{ij} < 2^{1/6}\sigma$ the interaction is repulsive; for $r_{ij} > 2^{1/6}\sigma$ it is attractive. Whereas the r^{-6}-term may be justified based on quantum perturbation theory, the r^{-12} term is just an ad hoc approximation of repulsive interactions preventing the molecules from simultaneously occupying the same space (cf. the b-parameter in the van der Waals equation). Because a computer can handle only a finite system, we use periodic boundary conditions parallel to the bottom face of the box (the surface), i.e. a molecule leaving the box through one of the side walls re-enters the box through the adjacent side. The top here is merely a reflective wall (We do not use reflective side walls, because reflective walls induce pronounced effects on the structure. Periodic boundaries are better in this respect, but since top and bottom of the box are not equivalent we cannot use them in this case.). The interaction between gas molecules and the surface are constructed similarly. In some simple cases we may use the above LJ potential to also describe the interaction between a gas molecule and an atom in the surface! Because the parameters ϵ and σ are characteristic for the two interacting species, we do have two sets of them: (ϵ_g, σ_g) and (ϵ_s, σ_s)—the indices distinguish gas and surface interactions. If our box contains N gas molecules their total potential energy can be written as

$$U = \frac{1}{2} \sum_{i=1,j=1}^{N} u_{LJ,ij} + 2\pi n_s \epsilon_s'(\sigma_s')^2 \sum_{i=1}^{N} \left[\frac{2}{5} \left(\frac{\sigma_s'}{z_i} \right)^{10} - \left(\frac{\sigma_s'}{z_i} \right)^4 \right]. \qquad (5.138)$$

Here the first term is a sum over all distinct pairs of gas molecules. The second term is a sum over the individual interactions of the gas molecules with the surface depending on their distance, z_i, from the latter. This expression includes the interaction with the topmost atom layer in the surface only. In addition there is just one type of atom in this surface. Another simplification is that the surface atoms are "smeared out" continuously inside the layer. The quantity n_s is the number density of surface atoms per area in the layer. The neglect of atoms below this first layer may be partially compensated by scaling the interaction parameters, which is indicated by the primes. All in all this is a simple and yet quite accurate surface potential for the system we have in mind—the adsorption equilibrium of methane on the graphite basal plane. The parameter values we use here are $\epsilon_g/R = 148.7$ K / mol, $\sigma_g = 3.79$ Å, $n_s = 0.382$ Å^{-2}, $\epsilon_s'/R = 72.2$ K/mol, $\sigma_s' = 3.92$ Å (taken from Aydt and Hentschke 1997). We will discuss computer simulation algorithms, especially the algorithm used in this example, in the next chapter. At this point we merely state that the following result is computed via grand-canonical Metropolis Monte Carlo using Eq. (5.135) as well as Eq. (5.136) for comparison.

Figure 5.15 shows isosteric heats of adsorption vs. pressure. Open symbols are based on Eq. (5.135); closed symbols are based on the approximation

Fig. 5.14 Computer simulation snapshot of gas particles near an adsorbing surface

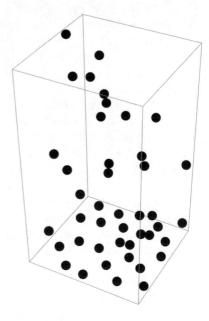

(5.136). The lines are quadratic fits. In the limit of vanishing coverage, which here means $P \to 0$, we obtain $q_{st}^{(o)} \approx 11.5$ kJ/mol at a temperature close to $40\,^{\circ}C$ in both cases. This is somewhat below the experimental values in the literature (e.g., $q_{st}^{(o)} \approx 14.6$ kJ/mol in Specovious and Findenegg 1978). There are a number of possible reasons. While we consider a perfectly smooth surface, the experimental systems are much less perfect and surface defects (like steps or corners), other adsorbates etc. may lead to an increase of $q_{st}^{(o)}$. On the theoretical side we must be critical with respect to the parameters as well as finite size effects due to the smallness of the system. Nevertheless, we learn that the isosteric heat of adsorption yields useful information on the microscopic gas-surface interaction.

5.3.3 Bosons and Fermions

The indistinguishability of elementary particles (such as photons or electrons) is an important concept suggesting the division of all known elementary particles into two classes: Bosons and Fermions. Indistinguishable means that the Hamiltonian operator commutes with another operator that interchanges two particles in a system. This leads to the conclusion that elementary particles in nature may simultaneously occupy the same quantum state in arbitrary number or just once. The former are Bosons and the latter are Fermions. Here we may introduce this distinction via the so called occupation number, n_i, of the one-particle quantum state i. Thus we may write

$$n_i = \begin{cases} 0, 1, 2, \ldots & \text{Bosons} \\ 0, 1 & \text{Fermions} \end{cases}. \tag{5.139}$$

Consequently if we consider a system in state ν, we may express the attendant total particle number, N_ν, via

$$N_\nu = \sum_i n_i \tag{5.140}$$

and the attendant total energy, E_ν, via

$$E_\nu = \sum_i \epsilon_i n_i, \tag{5.141}$$

where ϵ_i is the energy of the one-particle quantum state i. In particular different ν correspond to different sets of occupation numbers (for example: $\{n_1, n_2, n_3, \ldots\} = \{0, 1, 1, 0, \ldots\}$ or $\{1, 1, 1, 0, \ldots\}$). This means that Eq. (5.127) becomes

$$Q_{\mu VT} = \sum_{n_1, n_2, \ldots, n_i, \ldots} \exp\left[-\beta \sum_i (\epsilon_i - \mu)n_i\right]. \tag{5.142}$$

We want to combine this equation with the Eqs. (5.140) and (5.141) to work out its specific form for Bosons and Fermions. Notice that we may reshuffle the sums as follows[7]

$$\sum_{n_1, n_2, \ldots, n_i, \ldots} \exp[-\beta \sum_i \ldots] = \sum_{n_1, n_2, \ldots, n_i, \ldots} \prod_i \exp[-\beta \ldots]$$
$$= \sum_{n_1} e^{-\beta(..1..)} \sum_{n_2} e^{-\beta(..2..)} \ldots = \prod_i \sum_{n_i} \exp[-\beta(..i..)].$$

In the case of Bosons we find

$$Q_{\mu VT}^{(B)} = \prod_i \sum_{n_i=0}^{\infty} \exp[-\beta(\epsilon_i - \mu)n_i]$$
$$= \prod_i (1 - \exp[-\beta(\epsilon_i - \mu)])^{-1}, \tag{5.143}$$

whereas in the case of Fermions

$$Q_{\mu VT}^{(F)} = \prod_i \sum_{n_i=0}^{1} \exp[-\beta(\epsilon_i - \mu)n_i]$$
$$= \prod_i (1 + \exp[-\beta(\epsilon_i - \mu)]). \tag{5.144}$$

In both cases we may compute the average occupation number via

$$\langle n_j \rangle = Q_{\mu VT}^{-1} \sum_\nu n_j \exp[-\beta(E_\nu - \mu N_\nu)] = \frac{\partial \ln Q_{\mu VT}}{\partial(-\beta\epsilon_j)}. \tag{5.145}$$

This means for Bosons

$$\langle n_j \rangle^{(B)} = \left(\exp[\beta(\epsilon_j - \mu)] - 1\right)^{-1} \tag{5.146}$$

---

[7] via $\sum_{n=0}^{\infty} q^n = (1 - q)^{-1}$ for $q < 1$.

and for Fermions

$$\langle n_j \rangle^{(F)} = \left(\exp\left[\beta(\epsilon_j - \mu)\right] + 1\right)^{-1}. \tag{5.147}$$

Equation (5.146) imposes the condition $\mu < \epsilon_0$, where ϵ_0 is the one-particle ground state energy, because otherwise unphysical negative occupation is possible. No such restriction applies in the case of Fermions.

Particles in nature posses a property called spin. According to Pauli's famous Spin-Statistics-Theorem all particles possessing integer spin values (e.g. photons whose spin is one) are Bosons, whereas particles possessing half-integer spin values (e.g. the electron has spin one half) are Fermions.[8]

5.3.4 High Temperature Limit

With increasing temperature the particles may access higher and higher energies. At not too high densities this implies that there are more one particle energy states partially occupied than there are particles. Consequently the average occupation number, $\langle n_j \rangle$, is small, i.e.

$$\exp\left[\beta(\epsilon_j - \mu)\right] \gg 1 .$$

For both Bosons and Fermions we therefore have

$$\langle n_j \rangle \approx \exp\left[-\beta(\epsilon_j - \mu)\right]. \tag{5.148}$$

Insertion of this approximation into

$$\langle N \rangle = \sum_j \langle n_j \rangle \tag{5.149}$$

yields the useful relation

$$\frac{\langle n_j \rangle}{\langle N \rangle} \propto \exp\left[-\beta\epsilon_j\right] . \tag{5.150}$$

This is the probability for a particle to be in the state j – independent of the particle's type.

---

[8] Wolfgang Pauli, Nobel prize in physics for his discovery of the exclusion principle, 1945.

5.3.5 Two Special Cases—$\epsilon \propto k^2$ and $\epsilon \propto k$

The energy values of a free particle are $\epsilon = \hbar^2 k^2/(2m)$. This is the 3D version of Eq. (5.57) (omitting the index v). In addition the 3D version of Eq. (5.58) is

$$\sum_j \propto \int d^3 k \propto \int_0^\infty d\epsilon D(\epsilon). \tag{5.151}$$

The quantity $D(\epsilon)$ is the density of energy states, and because of $d^3 k \propto dk k^2$ we find $D(\epsilon) \propto \epsilon^{1/2}$. The second case we study here is $\epsilon \propto k$. For instance in the case of photons $\epsilon = \hbar\omega$ and $\omega = ck$, where c is the velocity of light and k is the magnitude of the wave vector. Eq. (5.151) still remains valid except now of course $D(\epsilon) \propto \epsilon^2$. Thus all in all we consider the following cases

$$\epsilon \propto k^2 \quad \text{with} \quad D(\epsilon) \propto \epsilon^{1/2}$$
$$\epsilon \propto k \quad \text{with} \quad D(\epsilon) \propto \epsilon^2.$$

First we compute the average energy of a systems of bosons in these two cases

$$\langle E \rangle = \begin{cases} C \int_0^\infty \dfrac{d\epsilon \epsilon^{3/2}}{z^{-1} e^{\beta\epsilon} - 1} = -\dfrac{3C}{2\beta} \int_0^\infty d\epsilon \epsilon^{1/2} \ln[1 - z e^{-\beta\epsilon}] & (\epsilon \propto k^2) \\[4mm] C' \int_0^\infty \dfrac{d\epsilon \epsilon^3}{z^{-1} e^{\beta\epsilon} - 1} = -\dfrac{3C'}{\beta} \int_0^\infty d\epsilon \epsilon^2 \ln[1 - z e^{-\beta\epsilon}] & (\epsilon \propto k), \end{cases} \tag{5.152}$$

where $z = \exp[\beta\mu]$ and C, C' are constants. Note that we have used partial integration, i.e.

$$\int_0^\infty \frac{d\epsilon \epsilon^x}{z^{-1} e^{\beta\epsilon} - 1} = \frac{1}{\beta} \underbrace{\Big|_0^\infty \epsilon^x \ln[1 - z e^{-\beta\epsilon}]}_{=0} - \frac{x}{\beta} \int_0^\infty d\epsilon \epsilon^{x-1} \ln[1 - z e^{-\beta\epsilon}].$$

Comparing the right sides in Eq. (5.152) to the pressure, i.e. $\beta V P^{(B)} = -\sum_i \ln[1 - z \exp[-\beta\epsilon_i]]$, we find

$$\frac{1}{V}\langle E \rangle = \frac{3}{2} P \quad (\epsilon \propto k^2)$$
$$\frac{1}{V}\langle E \rangle = 3P \quad (\epsilon \propto k). \tag{5.153}$$

Even though we have obtained this result in the case of Bosons an analogous calculation for Fermions yields identical formulas. Notice that these results confirm our previous findings for the energy density expressed in Eqs. (2.27) and (2.35). Notice also that in order for Eq. (5.152) ($\epsilon \propto k$-case) to yield agreement with the black body radiation energy density (cf. Remark 2 on page 239) we must require $z = 1$ or $\mu = 0$ for photons.

It is interesting to similarly relate the density, N/V, and the pressure, P. Unfortunately the result is not obtained simply via partial integration. We need to solve the integrals explicitly, i.e.

$$\frac{N}{V} = \begin{cases} \dfrac{C}{V} \displaystyle\int_0^\infty \dfrac{d\epsilon\,\epsilon^{1/2}}{z^{-1}e^{\beta\epsilon} \mp 1} = \pm \dfrac{C}{V} \dfrac{\Gamma[3/2]}{\beta^{3/2}} \sum_{n=0}^\infty \dfrac{(\pm z)^{n+1}}{(n+1)^{3/2}} & (\epsilon \propto k^2) \\[4mm] \dfrac{C'}{V} \displaystyle\int_0^\infty \dfrac{d\epsilon\,\epsilon^2}{z^{-1}e^{\beta\epsilon} \mp 1} = \pm \dfrac{C'}{V} \dfrac{\Gamma[3]}{\beta^3} \sum_{n=0}^\infty \dfrac{(\pm z)^{n+1}}{(n+1)^3} & (\epsilon \propto k) \end{cases}, \tag{5.154}$$

and

$$\beta P = \begin{cases} \dfrac{2C\beta}{3V} \displaystyle\int_0^\infty \dfrac{d\epsilon\,\epsilon^{3/2}}{z^{-1}e^{\beta\epsilon} \mp 1} = \pm \dfrac{C}{V} \dfrac{\Gamma[3/2]}{\beta^{3/2}} \sum_{n=0}^\infty \dfrac{(\pm z)^{n+1}}{(n+1)^{5/2}} & (\epsilon \propto k^2) \\[4mm] \dfrac{C'\beta}{3V} \displaystyle\int_0^\infty \dfrac{d\epsilon\,\epsilon^3}{z^{-1}e^{\beta\epsilon} \mp 1} = \pm \dfrac{C'}{V} \dfrac{\Gamma[3]}{\beta^3} \sum_{n=0}^\infty \dfrac{(\pm z)^{n+1}}{(n+1)^4} & (\epsilon \propto k) \end{cases}. \tag{5.155}$$

Here we have used $(1 - z\exp[-\beta\epsilon])^{-1} = \sum_{n=0}^\infty (z\exp[-\beta\epsilon])^n$ and

$$\int_0^\infty d\epsilon\,\epsilon^x \exp[-\beta\epsilon(n+1)] = \frac{\Gamma[x+1]}{(\beta(n+1))^{x+1}} \tag{5.156}$$

(Note: $\Gamma[x+1] = x\Gamma[x]$, where $\Gamma[x]$ is the Gamma-function (Abramowitz and Stegun 1972). Attention must be paid to the radius of convergence of the above sums, i.e. in cases when $z > 1$ (Fermions) the integrals must be evaluated by other means.

Both N/V and P are power series in z. However, we may conversely assume that z can be written as a power series of for instance $\rho = N/V$, i.e. $z = z(\rho) = a_0 + a_1\rho + a_2\rho^2 + \dots$. By inserting this into our above expansion of $\rho(z)$ we eliminate z from this equation and we may work out the coefficients a_i by equating coefficients multiplying the same power of ρ. If we now insert the power series of $z(\rho)$ with the known coefficients a_i into the expansion of $P = P(z)$ we obtain a power series $P = P(\rho)$.

Figure 5.16 shows the results (for the case $\epsilon \propto k^2$). The stars indicate reduced quantities defined via

$$\rho = \frac{C\Gamma[3/2]}{V\beta^{3/2}} \rho^* \quad P = \frac{C\Gamma[3/2]}{V\beta^{5/2}} P^*$$

$$\rho = \frac{C'\Gamma[3]}{V\beta^3} \rho^* \quad P = \frac{C'\Gamma[3]}{V\beta^4} P^* .$$

Fig. 5.15 Isosteric heat of adsorption of methane on graphite

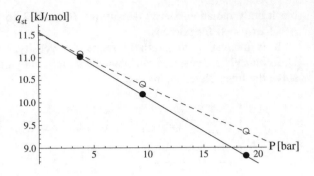

Fig. 5.16 Reduced pressure versus reduced density for particles obeying Fermi-Dirac statistics, Bose-Einstein statistics, or the ideal gas law

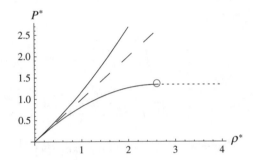

Of the two solid lines the lower one is for Bosons and the upper one for Fermions. The dashed line is the ideal gas law, i.e. $P = \beta\rho$. We note that in the case of Bosons we find a quantum statistical attraction leading to a lower pressure than in the ideal gas, whereas in the case of Fermions there is a quantum statistical repulsion leading to a higher pressure than in the ideal gas. The open circle indicates the Boson pressure for $z = 1$. At higher densities the pressure remains constant as shown by the dotted line.

To understand what is happening here we look at Fig. 5.17 showing z as function of reduced density, ρ^*. The upper line is for Fermions and the lower one for Bosons. In case of the former nothing special happens as z approaches or exceeds unity. In the Boson case the reduced density, ρ^*, approaches a finite value, $\rho^* = \zeta[3/2] = 2.61238$,[9] again indicated by a circle. This circle corresponds to the circle in the previous figure. Because we may increase ρ^* beyond the value marked by the circle we may ask what the corresponding z-values are. Clearly, $z \leq 1$ as pointed out above. But decreasing z means $\partial z/\partial\rho^* = \beta z \partial\mu/\partial\rho^* < 0$ or $\partial\mu/\partial\rho^* < 0$, which is thermodynamically unstable. Thus z must remain equal to unity as indicated by the dotted line.

---

[9]

$\zeta[s] = \sum_{k=1}^{\infty} k^{-s}$ is the Riemann Zeta-function (Abramowitz and Stegun 1972).

But why does our current approach not show the divergence of $\rho^*(z)$ in the limit $z = 1$ – or in other words: where do the particles go? There is a mathematical problem here, which we have overlooked thus far. Notice that in the Boson case

$$\frac{N}{V} \propto \int_0^\infty \frac{d\epsilon \, \epsilon^{1/2}}{z^{-1} e^{\beta \epsilon} - 1} \stackrel{z=1}{=} \frac{\sqrt{\pi} \zeta[3/2]}{2 \beta^{3/2}} \quad \text{(note: } C, C' \propto V\text{).} \qquad (5.157)$$

The integral obviously is finite. On the other hand, the average Boson occupation number of the energy level $\epsilon_0 = 0$ in the case $z = 1$ is infinity according to the original Eq. (5.146)[10]! Thus our conversion of the summation into an integration fails utterly—at least when $z = 1$.

A simple example may illustrate this point. Consider the following sums

$$S_0(m, k) = \sum_{n=0}^m \frac{1}{\sqrt{n+k}} \quad \text{and} \quad S_1(m, k) = \sum_{n=1}^m \frac{1}{\sqrt{n+k}}$$

compared to the integral

$$I(m, k) = \int_0^m \frac{dn}{\sqrt{n+k}} = 2(\sqrt{m+k} - \sqrt{k}).$$

Putting in some numbers we obtain

		$I(m,k)$	$S_1(m,k)$	$S_0(m,k)$
$m = 1$	$k = 10^{-3}$	1.94	1.0	32.6
$m = 100$	$k = 10^{-3}$	19.9	18.6	50.2
$m = 100$	$k = 10^{-5}$	20.0	18.6	335

We observe that summation and integration are in rather good agreement for large m independent of k. Including the $n = 0$-Term in the sum, especially in the limit of small k, spoils the agreement however. Therefore we may write $S_0(m, k) = S_1(m, k) + k^{-1/2} \approx I(m, k) + k^{-1/2}$.

Note that in the above example $k \to 0$ corresponds to $z \to 1$ and the $n = 0$-term corresponds to the ground state contribution ($\epsilon_0 = 0$) in Eq. (5.146). Here we may write

---

[10] Somebody may object that $\epsilon_0 = 0$ is not really possible due to the zero-point energy. But note that for a particle trapped in a cubic box one finds $\beta \epsilon_0 \sim (\Lambda_T / L)^2$. Here $V = L^3$ is the box volume. The thermal wavelength, Λ_T, is on the order of Å, so that for every macroscopic L we find that $\beta \epsilon_0$ is vanishingly small. This also is the reason why we consider the limit $z = 1$ rather than $\exp[\beta \epsilon_0]$.

$$\rho = \frac{C}{V} \frac{\sqrt{\pi}\zeta[3/2]}{2\beta^{3/2}} + \frac{1}{V}\frac{z}{1-z}. \tag{5.158}$$

We use the $=-$sign rather than $\approx$, because in this case the number of terms in the sum are so large. Now we see that in the limit $z = 1$ there is a condensation of particles into single particle states $\epsilon_0 = 0$. This phenomenon is called Bose condensation[11]!

Equation (5.158) also explains the pressure plateau (dotted line in Fig. 5.16). Assuming that ρ^* is large and fixed we conclude that $V \sim (1 - z)^{-1}$. If we now consider the ground state contribution to the pressure given by $\beta P|_o = -V^{-1}\ln[1 - z]$ we immediately obtain $\beta P|_o \sim V^{-1}\ln V \to 0$ for $V \to \infty$ at constant density. This means that $P|_o$ does not contribute to the pressure, which therefore remains constant. The same is true for the energy density, as one can easily find out by considering the limit $\epsilon_j \to 0$ of $\epsilon_j \langle n_j \rangle$.

Before leaving this subject, we want to also calculate the transition enthalpy predicted by our model. We estimate the pressure corresponding to the circle in Fig. 5.16 via Eq. (5.155) setting $z = 1$, i.e.

$$P = \frac{\sqrt{\pi}C}{2V}\frac{\zeta[5/2]}{\beta^{5/2}}. \tag{5.159}$$

Taking the derivative with respect to T we obtain

$$\frac{dP}{dT} = k_B \frac{5\sqrt{\pi}C}{4V}\frac{\zeta[5/2]}{\beta^{3/2}}. \tag{5.160}$$

Notice that this derivative is along the coexistence line in the T-P-plane. Therefore we can apply the Clapeyron equation, i.e.

$$\left.\frac{dP}{dT}\right|_{coex} = \frac{1}{T}\frac{\Delta h}{\Delta \rho^{-1}}. \tag{5.161}$$

Here Δh is the transition enthalpy per Boson, and $\Delta \rho^{-1} \approx \rho^{-1}$ is the corresponding inverse density difference between the inverse density given by the first term in Eq. (5.158) and the corresponding inverse ground state density. The latter

---

[11] The superfluid behavior exhibited by the helium isotope ^4He below 2.1768 K, the so-called lambda point, is a manifestation of Bose condensation. The mass density at this temperature is about 145 kg/m³. If we insert this number into Eq. (5.158) using $C = (gV/(4\pi^2))(2m_{He}/\hbar^2)^{3/2}$, where $g = 2s + 1$ and s is the boson's spin (in this case $s = 0$), we obtain a transition temperature of about 3.1 K, which, despite the ideality assumption, is rather close to the above value (an in depth discussion is given in Feynman (1972) ; Nobel prize in physics for his contributions to quantum electrodynamics, 1965). Notice that we have neglected the second term in Eq. (5.158), because for fixed z just slightly less than 1 the factor $1/V$ dominates and this term is vanishingly small.

density is much larger and its inverse is therefore neglected. Thus Δh is the (vaporization) enthalpy difference going from the condensed phase to the ideal gas. After some algebra we find[12]

$$\Delta h = \frac{5}{2} \frac{\zeta[5/2]}{\zeta[3/2]} k_B T \approx 1.28 k_B T.$$
(5.162)

5.4   The Third Law of Thermodynamics

This law, which does not introduce new functions of state, is about entropy in the limit of vanishing temperature. Its most common form is the Nernst heat theorem[13]:

$$\lim_{T \to 0} \Delta S = 0.$$
(5.163)

This means that all entropy changes are zero at absolute zero. In a generalization due to Planck the Δ is omitted and thus

$$\lim_{T \to 0} S = 0.$$
(5.164)

This follows from (5.163) if the constant entropy at $T = 0$ implied by the Nernst heat theorem is universal and finite and thus may be set to zero.

Equation (5.163) implies that partial derivatives of S like $\partial S/\partial V|_T$ vanish in the limit of vanishing temperature, i.e.

$$\lim_{T \to 0} \frac{\partial S}{\partial V}\bigg|_T = 0.$$
(5.165)

Because of $\partial^2 F/\partial V \partial T|_T|_V = \partial^2 F/\partial T \partial V|_V|_T$, the above equation implies

$$\lim_{T \to 0} \frac{\partial P}{\partial T}\bigg|_V = 0.$$
(5.166)

If we apply this to the classical ideal gas, i.e. $PV = Nk_B T$, we find that this gas does not obey the Nernst heat theorem. Photons on the other hand fulfill $S \propto T^3$ (cf. Eq. 2.41) or $P \propto T^4$ (cf. Eq. (2.38)) and satisfy Eq. (5.164) as well as Eq. (5.166).

Figure 5.16 shows that the classical ideal gas law at finite densities is not followed by either Fermions or Bosons in the limit of low temperatures. Only in the

---

[12] In the case of ^4He, i.e. using $T = 3.2$ K, we obtain $\Delta H \approx 34 J/mol$. This is about three times less than the experimental value.

[13] Walther Hermann Nernst, Nobel prize in chemistry for his contributions to thermodynamics, 1920.

high temperature limit, i.e. the origin of the graph, both quantum laws merge with the ideal gas law. We therefore want to study the above derivative for the quantum laws in Fig. 5.16. In the case of Fermions, the leading dependence of the pressure on temperature is

$$P \approx \frac{2}{d+2} \epsilon_F \rho + \frac{\pi^2 \rho}{\epsilon_F} (k_B T)^2. \tag{5.167}$$

Here d is the space dimension, ρ is the number density, and ϵ_F, the Fermi energy, is the energy of the highest occupied level at zero temperature. We omit the calculation of this formula, which may be found in the context of the Fermi gas model of electrons in solids in most solid state textbooks. Obviously the pressure this time satisfies Eq. (5.166). But what about the Bosons? The answer is contained in Fig. 5.16. At finite density $P \propto T^{5/2}$ and thus $\partial P/\partial T|_V \to 0$ for $T \to 0$. Again Nernst's theorem is satisfied. Let us therefore look at the relation between the Nernst theorem and quantum theory.

We start from the partition function

$$Q = \sum_s g_s \exp[-\beta E_s]. \tag{5.168}$$

Here all E_s are distinct and discrete $(E_o < E_1 < E_2 \cdots)$. The degeneracy of each level s is g_s. Factorizing the ground state we obtain

$$Q = g_o \exp[-\beta E_o] \left(1 + \frac{g_1}{g_0} \exp[-\beta(E_1 - E_o] + \cdots \right). \tag{5.169}$$

As the temperature approaches zero we can satisfy $\beta(E_1 - E_o) \ll 1$ and it becomes sufficient to retain the first two terms in the above sum. Using $S = \partial(k_B T \ln Q)/\partial T|_V$ we obtain

$$S/k_B \approx \ln g_o + (1 + \beta(E_1 - E_o)) \frac{g_1}{g_o} \exp[-\beta(E_1 - E_o)]. \tag{5.170}$$

Letting $T \to 0$, we get

$$S/k_B = \ln g_o. \tag{5.171}$$

Therefore $S = 0$ requires that the ground state is not degenerate and sufficiently separated from the next state. However, notice that the degeneracy of this state may already be large. To see this we imagine N independent harmonic oscillators. The state o corresponds to all oscillators in their ground state. The next state corresponds to one oscillator in its first excited state whereas all others are still in the ground state. There are N distinct ways to accomplish this. Again the next state corresponds to one oscillator in its second excited state and a second one in its first excited state.

Fig. 5.17 Fugacity versus reduced density for particles obeying Fermi-Dirac statistics and particles obeying Bose-Einstein statistics

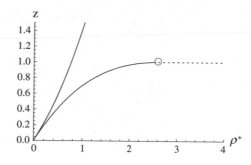

Fig. 5.18 Entropy per spin, S/n of a periodic 1D Ising chain with 10 spins ($J = 1$, $B = 10^{-4}$) versus temperature, T (here: $k_B = 1$)

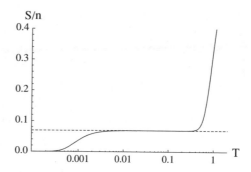

This "pair of excited oscillators" may be realized in $N(N-1)/2$ different ways. And so on.

Figure 5.18 shows an illustrative example of a model system—the 1D Ising chain (cf. the example on page 210). Without an external field the ground state is degenerate and $S/n > 0$ (here $n = 10$ and $S/n \to (\ln 2)/10$—dashed line) for $T \to 0$. But an external field, B, coupling to the spins (here via an additional term $-B \sum_{i=1}^{n} s_i$ in the energy (Eq. 5.17)) no matter how small, breaks the symmetry and "eventually" $S/n \to 0$ for $T \to 0$. In reality such symmetry-breaking fields should always be present.

Chapter 6
Thermodynamics and Molecular Simulation

In the previous chapter we have learned that analytic calculations on the basis of microscopic interactions can become very difficult or even impossible. In such cases computer simulations are helpful. And even though thermodynamics is not the theory of many particle systems based on microscopic interactions, Statistical Mechanics is this theory, it possesses noteworthy ties to computer simulation. In the following we want to discuss some of them.

There are two main methods in this field. One is Molecular Dynamics and the other is Monte Carlo. Additional simulation methods are either closely related to one or the other aforementioned methods or they apply on spatial scales far beyond the molecular scale. Molecular Dynamics techniques model a small amount of material (system sizes usually are on the nm-scale) based on the actual equations of motion of the atoms or molecules in this system. Usually this is done on the basis of mechanical inter- and intra-particle potential functions. In certain cases however quantum mechanics in needed. Monte Carlo differs from Molecular Dynamics in that its systems do not follow their physical dynamics. Monte Carlo estimates thermodynamic quantities via intelligent statistical sampling of (micro)states. Capabilities and applications of both methods overlap widely. But they both also have distinct advantages depending on the problem at hand. Here we concentrate on Monte Carlo—which is the "more thermodynamic" method of the two.

6.1 Metropolis Sampling

Consider a gas at constant density and constant temperature. Suppose we are interested in the internal energy, E, of the gas. In principle we can try to evaluate Eq. (5.8). Instead of doing this by an analytic method, we use a "device" which

© Springer Nature Switzerland AG 2022
R. Hentschke, *Thermodynamics*, Undergraduate Lecture Notes in Physics,
https://doi.org/10.1007/978-3-030-93879-6_6

supplies us with a sample of E_v-values distributed according to p_v as defined in Eq. (5.7). Our estimate for E would then be $\bar{E}$, where the bar indicates the average over our sample. In the limit of an infinite sample we have $E = \langle E \rangle = \bar{E}$. But how would such a "device" look like?

We consider a simple example. Rather than using an infinite set of E_v-values from which we generate our sample, we just work with four values. They are not called energy—we just call them 1, 2, 3, and 4. A possible sample might look like this:

$$234334412223\ldots \tag{6.1}$$

But what is the underlying probability distribution, p_v, in this example? For the moment we decide to invent a distribution, i.e. we require that the even digits, 2 and 4, are twice as probable as the odd ones, 1 and 3.[1]

A computer algorithm generating a series of digits possessing this distribution is the following:

(i)   choose a new digit, d_{new}, from the set $\{1, 2, 3, 4\}$ at random
(ii) IF

$$min[1, \frac{p(d_{new})}{p(d_{old})}] \geq \xi \tag{6.2}$$

  THEN d_{new} becomes the next digit in the series
  ELSE d_{old} becomes the next digit in the series
(iii) if the series contains less than M digits continue with (i)

Here *old* refers to the last already existing digit in the series. The next digit is *new*. Step (i) should be clear. But step (ii) requires explanation. The function $min[a, b]$ returns the smaller of the two arguments a and b. The quantity $p(d)$ denotes the probability of occurrence of digit d in the series, i.e. $p(1)/p(2) = 1/2$, $p(1)/p(3) = 1$, $p(1)/p(4) = 1/2$, ... The quantity ξ is a random number between 0 and 1. The condition (6.2) is the Metropolis criterion.

We can check this algorithm in two ways—we can implement it and count the occurrence of the individual digits in the series—we can study the attendant transfer matrix, $\underline{\pi}$, whose elements $\pi_{ij} \equiv \pi_{i \to j}$ are the probabilities that digit i is followed by digit j. At this point it is most instructive to do the latter.

---

[1] Every other conceivable distribution can replace this choice if desired.

The transfer matrix is

$$
\underline{\pi} = \frac{1}{4}
\begin{pmatrix}
1 & 1 & 1 & 1 \\
1/2 & 2 & 1/2 & 1 \\
1 & 1 & 1 & 1 \\
1/2 & 1 & 1/2 & 2
\end{pmatrix}
\quad \downarrow i.
$$

$$j \rightarrow$$

The factor $1/4$ is the statistical weight of any *new* digit according to step (i). The matrix elements on the other hand are the statistical weights of attempted transitions generated by the Metropolis criterion in step (ii). In particular the first row elements are the step (ii)-statistical weights for transitions from the *old* digit $i = 1$ to any *new* digit j. In this case the Metropolis criterion (6.2) is always fulfilled independent of the *new* digit that follows. The second row corresponds to the case when the *old* digit $i = 2$. The first element, i.e. $1/2$, is the statistical weight due (6.2) for the transition from 2 to 1, which is rejected half of the time. The same applies to the third entry, i.e. the transition from 2 to 3. The fourth element describes the transition from 2 to 4, which is always accepted. The entry 2 in this row is more difficult to understand. This is because the transition from 2 to 2 has three contributions, i.e. $2 = 1 + 1/2 + 1/2$. The criterion is always fulfilled when 2 is followed by 2, which contributes the one. In addition there is a $1/2$ from the rejection of transitions from 2 to 1. In this case the *old* digit 2 also will be the accepted *new* digit. The same applies when the criterion rejects transitions from 2 to 3. In this case the *old* digit 2 again will be the *new* digit. All in all the result from step (ii) is two. This discussion of the first and the second row may be carried over to explain rows three and four.

At this point we are able to answer to the following question: What is the probability that if the first digit in the series is i the 3rd digit will be j? The answer is obtained via simple matrix multiplication:

$$
\sum_{k=1}^{4} \pi_{ik} \pi_{kj} = \pi_{i1}\pi_{1j} + \pi_{i2}\pi_{2j} + \pi_{i3}\pi_{3j} + \pi_{i,4}\pi_{4j}, \tag{6.3}
$$

i.e. the total transition probability is the sum over all independent possibilities (or paths) to get from i to j. This easily is generalized to longer paths extending over n digits

$$\sum_{i,k,l,\ldots j} \underbrace{\pi_{ik}\pi_{kl}\pi_{l..}\cdots\pi_{..j}}_{n-1\,\text{factors}}. \tag{6.4}$$

Provided n is sufficiently large the result should be $2/6$ if j is even and $1/6$ if j is odd. Notice that the denominator 6 is due to the overall normalization, i.e. $2\cdot 2/6 + 2\cdot 1/6 = 1$. For $n = 3$ the explicit numerical result is

$$\begin{pmatrix} 0.1875 & 0.3125 & 0.1875 & 0.3125 \\ 0.15625 & 0.375 & 0.15625 & 0.3125 \\ 0.1875 & 0.3125 & 0.1875 & 0.3125 \\ 0.15625 & 0.3125 & 0.15625 & 0.375 \end{pmatrix}. \tag{6.5}$$

But already for $n = 5$ we obtain

$$\begin{pmatrix} 0.167969 & 0.332031 & 0.167969 & 0.332031 \\ 0.166016 & 0.335938 & 0.166016 & 0.332031 \\ 0.167969 & 0.332031 & 0.167969 & 0.332031 \\ 0.166016 & 0.332031 & 0.166016 & 0.335938 \end{pmatrix}. \tag{6.6}$$

This shows that our algorithm indeed produces the target distribution of digits in the series. The closeness of the numbers in the same column, which means the same final digit j, in the last matrix also shows that the probability of j is almost independent of the first digit i after only five steps.

Example : Metropolis MC. The following is a *Mathematica* program which estimates

$$\langle x^2 \rangle := \int_{-\infty}^{\infty} dx\, x^2 \exp[-x^2] \Big/ \int_{-\infty}^{\infty} dx \exp[-x^2] = 0.5 \;.$$

Here $p \propto \exp[-x^2]$ and $\{1, 2, 3, 4\}$ is now replaced by $x \in (-\infty, \infty)$. We seek $\overline{x^2} = (1/M) \sum_{i=1}^{M} x_i^2 |_{MMC}$. $|_{MMC}$ means that the x_i are weighted according to p via Metropolis MC.

```
"Metropolis Monte Carlo";
"total number of MC steps"; M = 100000;
"acceptance counter"; ac = 0;
"initial x"; xold = 1;
"maximum step size"; δx = 1;
"results storage"; x2array = {}; sx2 = 0;
Do[
   xnew = xold + δxRandom[Real, {−1, 1}];
   If[Min[1, Exp[−(xnew^2 − xold^2)]]>=Random[],
      {sx2+=xnew^2, xold = xnew, ac+=1},
      {sx2+=xold^2}];
   x2array = Append[x2array, sx2/i],
{i, 1, M}];

"output:";
p1 = ListPlot[x2array, Joined → True,
AxesLabel → {"MC step", "⟨x²⟩cave"},
PlotRange → {0.4, 0.6}, PlotStyle → {Black}];
p2 = ListPlot[{{0, 0.5}, {M, 0.5}}, Joined → True,
PlotStyle → {Dashed, Black}];
Show[p1, p2]
N[ac/M]
```

0.72954

A few remarks: Step (i) of the MC algorithm is xnew=xold + δx *Random*[Real, $\{-1, 1\}$]. We define a maximum step size, δx, and generate the new x near the old x. It is important that this step does not introduce a bias, i.e. all possible x-values do have equal probability in step (i). It does not matter that it may take many steps to reach a certain x-value. Notice that in contrast to our initial example, with merely 4 digits to choose from, x can be any real number. In this case it is important to build the new x from the old one in order to play out the particular advantage of the Metropolis criterion. It preferentially samples the important x-values, i.e. the x-values with high probability. Random "jumping around" between $-\infty$ and ∞ would not be efficient. The above graph shows the cumulative average, i.e. at every MC-step the current average of x^2 computed from the thus far accumulated x-values. The final result obviously is close to the exact solution (dashed line). The shown example yields $\overline{x^2} = 0.498$. However, the question is—what is the error? This is a computer experiment and as for real experimental data we may compute the standard error of the mean via $\sigma_{x^2}/\sqrt{M'}$. Here σ_{x^2} is the standard deviation of the x_i^2-values and M' is the number of independent values ($M' \ll M$!). We can estimate M' via the autocorrelation function of the data. But this is described in every good text on computer simulation methods (e.g., Frenkel and Smit 1996). Here we want to concentrate on our main goal—the ties between molecular simulation and thermodynamics.

6.2   Sampling Different Ensembles

In Chap. 5 we discuss the probabilities of state v given by

$$p_v \propto \exp[-S_v/k_B], \tag{6.7}$$

where

$$S_v = \frac{1}{T}E_v + \frac{P}{T}V_v - \frac{\mu}{T}N_v\ldots \tag{6.8}$$

In the following we are interested in the classical approximation of p_v, which means

$$p \propto g\exp[-\beta\mathcal{H} - \beta PV + \beta\mu N\ldots]. \tag{6.9}$$

Here $\mathcal{H}$ is the Hamilton function of the system and $\beta = (k_B T)^{-1}$. The factor g arises from the "translation" of $\sum_v$ to a corresponding integration over classical phase space, i.e.

$$\sum_v \approx \frac{1}{N!h^{3N}} \int d^{3N}p \, d^{3N}r \rightarrow \frac{1}{N!} \left(\frac{V}{\Lambda_T^3}\right)^N \int d^{3N}s. \qquad (6.10)$$

This formula applies to particles completely characterizable through their position in space $\vec{r} = V^{1/3}\vec{s}$, where $\vec{s}$ is a relative coordinate independent of the size of the (cubic) volume, and their translational momentum $\vec{p}$. We can separate a factor $\exp[-\beta\mathcal{K}]$ from $\exp[-\beta\mathcal{H}]$, where $\mathcal{K}$ is the kinetic energy of the system of particles, and the arrow indicates what we get after integrating out the momenta. Thus in the present case we may write

$$p(\{\vec{s}\}, V, N) \propto \frac{1}{N!} \left(\frac{V}{\Lambda_T^3}\right)^N \exp[-\beta\mathcal{U} - \beta PV + \beta\mu N \ldots]. \qquad (6.11)$$

Here $\mathcal{U}$ is the potential energy of the system. All in all this probability describes classical systems with variable volume, and particle number.

How do we apply this? First we must decide which ensemble to use. Is it sufficient to just translate the particles at constant temperature, volume and particle number? This would be the canonical ensemble. Or do we model an open system with variable particle number at constant chemical potential, volume, and temperature? This would be the grand-canonical ensemble. Remember our discussion of the isosteric heat of adsorption for methane on graphite in Sect. 5.3.2. In this example methane is well represented as a point particle. Here step (i) of a MC procedure consists in a random change of $(\{\vec{s}\}, V, N)$. We can select a methane molecule at random and move it a random distance in a random direction. Volume and particle number would be constant. But we can also decide to just change the particle number. We must decide whether to insert or remove a particle from the system. The following algorithm, used to generate the simulation results in the aforementioned example, alternates between these two MC "moves". The volume is kept constant all the time. Insertion and removal of particles makes additional translation of existent particles obsolete in this case.[2]

1. randomly select a position at which to insert a new particle into the gas
2. evaluate the condition

$$\min\left(1, \exp\left[-T^{-1}\left(\mathcal{U}_{N+1}^{(new)} - \mathcal{U}_N^{(old)}\right) + \ln\frac{aV}{N+1}\right]\right) \geq \xi_i.$$

---

[2] This works and is simple, but not necessarily efficient. It bypasses the importance sampling capability of the Metropolis MC mentioned above.

3. TRUE: Insert the particle and append the new configuration to the configuration list;
   FALSE: Do not insert the particle and append the old configuration to the configuration list.
4. select an already existing gas particle at random
5. evaluate the condition

$$\min\left(1, \exp\left[-T^{-1}\left(\mathcal{U}_{N-1}^{(new)} - \mathcal{U}_{N}^{(old)}\right) + \ln\frac{N}{aV}\right]\right) \geq \xi_r.$$

6. TRUE: Remove the selected particle and append the new configuration to the configuration list;
   FALSE: Do not remove the selected particle and append the old configuration to the configuration list.

Here ξ_i and ξ_r are random numbers on $[0, 1]$. The quantity a is $a = \Lambda_T^{-3}\exp[T^{-1}\mu]$. Notice that the second argument in $\min(1, \ldots)$ is just the ratio p_{new}/p_{old} (with $k_B = 1$). We emphasize that we do not have to know the proportionality constant in (6.11)! This would mean that we have to compute the full partition function, which in general is impossible. After having compiled a large number of configurations generated according to this algorithm the desired averages may be calculated. However, because MC does not yield information on the molecular dynamics, it is limited to configurational averages, i.e. the quantities of interest depend exclusively on coordinates and not on momenta.

For the sake of completeness we want to write down the equivalent to relation (6.11) for the case of small rigid molecules (e.g. water):

$$p(\{\vec{s}\}, \{\phi, \theta, \psi\}, V, N) \propto$$
$$\frac{1}{N!}\left(\frac{VQ_{cl}^{rot}}{\Lambda_T^3}\right)^N \Pi_{i=1}^N \sin\theta_i \exp[-\beta\mathcal{U} - \beta PV + \beta\mu N\ldots]. \tag{6.12}$$

In this case the molecular orientation in space must be included. We have discussed molecular rotation in the context of Eq. (5.98). Here $\{\phi, \theta, \psi\}$ are the Euler angles of the molecules. The factors $\sin\theta_i$ arise from the Jacobian J (discussed in Sect. 5.1.8) when we integrate over the momenta conjugate to the Euler angles. In the case of a MC reorientation of a single randomly chosen molecule the attendant probability ratio becomes

$$\frac{p_{new}}{p_{old}} = \frac{\sin\theta_{new}}{\sin\theta_{old}}\exp[-\beta(\mathcal{U}_{new} - \mathcal{U}_{old})]. \tag{6.13}$$

In the case of molecule insertion or removal we can omit the factor $\sin\theta_{new}/\sin\theta_{old}$ assuming $\theta_{new} = \theta_{old}$.

6.3 Selected Applications

The applicability of molecular simulation to microscopic phenomena is subject to numerous constrains. This is not the place to describe these constraints and the strategies used to extend the limits of simulation methods. The following examples are for gases and liquids with simple, which mostly means short-ranged, radially symmetric interactions between particles.

6.3.1 Simple Thermodynamic Bulk Functions

It is useful to express the quantities V, Λ_T, T, $\mathcal{U}$, ... in (6.11), as well as in every possible Metropolis probability ratio, in a new set of units, i.e. energies are in units of ϵ and lengths are in units of σ. Table 6.1 compiles the explicit conversion between the new dimensionless quantities X^* and the original physical quantities. The quantities X^* are said to be in Lennard-Jones (LJ) units, because ϵ and σ usually are identical to the same quantities in the LJ potential in Eq. (5.137). For small, non-polar molecules and not too high densities this potential usually is a good approximation. This means that we can carry out simulations of gases and (to some extend) liquids of noble gas atoms and small, non-polar molecules using the LJ potential with $\epsilon = \sigma = 1$. Subsequently we convert the results from LJ units to SI units. If we do this using atom or molecule specific values for ϵ and σ, we obtain results specific for the system of interest (e.g. methane or argon). But how do we get $\epsilon_{methane}$ and $\sigma_{methane}$ or ϵ_{argon} and σ_{argon}? One possibility is to look up the values of these system's critical parameters T_c, ρ_c, and P_c. Assuming that we also have simulation results for the critical point of our LJ system (the one with $\epsilon = \sigma = 1$), i.e. T_c^*, ρ_c^*, and P_c^*, we can use the conversion formulas in the table, e.g. $\epsilon_{methane} = k_B T_{c,methane}/T_c^*$ and $\sigma_{methane} = (\rho_c^*/\rho_{c,methane})^{1/3}$. We notice immediately that this is not unique. We have three critical parameters and only two model parameters. Alternatively we could have used ρ_c and P_c or T_c and P_c to obtain ϵ and σ. However, provided that LJ interactions do describe the interactions in the physical system reasonably well, we shall find that the differences are small.[3] As soon as we have decided which values to use for ϵ and σ, we can begin to convert other quantities of interest from their LJ values provided by the simulation to the SI-values comparable to experimental data. Or we can convert experimental data to LJ units.

---

[3] Probably you have noticed the similarity to our discussion of the universal van der Waals equation (4.12), where we also use the gas-liquid critical point expressed via the parameters a and b to map the results of the universal theory onto specific systems. There as well as here we can also use experimental data for the second virial coefficient to fit a and b or ϵ and σ. These values again will differ to some extend from the ones obtained via the critical parameters.

Table 6.1 Lennard-Jones units

Quantity	LJ-quantity	Conversion
T	T^*	$T = \epsilon T^*/k_B$
V	V^*	$V = \sigma^3 V^*$
ρ	ρ^*	$\rho = \rho^*/\sigma^3$
P	P^*	$P = \epsilon P^*/\sigma^3$
μ	μ^*	$\mu = \epsilon \mu^*$

Fig. 6.1 Simulation results, showing the temperature dependence of selected thermodynamic quantities, compared to experimental data

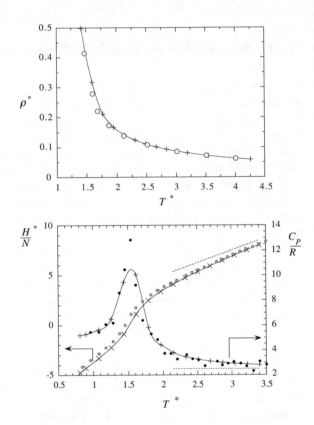

The top panel in Fig. 6.1 shows simulations results[4] (dotted circles) for the number density, ρ^*, versus temperature, T^*, for methane. Here $\epsilon_{methane}/k_B = 141.2$ K and $\sigma_{methane} = 3.688$ Å. The experimental results (plusses), converted to LJ units, are from the HCP. The experimental pressure is 10 MPa or 0.257 in LJ units. This pressure is roughly a factor of two above the critical pressure. The solid line is a fit

---

[4] These particular quantities were calculated for system of 108 particles using the Molecular Dynamics technique (R. Hentschke, E. M. Aydt, B. Fodi, E. Stöckelmann *Molekulares Modellieren mit Kraftfeldern.*, http://constanze.materials.uni-wuppertal.de), but Metropolis Monte Carlo could have used instead.

to the experimental data for the sake of easier comparability but without particular physical significance. Analogous curves obtain from the van der Waals equation are shown in Fig. 4.5. The above graph corresponds to the high temperature portion of the highest pressure isobar in Fig. 4.5. Notice that the critical temperature, which in case of the universal van der Waals equation is 1, is slightly above 1.3 in the case of the LJ system. The bottom panel in Fig. 6.1 shows the temperature dependence of the enthalpy per atom (dotted circles) and the isobaric heat capacity per particle (solid circles) for argon. Plusses and crosses are experimental results taken from HTTD. Again the pressure is 10 MPa, but this time ϵ and σ are different, i.e. $\epsilon_{argon}/k_B = 111.8$ K and $\sigma_{argon} = 3.369$ Å, and thus this pressure is 0.248 in LJ units. The dotted lines are the respective quantities calculated assuming an ideal gas. We note that the ideal gas behavior is approached on the high temperature side. The simulation yields the enthalpy and the heat capacity, $C_P = \partial H/\partial T|_P$, is obtained via simple numerical differentiation of the enthalpy data without prior smoothing. This relatively crude procedure is responsible for the larger deviations from the experimental results (plusses) in the vicinity of the peak. The peak in the heat capacity marks the Widom line, i.e. the smooth continuation of the gas-liquid saturation line (cf. the right panel in Fig. 4.4) above the critical point.

6.3.2 Phase Equilibria

We consider phase coexistence in a one-component system—for instance between gas (g) and liquid (l). We envision two coupled subsystems—one on either side of the saturation line. Each of the subsystems is represented by a simulation box containing N_g and N_l particles, respectively. By exchanging particles between the boxes and by varying their volumes, V_g and V_l, we attempt to generate the proper thermodynamic states for both gas and liquid at coexistence. Our result will be two densities, $\rho_g(T)$ and $\rho_l(T)$, at coexistence as functions of temperature.

The thermodynamic variables in the respective subsystem (simulation boxes) are T_g, T_l, P_g, P_l, N_g, and N_l. Phase equilibrium requires $T_g = T_l(= T)$, $P_g = P_l(= P)$, and $\mu_g = \mu_l(= \mu)$. In addition we require $N_g + N_l = constant$ and $V_g + V_l = constant$. This ensures that only one free variable remains in accordance with the phase rule, which will be temperature. The subsystem entropy changes compatible with the above conditions are $T\Delta S_g = \Delta E_g + P\Delta V_g - \mu\Delta N_g$ and $T\Delta S_l = \Delta E_l + P\Delta V_l - \mu\Delta N_l$. The resulting total entropy change is $\Delta S = (1/T)(\Delta E_g + \Delta E_l) + (P/T)\ (\Delta V_1 + \Delta V_2) - \mu(\Delta N_g + \Delta N_l) = (1/T)(\Delta E_g + \Delta E_l)$. From this we can read off the attendant phase space probability ratios, i.e.

$$\frac{P_{new}}{P_{old}} = \frac{g_{new}}{g_{old}} \exp[-\beta(\Delta \mathcal{U}_g + \Delta \mathcal{U}_l)], \tag{6.14}$$

where

$$g = \frac{1}{N_g!} \left(\frac{V_g}{\Lambda_T^3}\right)^{N_g} \frac{1}{N_l!} \left(\frac{V_l}{\Lambda_T^3}\right)^{N_l}, \tag{6.15}$$

and $\Delta \mathcal{U} = \mathcal{U}_{new} - \mathcal{U}_{old}$. Notice that the phase space integration factorizes into a product of integrations over the respective volumes. Notice also that the combinatorial factor $N!^{-1}$ becomes the product of two such factors (cf. Eq. 5.75 for the ideal gas mixture).

A suitable MC algorithm may be one that cycles through the steps—translation of a particle in the gas box, translation of a particle in the liquid box, volume change, transfer of a particle from gas to liquid, transfer from liquid to gas. The following are examples from which the missing cases can be worked out easily.

- Translation of particle i in the gas box $\vec{s}_{g,i} \to \vec{s}_{g,i} + \delta\vec{s}$:

$$\frac{p_{new}}{p_{old}} = \exp[-\beta \Delta \mathcal{U}_g(\delta\vec{s})], \tag{6.16}$$

where $\Delta \mathcal{U}_g = \mathcal{U}_{g,new} - \mathcal{U}_{g,old}$.
- Volume change, i.e. $V_g \to V_g - \delta V$ and $V_l \to V_l + \delta V$:

$$\frac{p_{new}}{p_{old}} = \left(1 - \frac{\delta V}{V_g}\right)^{N_g} \left(1 + \frac{\delta V}{V_l}\right)^{N_l} \exp[-\beta(\Delta \mathcal{U}_g(-\delta V) + \Delta \mathcal{U}_l(+\delta V))], \tag{6.17}$$

again $\Delta \mathcal{U}_g = \mathcal{U}_{g,new} - \mathcal{U}_{g,old}$ and $\Delta \mathcal{U}_l = \mathcal{U}_{l,new} - \mathcal{U}_{l,old}$.
- Particle transfer from gas to liquid, i.e. $N_g \to N_g - 1$ and $N_l \to N_l + 1$:

$$\frac{p_{new}}{p_{old}} = \frac{N_g}{N_l+1} \frac{V_l}{V_g} \exp[-\beta(\Delta \mathcal{U}_g(N_g - 1) + \Delta \mathcal{U}_l(N_l + 1))]. \tag{6.18}$$

Notice again that the selection of particle i for translation, or the selection of a particle for transfer, or the volume change δV must be completely random! The only restriction is that the translation step size δs as well as the volume change δV usually are small compared to the system size in order to take advantage of the importance sampling.[5]

There is one question though. How does the system decide which box contains the gas and which the liquid? First we must prepare initial conditions within the coexistence region, requiring some knowledge about its location and extend. Clearly, we can "bias" one box to be the gas box and the other one to be the liquid box by the initial distribution of particles from which we start. We shall find

---

[5] What we just have described is known as Gibbs-Ensemble Monte Carlo originally invented by Panagiotopoulos (1987).

Fig. 6.2 Gas-liquid phase separation via Gibbs-Ensemble Monte Carlo

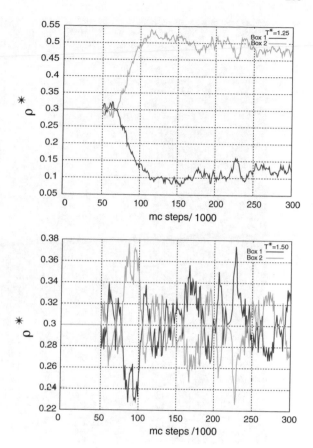

however that the particle number in the boxes fluctuates, particularly when we approach the critical temperature, T_c, and the identity of the boxes may switch. In fact, on approaching T_c the growing critical fluctuations render the distinction of gas and liquid meaningless in accordance with true physical systems. Figure 6.2 shows an example. The temperature in the top panel is below T_c and the densities in the boxes, initially kept at 0.3 by not allowing particles to transfer, subsequently develop into gas and liquid. The bottom panel shows the densities above the critical temperature (notice the change of scale of the ρ^*-axis). There is no distinction between the simulation boxes.

We note that the density fluctuations are quite large. This is due to the small size of these systems containing on the order of 100 particles. Close to the critical point there are critical density fluctuations on all length scales giving rise to critical opalescence (e.g. Stanley 1971). Even though this is a different matter, which we do not discuss here, it still is worth presenting a real experimental example of critical opalescence conducted by the author. The bottom inset in Fig. 6.3 shows a cylindrical pressure chamber with windows on opposite sides containing sulfurhexafluoride (SF_6) (cf. p. 166). We can see the meniscus of the liquid. Heating of

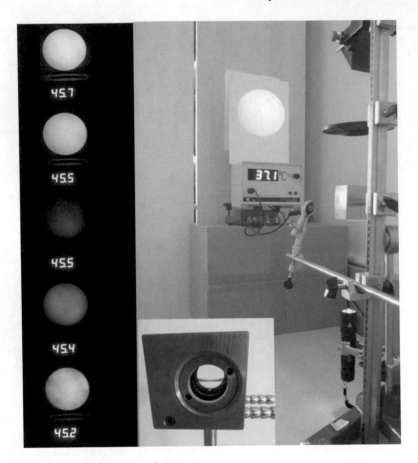

Fig. 6.3 Critical opalescence experiment

the chamber causes the SF_6 to pass through its critical point along the critical isochore. The critical temperature of SF_6 is $T_c = 318.7$ K or 45.5 °C. In the experimental setup the chamber is mounted so that the meniscus of the liquid is perpendicular to the gravitational field.[6] The flashlight mounted at the bottom illuminates the interior of the chamber and a corresponding bright spot of light can be seen on the white cardboard in the background. The instrument beneath the cardboard shows the chamber's temperature. The series of photographs on the left illustrates what happens when the temperature passes through T_c. At temperatures above and below T_c the light can pass through the chamber and the spot on the

---

[6] The effect is concentrated near the interface or, because above T_c the interface vanishes, where the interface develops upon cooling. The rotation of the pressure chamber used here ensures greater homogeneity.

Fig. 6.4 Gas-liquid
coexistence data obtained via
Gibbs-Ensemble Monte Carlo

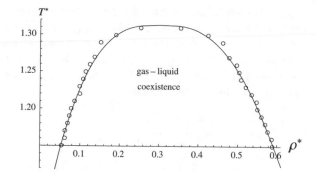

cardboard is rather bright. But very close and right at T_c a pronounced decrease in
the brightness is observed due to critical scattering.

Figure 6.4 shows a plot of the gas and liquid densities from the simulation
(circles). The solid line is a power law fit[7] yielding $T_c^* \approx 1.32$ and $\rho_c^* \approx 0.3$.[8]

We conclude our example with a remark. The simulation boxes posses no
physical connection, i.e. there is no continuous migration of molecules from one
region in space to another as in a real experiment. All that matters is that thermal,
mechanical, and chemical equilibrium is attained. Since the chemical potential is a
state function, it is not important which path we choose for this purpose. This
implies that unphysical but highly efficient Monte Carlo moves become possible! In
the present case this is the instantaneous transfer of particles between boxes.

6.3.3 Osmotic Equilibria

Another example allowing us to practice the above approach is the following.
A semi-permeable membrane divides a system into two compartments or subsystems—again represented by respective simulation boxes situated on opposite sides
of the membrane. Subsystem i ($= 1, 2$) contains $N_{l,i}$ solvent molecules and $N_{s,i}$
solute molecules, respectively. In particular we require $N_{s,1} = 0$, due to the membrane. The thermodynamic states of the two subsystems are defined through the
eight quantities T_1, T_2, P, $P + \Pi$, $N_{l,i}$, and $N_{s,i}$. At equilibrium we have the five
conditions $(T =)T_1 = T_2$, equality of the solvent chemical potentials, i.e.
$(\mu_l =)\mu_{l,1} = \mu_{l,2}$, as well as $N_{l,1} + N_{l,2} = N_l = constant$, $N_{s,1} = 0$, and

---

[7] We use $\rho_{liq}^* - \rho_{gas}^* = A_o t^\beta - A_1 t^{\beta + \Delta}$ and $\rho_{liq}^* + \rho_{gas}^* = 2\rho_c^* + D_o t^{1-\alpha}$ with $t = T_c^* - T^*$
(Ley-Koo and Green 1981). The (3D Ising) critical exponent values are $\beta = 0.326$, $\alpha = 0.11$,
and $\Delta = 0.52$ (cf. our discussion of critical exponents and scaling beginning on p. 162).

[8] These results were obtained during a student laboratory on computer simulation techniques—my
thanks to S. Reinecke and S. Mathys for Fig. 6.2 and to A. Obertacke and M. Götze for the data in
Fig. 6.4. The MC program was written by R. Kumar.

$N_{s,2} = constant$. The resulting $8 - 5 = 3$ degrees of freedom (phase rule!) are T, P, and Π. Furthermore we have $T\Delta S_1 = \Delta E_1 + P\Delta V_1 - \mu_l \Delta N_{l,1}$ and $T\Delta S_2 = \Delta E_2 + (P+\Pi)\Delta V_2 - \mu_l \Delta N_{l,2}$, where the E_i are the internal energies. The total entropy change in the system therefore is $\Delta S = (1/T)(\Delta E_1 + \Delta E_2) + (P/T)(\Delta V_1 + \Delta V_2) + (\Pi/T)\Delta V_2$. The complete phase space density entering into our Metropolis criteria thus becomes

$$p \propto \frac{V_1^{N_{l,1}} V_2^{N_{l,2}+N_{s,2}}}{\Lambda_{T,l}^{3N_l} \Lambda_{T,s}^{3N_{s,2}} N_{l,1}! N_{l,2}! N_{s,2}!} \tag{6.19}$$
$$\exp\left[-\beta(\mathcal{U}_1 + \mathcal{U}_2 + PV_1 + (P+\Pi)V_2 + \mu_l N_l + \mu_{s,2} N_{s,2})\right].$$

Our Monte Carlo algorithm consists of the following moves: molecule translation, volume change, and solvent transfer. The probability ratios entering into the Metropolis criterion are:

- for translation of a molecule in compartment i

$$\frac{p_{new}}{p_{old}} = \exp[-\beta \Delta \mathcal{U}_i], \tag{6.20}$$

where $\Delta \mathcal{U}_i = \mathcal{U}_{i,new} - \mathcal{U}_{i,old}$,
- for a volume change of compartment i,

$$\frac{p_{new}}{p_{old}} = \left(\frac{V_{i,new}}{V_{i,old}}\right)^{N_{l,i}+N_{s,i}} \exp[-\beta(\Delta \mathcal{U}_i + (P + \delta_{i2}\Pi)\Delta V_i)], \tag{6.21}$$

where $\Delta V_i = V_{i,new} - V_{i,old}$ and $\delta_{i2} = 1$ if $i = 2$ or zero otherwise, and
- for a molecule transfer from compartment 1 to compartment 2.

$$\frac{p_{new}}{p_{old}} = \frac{V_2 N_{l,1}}{V_1(N_{l,2}+1)} \exp[-\beta(\Delta \mathcal{U}_1 + \Delta \mathcal{U}_2)], \tag{6.22}$$

where $\Delta \mathcal{U}_1 = \mathcal{U}_1(N_{l,1}-1) - \mathcal{U}_1(N_{l,1})$ and $\Delta \mathcal{U}_2 = \mathcal{U}_2(N_{l,2}+1) - \mathcal{U}_2(N_{l,2})$.

Exchanging the compartment index yields the probability ratio governing the opposite transfer.

The following numerical data[9] were obtained with the LJ particle-particle interaction potential $u(r) = 4(r^{-12} - r^{-6})$ (cf. Eq. (5.137)). In this somewhat artificial model calculation all interactions are taken as identical. Figure 6.5 shows the osmotic pressure, Π, vs. solute mole fraction, $x_{s,2}$. The units are LJ units. Comparison of the above algorithm (up-triangles: $T^* = 1.15$, $P^* = 0.2$; down-triangles: $T^* = 1.5$, p as for the open circles) to data from the literature

---

[9] From Schreiber and Hentschke (2011).

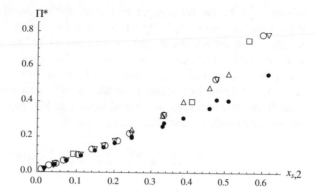

Fig. 6.5 Osmotic pressure versus solute mole fraction

(Panagiotopoulos et al. 1988): (open squares: $T^* = 1.15$, $P^* = 0.077$); Murad et al. (1995): (open circles: $T^* = 1.5$, $0.41 > P^* > 0.26$ from low to high $x_{s,2}$). The simulation data sets do agree closely even though the corresponding pressure, P, varies considerably. This is expected when the temperatures are low and the systems are rather dense. Notice that this example obeys van't Hoff's law (solid circles) over a wide concentration range. There is some scatter here because the solute concentration is determined from the simulation.

6.3.4 Chemical Potential

The chemical potential played an important role in the two preceding examples. But it did not enter explicitly into the calculations. How then can we obtain the chemical potential via computer simulation? Let us assume that we already have a value for the chemical potential, $\mu(T_o, P_o)$, in a one-component system. This particular phase point is indicated by the circle in Fig. 6.6. We may reach every other state point via a succession of steps along the directions indicated by the arrows. A path along which the temperature is constant may be followed by integrating Eq. (2.124), i.e.

$$\mu(T_o, P) = \mu(T_o, P_o) + \frac{1}{N} \int_{P_o}^{P} dP' V(T_o, P'), \tag{6.23}$$

Fig. 6.6 Simulation of the chemical potential on the basis of one known value

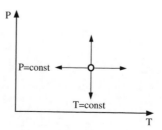

where V/N is the volume per molecule. Based on either Eq. (6.11) or (6.12), depending on the complexity of the molecule of interest, we may construct a suitable Metropolis criterion (MC moves would include translation, rotation (in the case of (6.12) only), and volume change). We obtain a series of values for $v(P)$ along the chosen path, which then can be used to numerically solve the integral (e.g. via a suitable interpolating function which is easy to integrate). An analogous procedure works for the $P = const$-direction in Fig. 6.6—this time however we use the Gibbs-Helmholtz equation (3.145), i.e.

$$\frac{\mu(T, P_o)}{T} = \frac{\mu(T_o, P_o)}{T_o} - \frac{1}{N} \int_{T_o}^{T} dT' \frac{H(T', P_o)}{T'^2}.$$

(6.24)

Here $H(T)/V$ is the enthalpy per molecule.

Figure 6.7 shows the chemical potential obtained for the so called TIP4P/2005 (Transferable Intermolecular Potential 4 Point) water model via thermodynamic integration at $P = 1$ bar (solid line) (Guse 2011). The crosses are corresponding experimental chemical potentials calculated as follows.

In Fig. 4.4, we had included the experimental saturation pressure, $P_{sat}(T)$, along the gas-liquid saturation line for water. In the temperature range of interest here we may consider water vapor as being ideal. The ideal water chemical potential was calculated before in the second example in Sect. 5.1.8 (vapor pressure of ice), i.e. $\mu_{H_2O,gas}(T, P_{sat}) \approx \mu_{H_2O}^{trans}(T, P_{sat}) + \mu_{H_2O}^{rot}(T, P_{sat})$. Along the saturation line we have $\mu_{H_2O,gas} = \mu_{H_2O,liq}$, the gas phase chemical potential is the same as the liquid phase chemical potential. The example on page 94 (relative humidity) has taught us that the chemical potential difference between a state point on the saturation line and the state point at the same temperature in the liquid at 1 bar is almost negligible. Thus we have $\mu_{H_2O,liq}(T, 1 \text{ bar}) \approx \mu_{H_2O,gas}(T, P_{sat})$ to very good approximation. The crosses in Fig. 6.7 show $\mu_{H_2O,liq}(T, 1 \text{ bar})$ computed in this fashion using the ideal gas expressions for $\mu_{H_2O}^{trans}$ and $\mu_{H_2O}^{rot}$ provided in our above calculation of the vapor pressure of ice (cf. Eqs. (5.101) and (5.102); note that the molecular vibrations may

Fig. 6.7 Water chemical potential versus temperature

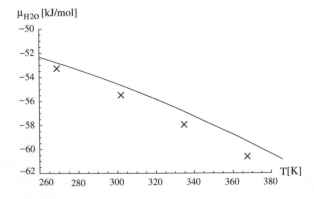

still be neglected despite the somewhat higher temperatures in the current case. The computer model for water also assumes rigid molecules.).

At this point we may ask whether there are direct methods for computing $\mu(T, P)$ avoiding lengthy paths? Yes, there are such methods. Because this is not a textbook on computer simulation methods, the reader is referred to the appropriate texts (e.g. Frenkel and Smit 1996 or Allen and Tildesley 1990). However, even without discussing these methods, we may suspect a certain difficulty. Measuring the chemical potential involves insertion of a molecule into a system. In dense systems it frequently happens that the new molecule is placed directly "on top of" an existing molecule. The attendant Metropolis MC ratio then is exceedingly small and the computer time needed to obtain sufficient successful insertions may be prohibitively large. Liquid water in the above example is such a system. Thermodynamic integration does not suffer from this problem and therefore may be preferable.[10] Two more questions remain. How must we proceed if more than one component is present? And what do we do when our path crosses a phase boundary?

Equations (6.23) and (6.23) for more than one component are

$$\mu_i(T_o, P) = \mu_i(T_o, P_o) + \frac{1}{N_A} \int_{P_o}^{P} dP' v_i(T_o, P') \tag{6.25}$$

and

$$\frac{\mu_i(T, P_o)}{T} = \frac{\mu_i(T_o, P_o)}{T_o} - \frac{1}{N_A} \int_{T_o}^{T} dT' \frac{h_i(T', P_o)}{T'^2}. \tag{6.26}$$

Here i denotes the component. The quantities v_i and h_i are the partial molar volume and enthalpy of component i, respectively. Equation (6.25) follows via differentiation of the free enthalpy in Eq. (2.124) with respect to n_i. Equation (6.26) follows directly via integration of (3.146). In principle the partial quantities may be determined for a set of given conditions, i.e. $T, P, n_{j(\neq i)}$, by simulating systems with different content of i, above and below the target mole fraction n_i. The derivatives may then be estimated via $v_i = \partial V / \partial n_i \approx \Delta V / \Delta n_i$ and $h_i = \partial H / \partial n_i \approx \Delta H / \Delta n_i$. Whether this additional effort is necessary again is a matter of experience.

The chemical potential is continuous at a phase boundary. But quantities like V and H are discontinuous (except at the critical point). The main practical problem upon crossing a phase boundary is equilibration of the respective phases. Usually it is a good idea to cycle through a path in both direction to check for hysteresis effects, e.g. supercooling (the liquid/gas is metastable inside the solid/liquid phase) or superheating (the liquid/gas is metastable inside the solid/liquid phase).

---

[10] In praxis choosing the method largely depends on the experience of the person doing the calculation.

Notice in this context that integration of the Clapeyron equation (4.46) can be used to trace out phase boundaries. For example, having obtained the location of the gas-liquid transition via the above method at a certain T_k and P_k, we may use the differences between enthalpy and density in the two boxes to obtain the slope of the saturation line, dP/dT, at this state point. With this information we can obtain an new state point, T_{k+1} and P_{k+1}, in the direction of the slope. This state point in turn may be used to find new values for the enthalpy and density in the two simulation volumes, which now do not trade molecules any more. This procedure is called Kofke integration in the literature on molecular simulation.

Chapter 7
Non-equilibrium Thermodynamics

In the preceding chapters with few exceptions we have studied systems at equilibrium. This means that the systems do not change or evolve over time. In Sect. 5.1.7 we have studied the likelihood of energy fluctuations. We found them to be exceedingly small in macroscopic systems. And yet our world is full of complexity and attendant order. Particularly "systems which do not change over time" are not what we observe. Here is a quote from Feynman (2003): "How then does thermodynamics work, if its postulates are misleading? The trick is that we always arrange things so that we do not experiments on things as we find them, but only after we have thrown out precisely all those situations which lead to undesirable orderings." Non-equilibrium thermodynamics is the part of thermodynamics where the undesirable orderings are not thrown out.

The usefulness of the fundamental laws and their consequences, as we have applied them, do require sufficiently distinct rates. Figure 7.1 shows the so-called H-function obtained from a Molecular Dynamics computer simulation of 108 LJ particles in an insulated, periodic box. The number density is $\rho^* = 0.05$. At time $t^* = 0$ the[1] simulation is started with the particles located on an fcc lattice. A random initial velocity is assigned to each particle based on a uniform distribution. Subsequently the function

$$H(t) = \int d^3v f(\vec{v}, t) \ln f(\vec{v}, t) \tag{7.1}$$

is obtained from the velocity distribution function $f(\vec{v}, t)$. If we describe $H(t)$ by a simple smooth function, it monotonously decreases—aside from short-lived fluctuations. In this particular example the characteristic time for the overall decrease is

---

[1] Here * means that time is in units of $\sqrt{m\sigma^2/\epsilon}$, where m is the particle mass and ϵ and σ are the parameters in the LJ-potential

© Springer Nature Switzerland AG 2022
R. Hentschke, *Thermodynamics*, Undergraduate Lecture Notes in Physics,
https://doi.org/10.1007/978-3-030-93879-6_7

about 2 time units. The "lifetime" of the fluctuations is comparable. In a real gas the above times are a few ps, i.e. a few 10^{-12} s. This may vary depending on density, initial velocities or microscopic interactions. But for ordinary gases or liquids these so-called relaxation times are very short compared to measurements of thermodynamic quantities. Temperature changes in the laboratory, deterioration of rubber gaskets, corrosion of the apparatus, etc., which eventually do affect our "experimental equilibrium", happen much more slowly.

Remark Ludwig Boltzmann[2] showed, based in particular on the assumption of molecular chaos (no correlations), that in the above sense

$$\frac{dH(t)}{dt} \leq 0. \tag{7.2}$$

It turns out that $-H(t)$ is essentially the entropy. This inequality, his famous H-theorem, opened the door to an understanding of the macroscopic world on the basis of molecular dynamics (Huang 1963).

We begin our discussion of non-equilibrium phenomena by exploiting the apparent similarity between the decay of spontaneous fluctuations and transport.

7.1 Linear Irreversible Transport

There are a number of well known equations describing irreversible transport processes. For instance Ohm's law

$$J_{q,\alpha} = \sum_{\beta} \sigma_{q,\alpha\beta} E_\beta. \tag{7.3}$$

Here $J_{q,i}$ is the charge current component α and E_β is the β-component of the electric field. The quantities $\sigma_{q,\alpha\beta}$ are the components of the conductivity tensor. Then there is Fourier's law of thermal conductivity

$$J_{Q,\alpha} = -\sum_{\beta} \lambda_{\alpha\beta} \frac{\partial T}{\partial x_\beta}. \tag{7.4}$$

Here $J_{Q,\alpha}$ is the local heat flux density component α due to the β-component of a temperature gradient. $\lambda_{\alpha\beta}$ are the components of the thermal conductivity tensor. Another example is Fick's law

---

[2] Ludwig Boltzmann, austrian physicist, *Wien 20.2.1844, †Duino 5.9.1906; fundamental contributions to Statistical Mechanics. His tombstone bears the inscription $S = k \log W$ (cf. Eq. 6.16).

$$J_{c,i} = -D_i \frac{\partial c_i}{\partial x}, \qquad (7.5)$$

where D_i is the diffusion coefficient of the diffusing component i and c_i is the concentration of i. The above transport flows are coupled. Two examples in the linear regime are

$$\vec{J}_q = A_{qq} \frac{\vec{E}}{T} - A_{qQ} \vec{\nabla} \frac{1}{T} \qquad (7.6)$$

and

$$\vec{J}_T = -A_{QQ} \vec{\nabla} \frac{1}{T} + A_{Qq} \frac{\vec{E}}{T}. \qquad (7.7)$$

More generally we may write

$$\Delta J_i = \sum_j L_{ij} \Delta X_j. \qquad (7.8)$$

Here the quantities ΔX_j are called generalized forces.[3] The prefix Δ reminds us that we stay close to the equilibrium state. At thermodynamic equilibrium, we have simultaneously for all irreversible processes $\Delta J_i = 0$ and $\Delta X_i = 0$. In the following we shall study the above linear relations and their coefficients in more detail.

Example—Insulation An important number for the quality of insulation material is its λ-value, i.e. the thermal conductivity. The unit of λ is $W/(mK)$. Table 7.1 lists some typical numbers (see also HCP). Notice that the thermal conductivity does depend on temperature and possibly pressure. The numbers given here correspond to "usual" ambient conditions.

Suppose the λ-value of an insulation material is 0.04 (This is rather typical for glass or mineral wools as well as for foams, because of their high air content.). How thick must the insulation layer be in order to maintain a temperature of 20°C in a garden shed, using a 500 W electric heater, when the outside temperature is 0 °C? We assume that the garden shed is a rectangular 3.5 by 4.0 by 2.2 m box and we neglect doors as well as windows. Thus the total area to be insulated is $A = 61 \, m^2$. The heat transport per unit time through the insulation, whose thickness is d, is

$$\frac{dQ}{dt} = \lambda \frac{A}{d} \Delta T. \qquad (7.9)$$

---

[3] For the sake of simplicity we treat the L_{ij} as scalar quantities.

Material	λ [W/(mK)]
Vacuum	0
Dry air	0.03
Wood	0.1...0.2
Snow	0.2
Water	0.6
Solid brick	0.5
Glass	1
Copper	400

Table 7.1 Thermal conductivity coefficients for selected materials

With $dQ/dt = 500$ W the result is $d \approx 10$ cm. The neglect of the timber siding and plaster is not serious. Their contribution to the walls thickness is comparatively minor and their λ-values are significantly higher than 0.04. This is not true for windows and doors. The main weakness of these construction elements is that they locally reduce the wall's thickness and strongly affect the thermal insulation of a building. For windows and doors, and for multilayered walls as well, λ usually is replaced by the more suitable U-value given by λ/d or by its inverse called thermal resistance.

7.1.1 Fluctuations Revisited

One approach to a deeper understanding of transport phenomena and in particular of the relations (7.8) exploits the analogy to fluctuations. The decay of a fluctuation involves irreversible transport. Notice that there is not much difference between the initial decrease of the H-function in Fig. 7.1 and the decay of a subsequent fluctuation. Thus we briefly recapitulate our previous discussion of small fluctuations in Sect. 5.1.

An isolated system possessing the internal energy E is divided into open subsystems. Its thermodynamic state is characterized by the thermodynamic quantities $x_j (j = 1, 2, \ldots, n)$. Examples for the x_i are the temperature or the mass density in one of the subsystems. The fluctuations of the x_j relative to their average values are Δx_j. The probability for a particular distribution of fluctuations throughout the collection of subsystems is[4]

---

[4] An explicit example in the case of energy fluctuations was worked out in Sect. 5.1.7.

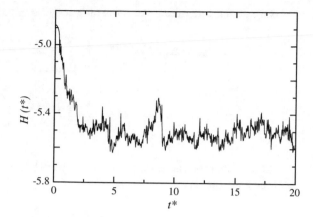

Fig. 7.1 H-function obtained from a Molecular Dynamics computer simulation

$$p(\{\Delta x\}) = \exp[(\Delta S(\{\Delta x\})/k_B)],\tag{7.10}$$

where

$$\Delta S(\{\Delta x\}) = -\frac{1}{2}\sum_{j,j'} g_{jj'}\Delta x_j \Delta x_{j'}\tag{7.11}$$

and

$$g_{jj'} \equiv -\frac{\partial^2}{\partial x_j \partial x_{j'}} S(E, x_1, \ldots, x_n)\Big|_{x_j^{(o)}, x_{j'}^{(o)}}.\tag{7.12}$$

The $x_j^{(o)}$ are the equilibrium values.
In the following

$$\Delta X_j = \frac{\partial \Delta S}{\partial \Delta x_j} = -\sum_{j'} g_{jj'}\Delta x_{j'}\tag{7.13}$$

are generalized forces and we shall need correlation functions $\langle \Delta x_j \Delta X_{j'}\rangle$, i.e.

$$\langle \Delta x_j \Delta X_{j'}\rangle = \frac{\int d\{\Delta x\}\Delta x_j \Delta X_{j'} \exp[(\Delta S(\{\Delta x\})/k_B)]}{\int d\{\Delta x\} \exp[(\Delta S(\{\Delta x\})/k_B)]}.\tag{7.14}$$

Using partial integration the numerator may be expressed as

$$\int d\{\Delta x\} \Delta x_j \Delta X_{j'} e^{\Delta S/k_B} = k_B \int d\{\Delta x\} \underbrace{\frac{\partial}{\partial \Delta x_{j'}} \left(\Delta x_j e^{\Delta S/k_B} \right)}_{=0}$$

$$- k_B \int d\{\Delta x\} \underbrace{\frac{\partial \Delta x_j}{\partial \Delta x_{j'}}}_{=\delta_{jj'}} e^{\Delta S/k_B}.$$

The first integral vanishes, because the probability of infinite fluctuations vanishes, and we obtain

$$\langle \Delta x_j \Delta X_{j'} \rangle = -k_B \delta_{jj'} \tag{7.15}$$

Remark We may use this result to derive other useful correlation functions. Inserting (7.13) into (7.15) yields

$$\sum_k g_{j'k} \langle \Delta x_k \Delta x_j \rangle = k_B \delta_{j'j}.$$

This shows that

$$\langle \Delta x_j \Delta x_{j'} \rangle = k_B (g^{-1})_{jj'}. \tag{7.16}$$

$(g^{-1})_{jj'}$ are the elements of the inverse of the (symmetric) matrix **g**.

Also of interest is the combination of this equation with (7.11) which yields

$$\langle \Delta S \rangle = \frac{-k_B}{2} \sum_{j j'}^m g_{jj'} (g^{-1})_{j'j} = -k_B \frac{m}{2}. \tag{7.17}$$

This result is analogous to the equipartition formula for the average thermal energy of a system with m degrees of freedom, whose Hamilton function is quadratic in the attendant momenta and coordinates (cf. Eq. 5.92). Equation (7.17) shows that the average entropy fluctuation in a system containing m fluctuating quantities is contributed in increments of $-k_B/2$ by each of these quantities.

7.1.2   Onsager's Reciprocity Relations

Let us assume that there are currents defined via

$$\Delta J_j = \frac{d\Delta x_j}{dt} \tag{7.18}$$

and that these currents are linearly coupled through Eq. (7.8) to the generalized forces defined above. One can then show that the coefficients obey Onsager's reciprocity relations (Onsager 1931):

$$\boxed{L_{jj'} = L_{j'j}}. \tag{7.19}$$

If an external magnetic field $\vec{B}$ is applied and the system of interest rotates with a constant angular velocity $\vec{\omega}$, then the coefficients L_{ij} do depend on these quantities and one finds

$$L_{jj'}(\vec{B}, \vec{\omega}) = L_{j'j}(-\vec{B}, -\vec{\omega}). \tag{7.20}$$

We show the validity of the reciprocity relations by going backwards starting from Eq. (7.19). Using (7.15) we have

$$\sum_k L_{jk}\langle \Delta x_{j'} \Delta X_k \rangle = \sum_k L_{j'k}\langle \Delta x_j \Delta X_k \rangle \tag{7.21}$$

or, using (7.8),

$$\langle \Delta x_{j'} \Delta J_j \rangle = \langle \Delta x_j \Delta J_{j'} \rangle. \tag{7.22}$$

Substituting (7.18) yields

$$\langle \Delta x_{j'} \frac{d\Delta x_j}{dt} \rangle = \langle \Delta x_j \frac{d\Delta x_{j'}}{dt} \rangle. \tag{7.23}$$

The time derivative is now expressed in terms of an explicit (infinitesimal) time difference, i.e.

$$\langle \Delta x_{j'}(t) \frac{\Delta x_j(t+\tau) - \Delta x_j(t)}{\tau} \rangle = \langle \Delta x_j(t) \frac{\Delta x_{j'}(t+\tau) - \Delta x_{j'}(t)}{\tau} \rangle, \tag{7.24}$$

which easily is reduced to

$$\langle \Delta x_{j'}(t)\Delta x_j(t+\tau) \rangle = \langle \Delta x_j(t)\Delta x_{j'}(t+\tau) \rangle. \tag{7.25}$$

For this equation to be valid we must first shift the time origin on the right side $(t \to t - \tau)$:

$$\langle \Delta x_{j'}(t) \Delta x_j(t + \tau) \rangle = \langle \Delta x_j(t - \tau) \Delta x_{j'}(t) \rangle. \tag{7.26}$$

In order for the resulting equation and therefore (7.19) to hold we must require microreversibility, i.e. the inversion $\tau \to -\tau$ does not affect the result of the ensemble averaging. But here we rely on the time reversibility of the microscopic equations of motion (at least for short times). This also is the reason for the requirements expressed in (7.20). In order for the time reversibility to hold, the Lorentz force, which is proportional to $d\vec{r}/dt \times \vec{B}$, must be invariant as well as the Coriolis force, proportional to $d\vec{r}/dt \times \vec{\omega}$, in the case of a rotating system. Therefore the magnetic field, $\vec{B}$, and the angular velocity, $\vec{\omega}$, appear with reversed signs on the right side of Eq. (7.20).

Remark Based on Eq. (7.11) the entropy production[5] is given by

$$\frac{d\Delta S}{dt} = -\frac{1}{2}\frac{d}{dt}\sum_{j,j'} g_{jj'} \Delta x_j \Delta x_{j'} = -\sum_{j,j'} g_{jj'} \Delta x_{j'} \frac{d\Delta x_j}{dt}. \tag{7.27}$$

Thus according to Eqs. (7.13) and (7.18)

$$\frac{d\Delta S}{dt} = \sum_j \Delta X_j \Delta J_j. \tag{7.28}$$

The entropy production is a bilinear function of the generalized forces and the currents. This is useful if we want to identify the correct form of the transport coefficients $L_{jj'}$ as we shall see in the following. Another point worth mentioning is illustrated in Fig. 7.2. A fluctuation will drive the entropy away from its equilibrium value. This means that

$$\Delta S \leq 0. \tag{7.29}$$

Stability then requires that the entropy production is positive:

$$\frac{d\Delta S}{dt} \geq 0. \tag{7.30}$$

---

[5] We discuss entropy production in more detail in the next section.

Fig. 7.2 Entropy change in response to a fluctuation

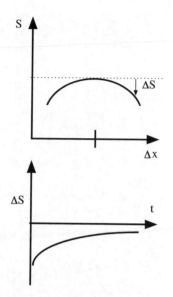

Example—Currents and Generalized Forces   Here we want to learn how to obtain specific relations according to Eq. (7.8) using Eq. (7.28). Our starting point is the first line in the expression (3.13) for ΔS. In this example we assume that all ΔV_v are zero. Instead we include charge fluctuations Δq_v. Thus we have

$$
\begin{aligned}
\Delta S = \frac{1}{2} \sum_v \Bigg(& \Delta E_v{}^2 \frac{\partial^2 S_v}{\partial E_v{}^2}\Bigg|^o_{q_v,n_v} + \Delta q_v{}^2 \frac{\partial^2 S_v}{\partial q_v{}^2}\Bigg|^o_{E_v,n_v} + \Delta n_v{}^2 \frac{\partial^2 S_v}{\partial n_v{}^2}\Bigg|^o_{E_v,q_v} \\
& + 2\Delta E_v \Delta q_v \frac{\partial^2 S_v}{\partial E_v \partial q_v}\Bigg|_{q_v,n_v}\Bigg|^o_{E_v,n_v} + 2\Delta E_v \Delta n_v \frac{\partial^2 S_v}{\partial E_v \partial n_v}\Bigg|_{q_v,n_v}\Bigg|^o_{E_v,q_v} \\
& + 2\Delta q_v \Delta n_v \frac{\partial_v^2}{\partial q_v \partial n_v}\Bigg|_{E_v,n_v}\Bigg|^o_{E_v,q_v} \Bigg).
\end{aligned}
\tag{7.31}
$$

Differentiating with respect to time this becomes

$$
\frac{d\Delta S}{dt} = \sum_v \Bigg(\underbrace{\frac{\partial \Delta S_v}{\partial E_v}}_{=\Delta(1/T)} \frac{d\Delta E_v}{dt} + \underbrace{\frac{\partial \Delta S_v}{\partial q_v}}_{=\Delta(-\phi/T)} \frac{d\Delta q_v}{dt} + \underbrace{\frac{\partial \Delta S_v}{\partial n_v}}_{=\Delta(-\mu/T)} \frac{d\Delta n_v}{dt} \Bigg).
$$

The following definitions of the currents

$$
\Delta J_E = \frac{d\Delta E}{dt} \qquad \Delta J_q = \frac{d\Delta q}{dt} \qquad \Delta J_n = \frac{d\Delta n}{dt}
\tag{7.32}
$$

determine, according to Eq. (7.28), the generalized forces:

$$\Delta X_E = \Delta\left(\frac{1}{T}\right) = -\frac{1}{T^2}\Delta T \tag{7.33}$$

$$\Delta X_q = \Delta\left(-\frac{\phi}{T}\right) \tag{7.34}$$

$$\Delta X_n = \Delta\left(-\frac{\mu}{T}\right). \tag{7.35}$$

Now we can write down the linear relations between the currents and the generalized forces in which the coefficients satisfy the reciprocity relations (7.19):

$$\Delta J_E = L_{EE}\Delta\left(\frac{1}{T}\right) + L_{Eq}\Delta\left(-\frac{\phi}{T}\right) + L_{En}\Delta\left(-\frac{\mu}{T}\right) \tag{7.36}$$

$$\Delta J_q = L_{qE}\Delta\left(\frac{1}{T}\right) + L_{qq}\Delta\left(-\frac{\phi}{T}\right) + L_{qn}\Delta\left(-\frac{\mu}{T}\right) \tag{7.37}$$

$$\Delta J_n = L_{nE}\Delta\left(\frac{1}{T}\right) + L_{nq}\Delta\left(-\frac{\phi}{T}\right) + L_{nn}\Delta\left(-\frac{\mu}{T}\right). \tag{7.38}$$

Remark The generalized force ΔX_n, for example, depends on a temperature difference and/or a chemical potential difference. If it is more convenient we can at this point express ΔX_n in terms of a temperature and a pressure difference. Once again we make use of $d\mu = -sdT + \rho^{-1}dP$, where s is the molar entropy and ρ is the density, i.e.

$$-\Delta X_n = \Delta\left(\frac{\mu}{T}\right) = \frac{1}{T}\Delta\mu - \frac{\mu}{T^2}\Delta T = \frac{1}{T}\left(-s\Delta T + \frac{1}{\rho}\Delta P\right) - \frac{\mu}{T^2}\Delta T. \tag{7.39}$$

Using $h = Ts + \mu$ this becomes

$$\Delta X_n = \frac{h}{T^2}\Delta T - \frac{1}{\rho T}\Delta P. \tag{7.40}$$

Fig. 7.3 Thermomolecular
pressure effect

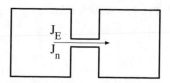

Example—Thermomolecular Pressure Effect A closed system consists of
two compartments joined by an aperture allowing the exchange of energy and
matter (cf. Fig. 7.3). The two compartments are kept at slightly different
temperatures. This situation can be described by the Eqs. (7.36) and (7.38)
with $\Delta\phi = \phi = 0$. With the help of Eqs. (7.33) and (7.40) we obtain

$$\Delta J_E = -L_{EE}\frac{1}{T^2}\Delta T + L_{En}\left(\frac{h}{T^2}\Delta T - \frac{1}{\rho T}\Delta P\right) \tag{7.41}$$

$$\Delta J_n = -L_{nE}\frac{1}{T^2}\Delta T + L_{nn}\left(\frac{h}{T^2}\Delta T - \frac{1}{\rho T}\Delta P\right). \tag{7.42}$$

Assuming for the moment that the temperatures in the compartments are
equal, i.e. $\Delta T = 0$, we deduce

$$\frac{\Delta J_E}{\Delta J_n} = \frac{L_{En}}{L_{nn}} \overset{(7.19)}{=} \frac{L_{nE}}{L_{nn}}. \tag{7.43}$$

The second equality is based on the validity of Onsager relations for the
transport coefficients L_{En} and L_{nE}.

Now we return to the situation when $\Delta T \neq 0$. Because the system is
closed, we expect $\Delta J_n = 0$ after some time. Inserting this into (7.42) the
equation yields

$$\frac{\Delta P}{\Delta T} = \frac{\rho}{T}\left(h - \frac{\Delta J_E}{\Delta J_n}\right). \tag{7.44}$$

If the compartments contain dilute gas, we may obtain $\Delta J_E/\Delta J_n$ from
kinetic gas theory as applied in "kinetic pressure" on p. 32:

$$\frac{\Delta J_E}{\Delta J_n} = \frac{\int' dN_{\vec{p}}\frac{\Delta z A}{V}\frac{\vec{v}_{\vec{p}}\cdot\vec{n}}{\Delta z}\frac{1}{2}mv_{\vec{p}}^2}{\int' dN_{\vec{p}}\frac{\Delta z A}{V}\frac{\vec{v}_{\vec{p}}\cdot\vec{n}}{\Delta z}m}N_A = 2RT. \tag{7.45}$$

In addition $h = (5/2)RT$ and thus

$$\frac{\Delta P}{\Delta T} = \frac{1}{2}\frac{P}{T} \tag{7.46}$$

Fig. 7.4 Seebeck effect

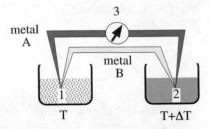

or, via $dP/P = d \ln P$ and $dT/T = d \ln T$,

$$\frac{P_1}{P_2} = \left(\frac{T_1}{T_2}\right)^{1/2}.$$ (7.47)

Example—Seebeck and Peltier Effect The Seebeck effect utilizes a temperature difference to generate a potential gradient as illustrated in Fig. 7.4. Two pieces of distinct metals, A and B (i.e. Cu/Al or Fe/Ni), are joined at 1 and 2. The junctions are exposed to different temperatures and this in turn gives rise to a (small) voltage drop, $\Delta\phi$, at 3. The Seebeck coefficient

$$\mathcal{S}_{AB} = \frac{\Delta\phi}{\Delta T}$$ (7.48)

is a material parameter.

Another experiment goes like this. Initially $\Delta T = 0$ and a current, I, flows through the same metal loop. The result is a heat current ΔJ_E such that T_2 and T_1 begin to differ. This is the Peltier effect. The quantity

$$\Pi_{AB} = \left.\frac{\Delta J_E}{I}\right|_{\Delta T \to 0}$$ (7.49)

is called Peltier coefficient.

We can connect the two coefficients using the first two terms in Eqs. (7.36) and (7.37):

$$\Delta J_E = -L_{EE}\frac{\Delta T}{T^2} - L_{Eq}\frac{\Delta\phi}{T}$$ (7.50)

$$\Delta J_q = -L_{qE}\frac{\Delta T}{T^2} - L_{qq}\frac{\Delta\phi}{T}.$$ (7.51)

In the case of the Seebeck effect we have $\Delta J_q = 0$ and therefore

$$S_{AB} = -\frac{1}{T}\frac{L_{qE}}{L_{qq}}. \tag{7.52}$$

In the case of the Peltier effect $\Delta T = 0$ and thus

$$\frac{\Delta J_E}{\Delta J_q} = \frac{L_{Eq}}{L_{qq}}. \tag{7.53}$$

With $\Delta J_q = I$ and using the Onsager relation, i.e. $L_{Eq} = L_{qE}$, we obtain the non-trivial result

$$-TS_{AB} = \Pi_{AB}. \tag{7.54}$$

7.2 Entropy Production

7.2.1 Entropy Production—Fluctuation Approach

Due to the local nature of the fluctuation approach, we did not pay much attention to the origin of the currents. We want to be more precise in this respect. Therefore we separate the total entropy change, dS, into two distinctly different contributions, i.e.

$$dS = d_i S + d_e S \tag{7.55}$$

(cf. Fig. 7.5). The quantity $d_i S$ is the entropy change inside our system of interest due to processes inside the system. The quantity $d_e S$ describes flow of entropy due to the interaction of the system with the outside. Notice that the entropy change $d_i S$ is never negative:

$$d_i S = 0 \quad \text{for reversible processes}$$
$$d_i S > 0 \quad \text{for irreversible processes.}$$

Fig. 7.5 Contributions to the total entropy change

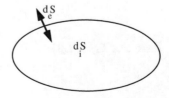

While d_iS can never be negative, d_eS does not have a definite sign and can be positive or negative.

As an example consider a closed system at constant temperature. For a reversible process we have learned that

$$dS = \frac{\delta q}{T}. \tag{7.56}$$

Now we have

$$d_eS = \frac{\delta q}{T}, \tag{7.57}$$

the entropy change is the heat flow across the system boundary divided by temperature. Without irreversible processes inside the system we can also write

$$dS = d_eS = \frac{\delta q}{T}. \tag{7.58}$$

Including irreversible processes inside the system this becomes

$$Td_iS = TdS - \delta q^{(1.1)}\underline{\underline{}}TdS - dE - PdV \geq 0. \tag{7.59}$$

We can combine (7.28) with (7.30) obtaining

$$\frac{d_i\Delta S}{dt} = \sum_j \Delta X_j \Delta J_j \geq 0. \tag{7.60}$$

Below we shall show that this relation remains valid outside the linear regime. Here we briefly mention a consequence of this for the Onsager coefficients. Inserting (7.8) into (7.60) yields

$$\frac{d_i\Delta S}{dt} = \sum_{jj'} L_{jj'} \Delta X_j \Delta X_{j'} \geq 0. \tag{7.61}$$

This means that the matrix of the coefficients $L_{jj'}$ is positive (semi-) definite, i.e. the eigenvalues of this matrix are all larger or equal to zero. If we apply this to the case when two currents are coupled linearly to two generalized forces, this conditions implies (as one can easily work out):

$$L_{11} \geq 0 \quad L_{22} \geq 0 \quad L_{12}^2 \geq L_{11}L_{22} \tag{7.62}$$

(note that $L_{12} = L_{21}$).

7.2.2   Theorem of Minimal Entropy Production

At this point it is useful to introduce the concept of steady states. Consider the example on p. 291—a system divided into two subsystems. The subsystems can exchange matter via for instance a membrane, capillary, or aperture (see Fig. 7.3). Additionally the two subsystems are kept at different temperatures. Thus there are two forces, ΔX_E and ΔX_n, due to the different temperatures and chemical potentials. Over time the system will reach a state in which the flow of matter vanishes, i.e. $\Delta J_n = 0$, whereas the transport of energy between the two subsystems continues. Likewise a non-zero production of entropy continues also. The state variables eventually become time independent. This non-equilibrium state is called steady state. This type of state is different from the equilibrium state of entropy continues also. The state variables eventually become time independent. This non-equilibrium state is called steady state. This type of state is different from the equilibrium state in which all forces, currents, and the entropy production vanish. Another example is a series of coupled chemical reactions. The start compounds are continuously supplied at a constant rate and the final products are removed likewise at a constant rate. A steady state is reached if the concentrations of all other intermediate compounds are constant. We can see that steady states require systems open to some type of transport.

For a steady state close to equilibrium the entropy production is at a minimum compatible with the imposed constraints. What does this mean? A steady state is characterized by constant generalized forces $\Delta X_j (j = 1, \ldots, n)$ and by non-vanishing currents ΔJ_j for some $j = 1, \ldots, m$ and vanishing currents for the remaining $j = m+1, \ldots, n$. If we take the derivative of $d_i \Delta S / dt$ with respect to the generalized force ΔX_k we obtain

$$\frac{\partial}{\partial \Delta X_k} \frac{d_i \Delta S}{dt} \overset{(7.28),(7.8)}{=} \frac{\partial}{\partial \Delta X_k} \sum_{j,j'} L_{jj'} \Delta X_j \Delta X_{j'} \overset{(7.19)}{=} 2 \sum_j L_{kj} \Delta X_j = 2 \Delta J_k.$$

For those k for which $\Delta J_k = 0$ we therefore find

$$\boxed{\frac{\partial}{\partial \Delta X_k} \frac{d_i \Delta S}{dt} = 0}. \tag{7.63}$$

This is the theorem of minimal entropy production (Prigogine 1947, Desoer[6]). With respect to the variation of ΔX_k the entropy production is at a minimum. We can see that this is indeed a minimum by reducing the forces $\Delta X_k (k = 1, \ldots, m)$ to zero, i.e. we approach the equilibrium state where $d_i \Delta S / dt = 0$. We can now reverse direction and conclude by reason of continuity that the entropy production in the steady state indeed is at a minimum compatible with the imposed constraints.

---

[6] Ilya Prigogine, Nobel Prize in chemistry for his contributions to non-equilibrium thermodynamics, 1977.

Example—Steady State and Minimal Entropy Production   We consider the monomolecular reaction

$$A \rightleftharpoons X \rightleftharpoons B \tag{7.64}$$

in a system with constant volume V. There is a constant flow of A into the system and a constant flow of B exiting the system. The entire process is in a steady state. What does the theorem of minimal entropy production tell us, when the process is perturbed by small fluctuations in the mass variables? We begin by writing down the rate equations for the individual reactions (We assume a certain familiarity with reaction kinetics on the level of Atkins (1986):

$$A \rightleftharpoons X \quad -\frac{dn_A}{dt} \overset{(3.70)}{=} \frac{d\xi^{(1)}}{dt} = k_1 n_A - k_{-1} n_X$$

$$\frac{dn_X}{dt} = k_1 n_A - k_{-1} n_X - k_2 n_X + k_{-2} n_B$$

$$X \rightleftharpoons B \quad \frac{dn_B}{dt} \overset{(3.70)}{=} \frac{d\xi^{(2)}}{dt} = k_2 n_X - k_{-2} n_B.$$

The $k_{\ldots}$ are rate constants, where the number refers to the reaction and the sign indicates the forward and reverse reactions, respectively. We assume that there is a small deviation from the steady state, i.e.

$$n_A = n_A^{(o)} + \Delta n_A \quad n_X = n_X^{(o)} + \Delta n_X \quad n_B = n_B^{(o)} + \Delta n_B. \tag{7.65}$$

The index (o) indicates steady state values. Insertion into the above rate equations yields

$$n_X^{(o)} = \frac{k_1}{k_{-1}} n_A^{(o)} = \frac{k_{-2}}{k_2} n_B^{(o)}. \tag{7.66}$$

The time dependence of the deviations from the steady state is

$$A \rightleftharpoons X \quad -\frac{d\Delta n_A}{dt} = \frac{d\Delta \xi^{(1)}}{dt} = k_1 \Delta n_A - k_{-1} \Delta n_X$$

$$\frac{d\Delta n_X}{dt} = k_1 \Delta n_A - (k_{-1} + k_2) \Delta n_X + k_{-2} \Delta n_B \tag{7.67}$$

$$X \rightleftharpoons B \quad \frac{d\Delta n_B}{dt} = \frac{d\Delta \xi^{(2)}}{dt} = k_2 \Delta n_X - k_{-2} \Delta n_B.$$

The temperature is assumed to be constant everywhere, and the entropy production due to the decay of the fluctuation is contributed by products $\Delta J_n \Delta X_n$ exclusively:

$$\frac{d_i \Delta S}{dt} = -\frac{d\Delta n_A}{dt}\frac{\Delta \mu_A}{T} - \frac{d\Delta n_X}{dt}\frac{\Delta \mu_X}{T} - \frac{d\Delta n_B}{dt}\frac{\Delta \mu_B}{T}. \tag{7.68}$$

We emphasize that this is the entropy production due to the decay of a small fluctuation. The entropy production associated with the steady state reaction as such is discussed below in a separate example. The chemical potential fluctuations $\Delta \mu_i$ $(i = A, X, B)$ are

$$\Delta \mu_i = RT\Delta \ln \frac{n_i}{n} = RT\left(\frac{\Delta n_i}{n_i^{(o)}} - \frac{\Delta n_A + \Delta n_X + \Delta n_B}{n^{(o)}}\right), \tag{7.69}$$

where $n = n_A + n_X + n_B$ and $n^{(o)} = n_A^{(o)} + n_X^{(o)} + n_B^{(o)}$. We insert this into Eq. (7.68) and subsequently set the derivative with respect to Δn_X equal to zero, i.e. $d_{\Delta n_X}(d_i\Delta S/dt) = 0$. Even though $-\Delta \mu_X/T$ is the generalized force, we may differentiate with respect to Δn_X instead, because of the linear relationship between the two. After some algebra we obtain

$$\Delta n_X = \frac{k_1 \Delta n_A + k_{-2}\Delta n_B}{k_{-1} + k_2}. \tag{7.70}$$

Notice that this result of the theorem of minimal entropy production is the same as what we obtain if we work from Eq. (7.67) requiring $d\Delta n_X/dt = 0$.

7.2.3 A Differential Relation in the Linear Regime

We consider the differential

$$d\frac{d_i \Delta S}{dt} = \sum_j \left(\underbrace{\Delta X_j d\Delta J_j}_{\equiv d_J \frac{d_i \Delta S}{dt}} + \underbrace{\Delta J_j d\Delta X_j}_{\equiv d_X \frac{d_i \Delta S}{dt}}\right). \tag{7.71}$$

Using Eq. (7.8) together with the reciprocity relation (7.19) it follows that

$$d_X \frac{d_i \Delta S}{dt} = d_J \frac{d_i \Delta S}{dt} \tag{7.72}$$

and

$$dx \frac{d_i \Delta S}{dt} = \frac{1}{2} d \frac{d_i \Delta S}{dt}. \tag{7.73}$$

Remark There exists a general relation, the so called evolution criterion,[7] valid also beyond the linear regime, $dx \frac{d_i S}{dt} \leq 0$. If we apply this to (7.73) we conclude that

$$\frac{d_i^2 \Delta S}{dt^2} \leq 0. \tag{7.74}$$

In accord with the theorem of minimal entropy production we see that the entropy production (following a perturbation) continuously decreases reaching a minimum in the steady state.

7.2.4 Entropy Production—Balance Equation Approach

Here we follow Glansdorff and Prigogine (1971). Our goal is the calculation of entropy production beyond the linear regime.

7.2.5 General Form of a Balance Equation

We wish to follow the time evolution of the scalar quantity

$$I(t) = \int_V \rho[I] dV. \tag{7.75}$$

V is a certain volume at rest, and $\rho[I]$ is the local density of I inside this volume. The change of I per unit of time is given by

$$\frac{\partial}{\partial t} I(t) = \int_V \sigma[I] dV + \int_{\partial V} \vec{j}[I] d\vec{A}. \tag{7.76}$$

The first term is a source term corresponding to the production or elimination of I inside V. The second term describes the change of I due to flow across the surface of V, i.e. $\vec{j}[I]$ is a current density and $d\vec{A}$, pointing towards the inside of V, is a

---

[7] Which we discuss in more detail below.

surface element. Using Green's theorem we may also write the balance equation in differential form

$$\frac{\partial}{\partial t} \rho[I] = \sigma[I] - \partial_\alpha j_\alpha[I]. \tag{7.77}$$

The minus sign results from the orientation of the area element. Because it is useful, we have introduced the shorthand notation $\vec{\nabla} \cdot \vec{j}[I] = \sum_{\alpha=1}^{3} dj_\alpha /dx_\alpha \equiv \partial_\alpha j_\alpha[I]$. If I is a conserved quantity, then $\sigma[I] = 0$ and (7.77) is the usual continuity equation.

In particular we may write

$$\frac{d_i S}{dt} = \int_V \sigma[S] dV, \tag{7.78}$$

where $\sigma[S]$ is the entropy production per unit time and volume, and

$$\frac{d_e S}{dt} = \int_{\partial V} j_\alpha[S] dA_\alpha. \tag{7.79}$$

Because the macroscopic integration volume (system volume) in Eq. (7.78) is arbitrary, we conclude that

$$\sigma[S] \geq 0. \tag{7.80}$$

Our goal is it to derive $\sigma[S]$ expressed in terms of generalized forces and attendant currents, without necessarily being close to equilibrium. Before we can do this, however, we must go through a list of ingredients.

7.2.6 A Useful Formula

First we derive a useful formula. The total derivative of the function $\phi(\vec{r}, t)$ with respect to time is

$$\frac{d}{dt} \phi(\vec{r}, t) = \left(\frac{\partial}{\partial t} + v_\alpha \partial_\alpha\right) \phi(\vec{r}, t), \tag{7.81}$$

where $\vec{v} = d\vec{r}(t)/dt$. Multiplication with the density of I, $\rho[I](\vec{r}, t)$, yields

$$\rho[I] \frac{d}{dt} \phi = \rho[I] \left(\frac{\partial}{\partial t} + v_\alpha \partial_\alpha\right) \phi + \phi \underbrace{\left(\frac{\partial}{\partial t} + \partial_\alpha v_\alpha\right) \rho[I]}_{=0}, \tag{7.82}$$

where a "zero" in the form of the continuity equation has been added—provided that $\sigma[I] = 0$ for this I. Using $\partial_\alpha(ap_\alpha) = (p_\alpha\partial_\alpha)a + a(\partial_\alpha p_\alpha)$ the equation assumes its final form

$$\rho[I]\frac{d}{dt}\phi = \frac{\partial}{\partial t}(\rho[I]\phi) + \partial_\alpha(\rho[I]v_\alpha\phi). \tag{7.83}$$

7.2.7   Mass Balance

In the case of mass we can write down the source term quite easily (cf. Eq. 3.70):

$$\sigma[m_i] = v_i m_i \frac{d\xi'}{dt}. \tag{7.84}$$

Here m_i is the molar mass of component i in a reaction and $d\xi'/dt$ is the reaction rate in moles per unit time and unit volume (indicated by the prime). By convention $v_i < 0$ for reactants and $v_i > 0$ for products. If there are several coupled reactions taking place, then the attendant generalization is

$$\sigma[m_i] = \sum_r v_i^{(r)} m_i \frac{d\xi'^{(r)}}{dt}, \tag{7.85}$$

where $d\xi'^{(r)}/dt$ is the reaction rate in the r-th reaction. The mass current associated with component i is

$$\vec{j}[m_i] = \rho[m_i]\vec{v}_i = \rho[m_i](\vec{\Delta}_i + \vec{v}). \tag{7.86}$$

Here $\vec{\Delta}_i = \vec{v}_i - \vec{v}$ is the velocity of component i relative to the center of mass velocity $\vec{v} = \sum_i \rho[m_i]\vec{v}_i / \sum_i \rho[m_i]$ taken over all components. The resulting mass balance equation is

$$\frac{\partial}{\partial t}\rho[m_i] = \sum_r v_i^{(r)} m_i \frac{d\xi'^{(r)}}{dt} - \partial_\alpha\rho[m_i](\Delta_{i,\alpha} + v_\alpha). \tag{7.87}$$

7.2.8 Internal Energy Balance

Due to conservation of the overall energy we have

$$\sigma[E] + \sigma[K] + \sigma[U] = 0, \qquad (7.88)$$

where $\sigma[E]$ is the internal energy source per unit volume in V, $\sigma[K]$ is the macroscopic kinetic energy source of the same unit volume, and $\sigma[U]$ is the potential energy source due to external forces. We obtain $\sigma[K]$ via the equation of motion for the mass density $\rho[m]$ in a continuous medium:

$$\rho[m]\frac{dv_\alpha}{dt} = \rho[m]g_\alpha - \partial_\beta p_{\alpha\beta}. \qquad (7.89)$$

Here g_α is the respective component of an external force (per unit mass) acting on the volume element. The second term on the right is the same force density component due to internal forces, where the stress tensor introduced in Eq. (1.3) is replaced by its negative—the pressure tensor. Multiplication with v_α (including summation over α) and application of Eq. (7.83) yields

$$\frac{\partial}{\partial t}\frac{1}{2}\rho[m]v^2 = \underbrace{\rho[m]v_\alpha g_\alpha + p_{\alpha\beta}\partial_\beta v_\alpha}_{=\sigma[K]} - \partial_\alpha\left(\frac{1}{2}\rho[m]v^2 v_\alpha + v_\beta p_{\beta\alpha}\right). \qquad (7.90)$$

An analogous equation for the potential energy follows via $dU = -dr_\alpha g_\alpha$, i.e. the force $\vec{g}$ is the negative gradient of U, and $dU/dt = -v_\alpha g_\alpha$. Multiplication of this equation by $\rho[m]$ and subsequent application of (7.83) yields

$$\frac{\partial}{\partial t}\rho[m]U = \underbrace{-\rho[m]v_\alpha g_\alpha}_{=\sigma[U]} - \partial_\alpha(\rho[m]v_\alpha U). \qquad (7.91)$$

If there is more than one component we can write for the source of component i:

$$\sigma[U_i] = -\sum_i \rho[m_i]v_{i,\alpha}g_{i,\alpha}. \qquad (7.92)$$

The flow of internal energy is

$$j_\alpha[E] = \rho[E]v_\alpha + J_{Q,\alpha}. \qquad (7.93)$$

The first term on the right is convection, whereas the second is heat flow. Combination of these results yields the following balance equation for the internal energy

$$\frac{\partial}{\partial t}\rho[E] = \sum_i \rho[m_i]\Delta_{i,\alpha}g_{i,\alpha} - p_{\alpha\beta}\partial_\beta v_\alpha - \partial_\alpha\left(\rho[E]v_\alpha + J_{Q,\alpha}\right). \tag{7.94}$$

7.2.9 Affinity

One more ingredient is the affinity. We consider a process consisting of r coupled chemical reactions. It is then useful to rewrite $\sum_{i=1}^{K} \mu_i dn_i$ as follows

$$\sum_i \mu_i dn_i \overset{(3.70)}{=} \sum_i \mu_i v_i d\xi = \sum_j \mu_j \sum_r v_j^{(r)} d\xi^{(r)} = -\sum_r A^{(r)} d\xi^{(r)} \tag{7.95}$$

(the index j indicates the components in a particular reaction), where

$$A^{(r)} = -\sum_j v_j^{(r)} \mu_j \tag{7.96}$$

defines the affinity. Notice that a non-vanishing affinity means that the system is not at equilibrium.

7.2.10 Entropy Balance Equation

The last ingredient is the assumption of local equilibrium even if the system as a whole is not at equilibrium. We can always express the extensive quantities in the form of local densities. Examples are

$$m = \int_V \rho[m]dV \tag{7.97}$$

$$E = \int_V \rho[m]edV \qquad (\rho[E] = \rho[m]e) \tag{7.98}$$

$$G = \int_V \rho[m] \sum_i \frac{N_i}{m_i}\mu_i dV \tag{7.99}$$

$$S = \int_V \rho[m]s\,dV.$$

(7.100)

Notice that

$$N_i = \frac{\delta m_i}{\delta m}$$

(7.101)

is a mass fraction ($\sum_i N_i = 1$) and

$$\rho[m_i] = \rho[m]N_i.$$

(7.102)

Local equilibrium means that the entropy per unit mass inside a small volume element, s, is the same function of the local macroscopic variables as in a situation of global equilibrium. Consequently, the equations from equilibrium thermodynamics remain applicable on the local scale.

Using the above local quantities we write (cf. Eq. 1.51)

$$\frac{ds}{dt} = \frac{1}{T}\frac{de}{dt} + \frac{P}{T}\frac{d}{dt}\frac{1}{\rho[m]} - \sum_i \frac{\mu_i}{T}\frac{d(N_i/m_i)}{dt}.$$

(7.103)

We multiply this equation with $\rho[m]$. Term by term application of Eq. (7.83), together with the indicated previous results, yields

$$\rho[m]\frac{ds}{dt} = \frac{\partial}{\partial t}(\rho[m]s) + \partial_\alpha(\rho[m]v_\alpha s)$$

$$\frac{\rho[m]}{T}\frac{de}{dt} \overset{(7.94)}{=} \sum_i \rho[m_i]\Delta_{i,\alpha}\frac{g_{i,\alpha}}{T} - \frac{p_{\alpha\beta}}{T}\partial_\beta v_\alpha - \frac{1}{T}\partial_\alpha J_{Q,\alpha}$$

$$\frac{P}{T}\rho[m]\frac{d}{dt}\frac{1}{\rho[m]} = \frac{P\delta_{\alpha\beta}}{T}\partial_\beta v_\alpha$$

$$-\sum_i \frac{\mu_i/m_i}{T}\rho[m]\frac{dN_i}{dt} \overset{(7.87),(7.95),(7.99)}{=} \sum_r \frac{A^{(r)}}{T}\frac{d\xi^{\prime(r)}}{dt} + \sum_i \frac{\mu_i/m_i}{T}\partial_\alpha(\rho[m_i]\Delta_{i,\alpha}).$$

In collecting the right sides together according to (7.103) we also make use of

$$\frac{1}{T}\partial_\alpha J_{Q,\alpha} = \partial_\alpha\left(\frac{J_{Q,\alpha}}{T}\right) - J_{Q,\alpha}\partial_\alpha\left(\frac{1}{T}\right)$$

and

$$\sum_i \frac{\mu_i/m_i}{T}\partial_\alpha(\rho[m_i]\Delta_{i,\alpha}) = \partial_\alpha\left(\sum_i \frac{\mu_i/m_i}{T}\rho[m_i]\Delta_{i,\alpha}\right) - \sum_i \rho[m_i]\Delta_{i,\alpha}\partial_\alpha\left(\frac{\mu_i/m_i}{T}\right).$$

Table 7.2 Currents, forces, and the attendant type of transport

Current J	Force X	Transport
$\rho[m_i]\Delta_{i,\alpha}$	$g_{i,\alpha}/T - \partial_\alpha(\mu_i/m_iT)$	Matter
$p_{\alpha\beta} - P\delta_{\alpha\beta}$	$-\partial_\beta v_\alpha/T$	Momentum
$J_{Q,\alpha}$	$\partial_\alpha T^{-1}$	Heat
$d\xi'^{(r)}/dt$	$A^{(r)}/T$	Chemical reaction

Because

$$\sigma[S] - \partial_\alpha j_\alpha[S] \stackrel{(7.77),(7.100)}{=} \frac{\partial}{\partial t}\rho[m]s \stackrel{(7.83)}{=} \rho[m]\frac{ds}{dt} - \partial_\alpha(\rho[m]v_\alpha s), \qquad (7.104)$$

we can collect the appropriate terms from the above non-numbered equations, i.e.

$$\sigma[S] = \sum_i \rho[m_i]\Delta_{i,\alpha}\left(\frac{g_{i,\alpha}}{T} - \partial_\alpha\frac{\mu_i/m_i}{T}\right)$$

$$- \frac{p_{\alpha\beta} - P\delta_{\alpha\beta}}{T}\partial_\beta v_\alpha + J_{Q,\alpha}\partial_\alpha\left(\frac{1}{T}\right) + \sum_r \frac{d\xi'^{(r)}}{dt}\frac{A^{(r)}}{T} \qquad (7.105)$$

and

$$j_\alpha[S] = \frac{J_{Q,\alpha}}{T} - \sum_i \frac{\mu_i/m_i}{T}\rho[m_i]\Delta_{i,\alpha} + \rho[m]v_\alpha s. \qquad (7.106)$$

We notice that the entropy production has the bilinear form,

$$\boxed{\sigma[S] = \sum_j J_j X_j \geq 0}, \qquad (7.107)$$

encountered before using the fluctuation approach (cf. Eq. 7.28). Table 7.2 lists currents, forces, and the type of transport (cf. Eqs. 7.32–7.35).

Notice that Eq. (7.107) also holds beyond the linear regime, provided that the local entropy assumptions is valid. Notice also that it is possible to use different sets of generalized currents, J'_j, and generalized forces, X'_j. However, this should not change the entropy production (for details see again §3 in Glansdorff and Prigogine (1971)).

Example—Steady State Entropy Production in a Mono Molecular Reaction We return to our previous example of entropy production in a chemical system undergoing the monomolecular reaction

$$A \rightleftharpoons X \rightleftharpoons B. \qquad (7.108)$$

Fig. 7.6 A monomolecular steady state reaction

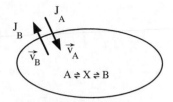

$$A \rightleftharpoons X \rightleftharpoons B$$

Figure 7.6 shows a cartoon of the system. A is supplied to the system, whereas B is leaving the system. In this example we calculate the steady state entropy production itself—not just the entropy production due to the decay of a fluctuation deviation from the steady state.

First we want to compute the internal entropy production d_iS/dt using Eqs. (7.78) and (7.105). There are no external forces, the interior of the system is homogeneous in the concentrations as well as temperature, and viscosity effects (off diagonal elements of the pressure tensor) are negligible. This means that only the last two terms in Eq. (7.105) must be included in the calculation. We begin with the activities for the two reactions:

$$A \rightleftharpoons X \qquad A^{(1)} = -v_A \mu_A - v_X^{(1)} \mu_X \qquad \frac{d\xi'^{(1)}}{dt} = v_A^{-1} \frac{dn_A/V}{dt}$$

$$X \rightleftharpoons B \qquad A^{(2)} = -v_X^{(2)} \mu_X - v_B \mu_B \qquad \frac{d\xi'^{(2)}}{dt} = v_B^{-1} \frac{dn_B/V}{dt}.$$

Thus

$$\frac{d_iS}{dt} = \int_V \sigma[S]dV = \int_V \left(J_{Q,\alpha} \partial_\alpha \left(\frac{1}{T} \right) - \frac{1}{T} (\mu_A + v_X^{(1)} v_A^{-1} \mu_X) \frac{dn_A/V}{dt} \right.$$
$$\left. - \frac{1}{T} (v_X^{(2)} v_B^{-1} \mu_X + \mu_B) \frac{dn_B/V}{dt} \right) dV.$$

With $\quad v_A = -1, \quad v_X^{(1)} = -v_X^{(2)} = 1, \quad v_B = 1, \quad d(n_B/V)/dt = -d(n_A/V)/dt > 0$ this becomes

$$\frac{d_iS}{dt} = -\frac{\mu_A}{T}\frac{dn_A}{dt} - \frac{\mu_B}{T}\frac{dn_B}{dt}. \tag{7.109}$$

The heat flow term has vanished because (by definition) there is no temperature gradient in V.

Now we compute the flow of entropy due to the interaction of the system with the outside using Eqs. (7.79) and (7.106), i.e.

$$\frac{d_e S}{dt} = \int_{\partial V} j_\alpha[S] dA_\alpha = \int_{\partial V} \left(\frac{J_{Q,\alpha}}{T} - \frac{\mu_A}{T} \rho[m_A] v_{A,\alpha} - \frac{\mu_B}{T} \rho[m_B] v_{B,\alpha} \right) dA_\alpha.$$

Here we have assumed that there is no center of mass motion and no motion of X across the surface of V. Now we use $(\rho[m_A]/m_A) v_{A,\alpha} dA_\alpha = \delta n_A/(A\delta s)(\delta s/\delta t) dA = (-dn_A/dt) dA/A$ and $(\rho[m_B]/m_B) v_{B,\alpha} dA_\alpha = -\delta n_B/(A\delta s)(\delta s/\delta t) dA = -(dn_B/dt) dA/A$. The quantities $\delta n_B/(A\delta s)$ and $\delta n_B/(A\delta s)$ are the molar amounts of A and B in a thin surface shell divided by the volume of this shell. Our final result is

$$\frac{d_e S}{dt} = \frac{\mu_A}{T} \frac{dn_A}{dt} + \frac{\mu_B}{T} \frac{dn_B}{dt}. \tag{7.110}$$

Once again does the assumed uniformity conditions permit no net heat flow across the boundary of the volume and we find

$$\frac{d_e S}{dt} = -\frac{d_i S}{dt}, \tag{7.111}$$

which means $dS/dt = d_i S/dt + d_e S/dt = 0$. This is correct, because under steady state conditions there is no overall entropy change inside the system. But notice that still $d_i S/dt > 0$ and therefore $d_e S/dt < 0$. The positive entropy production of the non equilibrium state inside the system is maintained by a flow of negative entropy into the system.

7.2.11 Evolution Criterion

Using the balance equation approach in conjunction with the local equilibrium assumption its is possible to show

$$\boxed{\frac{d_X}{dt} \frac{d_i S}{dt} = \int_V \sum_j J_j \frac{dX_j}{dt} dV \leq 0}, \tag{7.112}$$

where $=$ applies in the steady state (Chapter 9 in Glansdorff and Prigogine (1971)) This is the evolution criterion mentioned earlier (cf. 7.74). The evolution criterion is the most general relation of non-equilibrium thermodynamics. Therefore it is tempting to define the kinetic potential

$$d\Phi = Td_X \frac{d_iS}{dt}. \tag{7.113}$$

However, while it is possible to find a suitable integrating factor in some cases, $d\Phi$ in general is not an exact differential.

7.3   Complexity in Chemical Reactions

This section discusses entropy production in a specific context. Owing to the complexity of the matter we shall focus on processes describable entirely in terms of the last line in Table 7.2. This certainly is crude, because chemical reactions always involve other types of currents and forces as well.

Consider the reaction

$$X + Y \rightleftharpoons C + D. \tag{7.114}$$

Momentarily we are interested in the process far from equilibrium and we neglect the reverse reaction. Suppose that the reaction rate is given by

$$\frac{d\xi}{dt} = k_1 n_X n_Y, \tag{7.115}$$

where k_1 is a rate constant. For the affinity we have

$$A = RT \ln \frac{n_X n_Y}{n_C n_D} + const. \tag{7.116}$$

A fluctuation of the amount of X thus gives rise to the entropy production

$$\frac{d_iS}{dt} \propto \frac{n_Y}{n_X} (\Delta n_X)^2 > 0 \tag{7.117}$$

—in accord with thermodynamic stability. However, if instead we repeat this calculation for the autocatalytic reaction

$$X + Y \rightarrow 2X, \tag{7.118}$$

using again (7.115), the entropy production becomes

$$\frac{d_iS}{dt} \propto -\frac{n_Y}{n_X} (\Delta n_X)^2 < 0. \tag{7.119}$$

It looks as if in this case there is the danger of an unstable process. We shall see that things return to normal, i.e. stability, when we analyze this more carefully.

However, what is important to remember is that autocatalytic reactions are special and, as it turns out, are a key ingredient to the explanation of the creation of order not possible in systems remaining close to equilibrium.[8]

7.3.1 Bray Reaction

The following system of coupled reactions,[9] which has a realistic origin (Ebeling and Feistel 1986), here serves to illustrate a number of important aspects of non-linearity and autocatalysis. Most important, perhaps, is the possibility of bifurcation providing systems with the choice between different steady states. This in principle offers the possibility for competing alternative pathways along which chemical systems can evolve and (sometimes) compete along the way.

Our reaction schema is this:

$$X + Y \rightleftharpoons 2X$$
$$2X + Y \rightleftharpoons 3X$$
$$X \rightleftharpoons F \qquad (7.120)$$
$$B \rightarrow Y$$
$$F \rightarrow B.$$

The time dependence of the respective mole fractions[10] shall be the following

$$\frac{d}{dt} n_X = k_1 n_X n_Y - k_{-1} n_X^2 + k_2 n_X^2 n_Y - k_{-2} n_X^3 - k_3 n_X + k_{-3} n_F$$
$$\frac{d}{dt} n_Y = -k_1 n_X n_Y + k_{-1} n_X^2 - k_2 n_X^2 n_Y + k_{-2} n_X^3 + k_4 n_B$$
$$\frac{d}{dt} n_F = k_3 n_X - k_{-3} n_F - k_5 n_F \qquad (7.121)$$
$$\frac{d}{dt} n_B = -k_4 n_B + k_5 n_F.$$

Just as a reminder: (i) the k's are rate constants; (ii) negative indices indicate reverse reactions; (iii) a term like $n_X n_Y$ assumes that n_X reacts with n_Y in a two-molecule collision; (iv) in general, the powers indicate the number of

---

[8] A particular importance of autocatalytic reactions is their key role in models of prebiotic evolution—an idea that was developed quite a long time ago (Allen 1957). We return to this aspect in the next section.

[9] This is an early representative of the coupled reactions schemes discussed in the context of chemical oscillations. Perhaps the most famous experimental representative is the Belousov-Zhabotinsky reaction.

[10] We assume constant volume.

molecules of this type involved in the respective reaction; (vi) minus signs mean that this reaction reduces the mole fraction on the left side of the equation.

Example—Simple Reaction Kinetics We solve the following special case of (7.121): $n_Y \equiv n_Y(0) = const$, $B = const$, $k_2 = k_{-2} = k_{-3} = 0$. Insertion into (7.121) yields

$$\frac{d}{dt} n_X = (k_1 n_Y(0) - k_3) n_X - k_{-1} n_X^2$$

$$\frac{d}{dt} n_F = -\frac{d}{dt} n_X. \tag{7.122}$$

Integration of the first equation yields the solution

$$n_X(t) = \frac{(k_1 n_Y(0) - k_3) n_X(0)}{k_{-1} n_X(0) - [k_{-1} n_X(0) - k_1 n_Y(0) + k_3] \exp[(-k_1 n_Y(0) + k_3)t]} \tag{7.123}$$

Depending on whether $n_Y(0) < n_{Y,crit} = k_3/k_1$ or $n_Y(0) > n_{Y,crit}$ there are two steady state solutions $n_X(\infty) = 0$ or $n_X(\infty) = (k_1 n_Y(0) - k_3)/k_{-1}$ if $n_X(0) > 0$ If $n_X(0) = 0$ the only solution is $n_X(t) = 0$ (cf. Fig. 7.7). The existence of the two steady state solutions actually follows immediately by setting the right side of Eq. (7.122) equal to zero. However, the system's choice which of the two solutions it prefers, i.e. the stable solution, depends on the parameter $n_Y(0)/n_{Y,crit}$.

Fig. 7.7 Two solutions of Eq. (7.122)

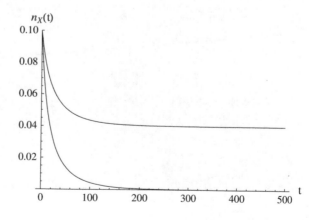

Example—Entropy Production   We briefly reconsider the potential stability problem expressed in (7.119), i.e. we study the first line in the reaction scheme (7.120) by itself (cf. 7.118) with $n_Y \equiv n_Y(0) = const$. The reaction rate is given by

$$\frac{d\xi(t)}{dt} = k_1 n_X(t) n_Y(0) - k_{-1} n_X^2(t). \tag{7.124}$$

In comparison to (7.115) we now include the reverse reaction. The activity is

$$A(t) = RT \ln \frac{n_X(t) n_Y(0)}{n_X^2(t)} + const. \tag{7.125}$$

The entropy production due to a small fluctuation $\Delta n_X = \Delta n_X(t)$ is given by

$$\frac{d_i \Delta S}{dt} = \frac{d\Delta \xi}{dt} \frac{\Delta A}{T} = -R \frac{\Delta n_X}{n_X(t)} \left(k_1 n_Y(0) \Delta n_X - 2k_{-1} n_X(t) \Delta n_x \right)$$
$$= R \left(2k_{-1} - k_1 \frac{n_Y(0)}{n_X(t)} \right) \Delta n_X^2. \tag{7.126}$$

Using $n_X(t) \approx n_X(\infty)$ and subsequent insertion of the (non-zero) steady state solution $d\xi/dt = 0$, i.e. $n_X(\infty) = (k_1/k_{-1}) n_Y(0)$ yields

$$\frac{d_i \Delta S}{dt} \approx R k_1 \Delta n_X^2 \geq 0. \tag{7.127}$$

We may include a check of the evolution criterion (7.112) for the present reaction, i.e.

$$\frac{d_A}{dt} \frac{d_i S}{dt} = \frac{1}{T} \frac{d\xi(t)}{dt} \frac{dA}{dt}. \tag{7.128}$$

We assume uniformity throughout the volume and thus omit the integration. The result is

$$\frac{d_A}{dt} \frac{d_i S}{dt} = -R \frac{1}{n_X(t)} \left(\frac{dn_X(t)}{dt} \right)^2 \leq 0. \tag{7.129}$$

7.3.2 Logistic Map

Here we return to the above example *simple reaction kinetics*, because we want to discuss the "stability issue" from another angle introducing the *logistic*-map:

$$x_{k+1} = 4rx_k(1 - x_k).$$ (7.130)

If we identify the index $k = 0, 1, 2, \ldots$ with time t, i.e. $x(k+1) - x(k) \sim dn_X(t)/dt$ (reasonable as long as $x(k+1) \approx x(k)$), we can express (7.122) by (7.130) if in addition $4r = k_1 n_Y(0) - k_3 + 1 = k_{-1}$.

The mapping (7.130) may be iterated graphically as shown in Fig. 7.8 (left: $r = 0.1$; right: $r = 0.6$). Starting from $x_o = 0.1$ (open circles) the value of x_1 is calculated (first arrow); the subsequent horizontal arrow is to (x_1, x_1); the following vertical arrow yields x_2; and so on. For $r = 0.1$ the mapping converges towards $x_\infty = 0$, while for $r = 0.6$ it converges on $x_\infty = 0.583333$. The two so called fixed points are indicated by the solid circles. Apparently $x = 0$ ceases to be a stable fixed point when the slope of $4rx(1 - x)$ at the origin is greater than one, which happens when $r > r_{crit} = 1/4$. Notice that r_{crit} corresponds to $n_{Y,crit}$! Figure 7.9 is a sketch illustrating the similarity between (7.122) and (7.130).

But there is more to discover here. Increasing the parameter r to 0.8 leads to the graph shown in Fig. 7.10. The final result is not one stable fixed point. The iteration yields a stable 2-cycle, i.e. asymptotically the mapping alternates between the two

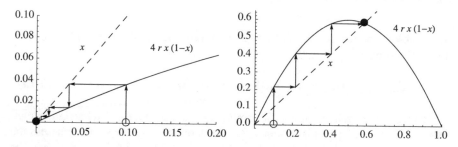

Fig. 7.8 Iterations of the logistic map

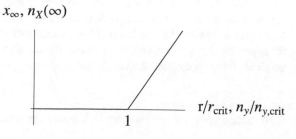

Fig. 7.9 A sketch illustrating the similarity between (7.122) and (7.130)

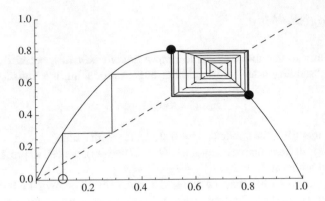

Fig. 7.10 Iteration of the logistic map with r = 0.8

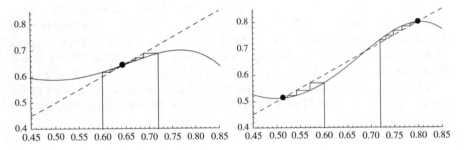

Fig. 7.11 A closer analysis of Fig. 7.10

solid circles. The reason why this happens is illustrated in the Fig. 7.11. The solid line is the square[11] of (7.130), i.e.

$$x_{k+2} = 4r(4rx_k[1 - x_k])(1 - 4rx_k[1 - x_k]). \tag{7.131}$$

The right side again is symmetric around $1/2$, but now there is a local minimum also at $1/2$. The left panel is obtained for this new mapping if $r = 0.7$, whereas the right panel is obtained when $r = 0.8$. Again it is the change of slope relative to the dotted line which makes the difference. This time however it is the slope at $1/2$ and not at the origin.

A particular interesting aspect of the right panel is that a "perturbation" of a "system" at the right attractive point (old steady state), i.e. a perturbative shift of x_n to a value below the central intercept with the dashed line, will cause the "system" to approach the lower fixed point (new steady state) rather than returning to its original fixed point. Analogously an opposite perturbation across the central

---

[11] We use the square because each particular fixed point is visited every second iteration.

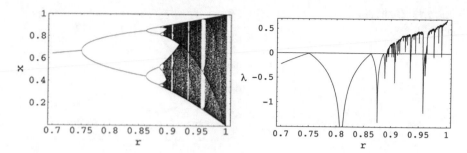

Fig. 7.12 Asymptotic x-values and stability analysis of the logistic map versus r

intercept will cause a transition to the upper fixed point. This of course is the reason why the stable 2-cycle emerges. Every iteration throws us to the opposite side of the intersection of the dotted with the solid line. On its respective side the governing fixed point attracts and continues to do so until the iteration produces the two fixed point values only.

Note in this context that the "transition" in Fig. 7.9 is similar to what happens upon cooling of a ferro-magnetic material below the Curie temperature T_c, i.e. increasing r means cooling. Above T_c there is only one stable state with zero magnetization. Below T_c the magnet is offered two stable states, whereas the zero-magnetization state becomes unstable. Fluctuations near T_c decide which magnetization direction is chosen. This is called spontaneous symmetry breaking (cf. the first example in Sect. 5.1 (model magnet)).

Finally, Fig. 7.12 is a demonstration of how complex the seemingly simple mapping (7.130) really is. The right graph shows the asymptotic x-values (and cycles) over a wide r-range. Notice that r_{crit} is outside the displayed range. Close to 0.75 occurs the bifurcation we have discussed. The bifurcation continues to repeat itself along the new branches until this becomes difficult to resolve. We do not want to discuss this graph further[12] and refer the interested reader to the original work by Feigenbaum (1983) or to Kadanoff (1993),[13] and particularly to the basic text by Gould and Tobochnik (1996). We also postpone the discussion of the right panel in Fig. 7.12 to p. 316 after the discussion of linear stability analysis.

What one should bear in mind, however, is that higher order non-linearity in chemical reactions, just as in the simplified example of the logistic map, may lead to bifurcations distinguishing chemical pathways involving different steady states.

---

[12] Not visible at this resolution are the self-similar copies of the original graph inside the "white gaps".

[13] Notice that this is the intercept of two lines of research. One objective is the understanding of the transition from order to chaos, whereas another group of researchers, Prigogine et al., peruse the opposite direction.

Fig. 7.13 Oscillating
chemical reaction

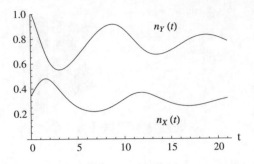

7.3.3   *Chemical Clocks*

Let us select another special variant of the above Bray reaction. We choose
$B = const.$, $k_{-1} = k_{-2} = k_{-3} = 0$. The system of rate equations (7.121) with this
choice becomes

$$\frac{d}{dt}n_X = k_1 n_X n_Y + k_2 n_X{}^2 n_Y - k_3 n_X$$

$$\frac{d}{dt}n_Y = -k_1 n_X n_Y - k_2 n_X{}^2 n_Y + k_4 n_B \qquad (7.132)$$

$$\frac{d}{dt}n_F = k_3 n_X - k_4 n_B.$$

Figure 7.13 shows a portion of the time evolution of $n_X(t)$ and $n_y(t)$. We do not
include $n_F(t)$, because it is not affecting the coupling between $n_X(t)$ and $n_y(t)$. The
initial values are close to the steady state solutions $n_X(\infty) = k_4 n_B/k_3$ and
$n_Y(\infty) = k_3^2/(k_1 k_3 + k_2 k_4 n_B)$. Here $k_1 = 0.5$, $k_2 = 1$, $k_3 = 0.9$, $k4 = 1$, and
$n_B = 0.3$. Thus $n_X(0) \approx 0.333$ and $n_Y(0) \approx 1.0125$. This variant of the Bray
reaction yields chemical oscillations.

The science of chemical oscillation is a wide field. The oscillations may be
oscillations in time, as in our example, or spatial oscillations.[14] The description of
real oscillation phenomena also requires to include diffusive or convective flows.
A rather detailed discussion of oscillation phenomena in the context of
non-equilibrium thermodynamics is given in Nicolis and Progogine (1977).

---

[14] Molecular pattern formation due to gradient induced gene transcription is part of the early
(Drosophila) embryo development (Nüsslein-Volhard 2006).

7.3.4 Linear Stability Analysis

We can solve the coupled system of the Bray reaction or similar coupled first-order differential equations on a computer quite easily. However, the answer is likely to be confusing, because we obtain a variation of distinct looking results depending on the rate constants or other conditions we impose on the system—why for instance do we choose $k_1 = 0.5$, $k_2 = 1$, $k_3 = 0.9$, $k_4 = 1$, and $n_B = 0.3$ in the above example?

A simple tool allowing a classification of our numerical solutions is the following. We consider a system of n coupled reactions described via

$$\frac{d}{dt} n_i = T_i(n_1, \ldots, n_n). \tag{7.133}$$

Notice that $T_i(n_1, \ldots, n_n)$ is an in general non-linear function of the n_j. A particular steady state solution of (7.133) is denoted $(n_1^{(o)}, \ldots, n_n^{(o)})$,[15] i.e.

$$0 = T_i(n_1^{(o)}, \ldots, n_n^{(o)}). \tag{7.134}$$

We now insert $n_i = n_i^{(o)} + \delta n_i$ into (7.133) and expand the right side around the steady state solution $(n_1^{(o)}, \ldots, n_n^{(o)})$ to linear order in the perturbations δn_i:

$$\frac{d}{dt} \delta n_i = \sum_{j=1}^{n} \frac{\partial T_i}{\partial n_j}\bigg|^{o} \delta n_j. \tag{7.135}$$

Or in matrix form:

$$\frac{d}{dt} \delta \vec{n} = A \delta \vec{n} \tag{7.136}$$

with $A_{ij} = \partial T_i / \partial n_j|^{o}$. Suppose the transformation SAS^{-1} diagonalizes A and therefore (7.136) becomes

$$\frac{d}{dt} \delta n_i' = \sum_{j=1}^{n} \lambda_i \delta_{ij} \delta n_j'. \tag{7.137}$$

Here $\delta_{ij} = 1$ if $i = j$ and zero otherwise and $\delta \vec{n}' = S \delta \vec{n}$. The now decoupled linear system (7.137) has the solution

$$\delta n_i'(t) \sim \exp[\lambda_i t]. \tag{7.138}$$

---

[15] There may be more than one steady state.

Fig. 7.14 Stability analysis
pertaining to the system in
Fig. 7.13

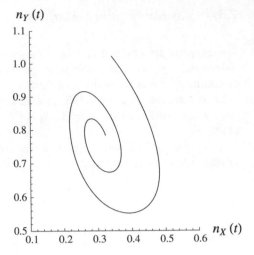

We see that the steady state solution $(n_1^{(o)}, \ldots, n_n^{(o)})$ is completely stable (unstable) with respect to the small perturbations if all eigenvalues λ_i of A are real and negative (positive).

In the case of Fig. 7.13 the two eigenvalues are $\lambda_1 = 0.0411 - i\,0.49831$ and $\lambda_2 = 0.0308333 - i\,0.37373$. Both eigenvalues posses positive real parts, i.e. the steady state close to which the trajectory is started is not stable. The imaginary parts give rise to the oscillatory behavior. Figure 7.14 illustrates how the solution spirals away from its starting point. We can use this type of analysis to find other types of trajectories depending on our choice of parameter values.

A similar type of analysis also explains the right panel in Fig. 7.12. Let us start the iteration of the logistic map from two x-values separated by the small distance $|\delta x_0|$. What happens to this distance after i iterations? Using $dx_{n+1}/dx_n = 4r(1 - 2x_n)$, we can work out the answer as follows:

$$
\begin{aligned}
|\delta x_i| &= |4r(1 - 2x_{i-1})||\delta x_{i-1}| \\
&= |4r(1 - 2x_{i-1})||4r(1 - 2x_{i-2})||\delta x_{i-2}| \\
&= \ldots \\
&= \prod_{n=0}^{i-1} |4r(1 - 2x_n)||\delta x_0|
\end{aligned}
$$

Assuming $|\delta x_i| = |\delta x_0| \exp[(i-1)\lambda]$ for large i, we may compute λ via $\lambda = \lim_{i \to \infty} i^{-1} \ln \sum_{n=0}^{i-1} |4r(1 - 2x_n)|$. Figure 7.12 shows $\lambda = \lambda(r)$. Negative values mean that the iteration approaches a stable fix point or limit cycle. Positive values mean that a small perturbation grows exponentially. We notice that the bifurcations are associated with $\lambda = 0$. λ is called Lyapunov-exponent.

7.4 Remarks on Evolution

If we think about evolution, what we usually have in mind are animal and plant populations adapting to changing environmental condition. The principle is seemingly easy to understand. Reproduction produces a new generation of individuals, who either by combination of their parents genetic information or by mutation develop an advantage over other individuals of their generation. This advantage acts as advantage due to changing environmental conditions and leads to an enhanced reproduction, perhaps because of a better chance for survival.[16]

Because evolution is the cause for the complexity thermodynamics is seemingly opposing (cf. the Feynman quote in the introduction to this chapter), it is interesting to trace evolution back to the beginnings of live itself and beyond. But the further one proceeds into the past, the more difficult it becomes to find the traces. Nevertheless, the question arises at what stage in our planet's development did evolution start? Even before "life" came into existence, there must have been a chemical evolution and chemical reactions can be described by thermodynamics.

Researchers have attempted to recreate the early chemical steps towards the development of life on the young earth in their laboratories, starting with the experiments of Stanley Miller and Harold Urey in 1953 extending ideas of A. I. Oparin and J. B. S. Haldane (Oparin 1964; Haldane 1990). The earth is an open system, and we have seen that there is an enormous flow of energy into this system primarily from the sun (Ebeling and Feistel 1986). Life and its development is possible in a thin shell on the surface of the earth. There also is heat and matter flowing into this shell from below. In addition the earth is hit by cosmic radiation and matter. Thus there is sufficient negative entropy production to allow ordering without causing a conflict with the second law. But the question still remains: at what point did "evolution" begin to drive things towards the development of "unlikely order?" Or is there the need for additional laws of nature—like an evolution law? This does not appear to be the case, even though it is probably fair to say that the very early evolution is still a matter of intense research at this time (e.g., Nowak and Ohtsuki 2008).

One may speculate that prebiotic evolution corresponds essentially to a sequence of instabilities bringing about increasing complexity (Nicolis and Progogine 1977). Significant insight along this line is due to M. Eigen[17] and coworkers (Eigen and Winkler 1993; Eigen 1996; Eigen et al. 1981; Eigen 1993). Eigen and coworkers have studied the autocatalytic synthesis or replication of RNA strands in a test tube. Their experiments were guided by the idea that the primeval broth constituted a suitable medium for Darwin's evolution acting on self-replicating molecular

---

[16] The principle may even be applied to the optimization of technical systems or material properties. So called genetic algorithms consist of a set of operators simulating reproduction, combination, and mutation applied to linear parameter sets defining the technical system (e.g. Goldberg 1989).

[17] Eigen 1967; he is perhaps better known for his work on prebiotic evolution.

species, i.e. RNA strands representing different nucleotide sequences. Strands with different sequences compete for the supply of monomers. Two key ingredients of the concept they develop for the RNA evolution are the quasispecies and the hypercycle. The former is the long-time result of the coupled reactions

$$\frac{dn_i}{dt} = (k_{ii} - \bar{k})n_i + \sum_{j(\neq i)} k_{ij} n_j. \tag{7.139}$$

Here n_i is the amount of sequences of type i. k_{ii} is the rate corresponding to a perfect replication of i. k_{ij} $(i \neq j)$ is the rate corresponding to a replication of j leading to the sequence i via sequence errors during replications. The quantity $\bar{k}$ is a mean excess productivity. The excess productivity of i is the difference between the rate of formation and the rate of decomposition of sequence i. This is adding the element of selective competition, because an increase of the mean excess productivity exerts a selective pressure on the individual sequence types.[18] The steady state solution consists of a core sequence m in constant competition with its own mutations. This distribution of sequences is called quasispecies.

However, the amount of information which can be stored by a quasispecies is limited. The longer the sequences becomes, the larger becomes the number of sequence errors during replication. Mathematically this summarized in the following criterion:

$$\bar{q}_i^{N_i} \mathcal{Y}_i \geq 1. \tag{7.140}$$

In this relation $\bar{q}_i$ is the (average) probability that a particular monomer in the replicated sequence is the correct one. The probability of a perfect replication of a sequence of length N_i therefore is $\bar{q}_i^{N_i}$. This number by itself is less than unity and therefore there must be another factor, $\mathcal{Y}_i$, outweighing replication errors. $\mathcal{Y}_i$ is a measure for the competitive advantage of sequence i. According to (7.140) the maximum possible length of a sequence satisfying this criterion is

$$N_{i,max} \approx \frac{\ln \mathcal{Y}_i}{1 - \bar{q}_i} \tag{7.141}$$

(using $\ln \bar{q}_i \approx 1 - q_i$ or $\bar{q}_i \approx 1$). This leads to the above conclusion that a quasispecies by itself can maintain only a very limited amount of information.[19] This information problem is improved via the hypercycle concept.

---

[18] The principle is analogous to a high jump competition. If a jumper clears the bar, the others must also clear this height in order to remain in the competition.

[19] The logarithm in the numerator does ensure that the latter will not be large. In addition, replication without additional mechanisms enhancing its precision limits the approach of $1 - \bar{q}_i$ towards zero.

Fig. 7.15 Hypothetical dynamic flow in a three-quasispecies-hypercycle

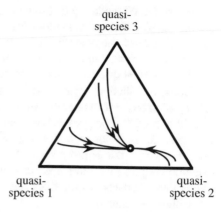

A hypercycle describes the autocatalytic coupling of a number of quasispecies. The requirements for the formation of a hypercycle are: (i) every quasispecies by itself must be stable; (ii) the quasispecies must tolerate each other; (iii) there must be however some kind of feed-back coupling between the sequences. Figure 7.15 is an illustration of a hypothetical dynamic flow due to condition (iii) in a three-quasispecies-hypercycle towards a fixed point (this triangular diagram is analogous to the ternary phase diagram in Fig. 4.28). The types of flows possible are similar to the dynamical flows discussed in the context of linear stability analysis. The type of coupling between the quasispecies populations is circular. One type of macromolecule catalyses the next and so forth leading to a closed loop—a hypercycle. Notice that autocatalysis enters as one necessary ingredient! In principle there are different types of circular couplings giving rise to competing hypercycles. Eigen and coworkers as well as others have analyzed the dynamics of hypercycles extensively on the basis of computer experiments combined with experimental observations on model systems. But the hypercycle is a qualitative concept and does not make concrete predictions. The hypercycle model also has been criticized because of stability problems (e.g., Dyson 1985). In his book Dyson constructs his own model of prebiotic evolution. His is a theoretical model based on polypeptides—an alternative to the polynucleotide based chemical evolution. It is not possible to discuss details here. However, it is interesting to mention that the main feature of the model is the possibility for switching between steady states (spontaneous symmetry breaking) akin to the mechanism depicted in Fig. 7.11.

Example—Hypercycle Game This example illustrates the idea of the hypercycle in the form a game or algorithm (taken from Eigen and Winkler (1993)). In this algorithm the reacting four quasispecies are represented by tiles of identical color. The quasi species are linked via the following circular sequential ordering of tile colors: blue $\rightarrow$ green $\rightarrow$ orange $\rightarrow$ red $\rightarrow$ blue.

Algorithm: (i) randomly distribute tiles of the four different colors on a periodic lattice; (ii) pick one tile at random (its color is "c"); (iii) if this tile has a common edge with at least one other tile of the preceding color in the sequence, then turn the color of another randomly chosen tile into "c"; (iii) goto (ii).

Figure 7.16 shows two 40×40 lattices (left: initial random distribution; right: particular distribution after $2 \cdot 10^5$ iteration steps of $2 \cdot 10^6$ total). The label "changes" stands for the number of actual tile replacements. Notice that the above algorithm produces pronounced oscillations. Notice also the pairing of next-nearest neighbors in the color sequence, e.g. when there is much blue there also is little orange. Clearly, this is just a game, but it provides a feeling for the type of auto-catalytic coupling we have been talking about and its consequences. It is worth noting that other sets of local rules can be invented giving rise to spatial structuring (cf. cellular automata).

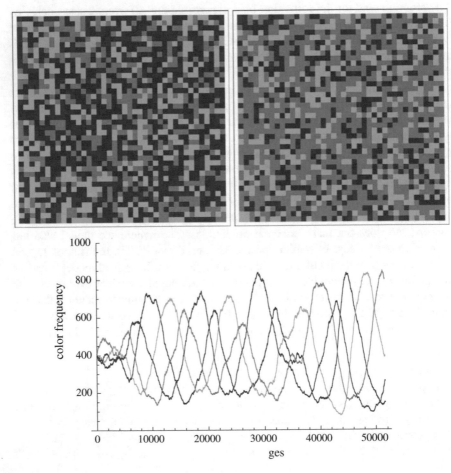

Fig. 7.16 Results of the hypercycle game

Experimental information on the early evolution and the beginning of live is extremely difficult to obtain; only the arrival of microbial genomics has allowed to reliably retrace subsequent developments. The earth was formed about $4.5 \cdot 10^9$ years ago. The earliest traces of primitive cells occurred possibly as early as $3.5 \cdot 10^9$ years ago (Schopf 2006). Nevertheless, this leaves several hundred million years for the actual chemical evolution. The evolution of cells (Woese 2002), however, has taken the major portion of the remaining time. The step from microorganisms (bacteria) to higher live forms occurred much later—less than $1 \cdot 10^9$ years ago. Even tough the very early steps towards cellular live are still in the dark, it is very likely that autocatalytic reactions in conjunction with steady state bifurcations, of which we have described the principles, did play a key role in the formation and maintenance of dissipative structures of increasing complexity.[20]

"Evolution is comparable to a soap box race. Entropy is the hill. Without the hill there is no race and thus no distinction between good or badly designed cars. Thermodynamics on the other hand describes the rules of the race."

A Final Remark—Mortality Due to Accumulation of Irreversible Defects In this final remark we briefly introduce the concept of percolation. There exists an interesting connection between homogeneity, criticality, scaling exponents, and non-linear mappings (logistic-map), all items we have talked about, and percolation. The connection is self-similarity or scale-invariance*An Introduction to Computer Simulation Methods*. Addison-Wesley. Here we use percolation as a model for a particular interaction between irreversible defects leading to the death of an organism.

We assume a population of organisms. Each organism is represented by a lattice of $L \times L$ tiles. Initially all tiles are white. Every organism may acquire irreversible defects, indicated by changing the color of a tile from white to black. An organism dies if the irreversible defects connect in a certain way, i.e. if they form a percolating cluster. Two adjacent tiles belong to the same cluster only if they have one common edge. Figure 7.17 shows an organism with 25 irreversible defects and 4 clusters. A cluster is a percolating cluster if it has at least one tile on the bottom and one tile on the top row of the lattice.

The attendant algorithm consists of the following steps: (i) generate a large number of blank lattices of size $L \times L$; (ii) with probability p every tile in every organism is changed from white to black; (iii) determine the number of surviving organisms, i.e. the number of lattices without percolating cluster(s); (iv) increase p and goto (i).

This is related to real live expectancy data as follows. We assume a constant average defect rate of n irreversible defects per year. Then $z = ny$ is the average

---

[20] Equilibrium structures—are formed and maintained through reversible transformations implying no appreciable deviations from equilibrium; dissipative structures—are formed and maintained through the effect of exchange of energy and matter in non-equilibrium conditions.

organism

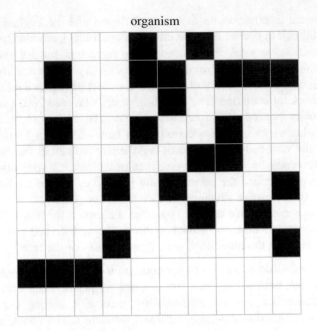

Fig. 7.17 Lattice representation of an organism

number of defects per organism after y years. We convert this into the above probability p via

$$y = \frac{pL^2}{n}. \tag{7.142}$$

The quantities n and L are parameters. Figure 7.18 shows a fit to experimental data showing the current survival probability in Germany (solid line; source: http://

survival probability

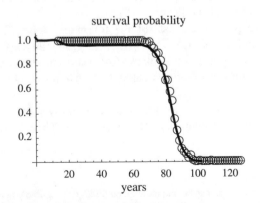

years

Fig. 7.18 Survival probability of the lattice organism and experimental data

www.uni-giessen.de/gi38/nublica/pharma/homepage.html) using $n = 4.5$, i.e. 1 defect every three months, and $L = 25$. Notice that the two initial "dips" of the solid line are due to early infant mortality as well as juvenile deaths due to traffic accidents—causes of death unrelated to irreversible defects. Notice also that increasing the size of the lattice increases the steepness of the drop of the survival probability.[21] The inflection point on the other hand is determined by L^2/n. The scatter of the model's results (open circles) is due to the relatively small population size of 200 organisms generated per p-value.

---

[21] On an infinite lattice the result will be a step function dropping to zero at $p_c = 0.5927$.

Appendix A
The Mathematics of Thermodynamics

A.1 Exact Differential and Integrating Factor

Consider a function of two independent variables

$$f = f(x, y).$$

The expression

$$df(x, y) = \frac{\partial f}{\partial x}\bigg|_y dx + \frac{\partial f}{\partial y}\bigg|_x dy$$

is called the differential of $f(x, y)$. Notice that the generalization to more than two variables is obvious. However, for most of our manipulations and transformations of thermodynamic relations this is the relevant case.

Now consider the expression

$$dg(x, y) = p\,dx + q\,dy.$$

Provided that

$$\frac{\partial p}{\partial y}\bigg|_x = \frac{\partial q}{\partial x}\bigg|_y$$

holds, then $dg(x, y)$ is an exact differential. An example of an exact differential is

$$dg(x, y) = \left(3x^2 + y\cos x\right)dx + \left(\sin x - 4y^3\right)dy,$$

© Springer Nature Switzerland AG 2022
R. Hentschke, *Thermodynamics*, Undergraduate Lecture Notes in Physics,
https://doi.org/10.1007/978-3-030-93879-6

because

$$\underbrace{\frac{\partial}{\partial y}\left(3x^2 + y\cos x\right)\Big|_x}_{=\cos x} = \underbrace{\frac{\partial}{\partial x}\left(\sin x - 4y^3\right)\Big|_y}_{=\cos x}.$$

An example of a differential which is not exact is

$$dg(x,y) = \left(3xy^2 + 2y\right)dx + \left(2x^2y + x\right)dy,$$

because

$$\underbrace{\frac{\partial}{\partial y}\left(3xy^2 + 2y\right)\Big|_x}_{=6xy+2} \neq \underbrace{\frac{\partial}{\partial x} + \left(2x^2y + x\right)\Big|_y}_{=4xy+1}.$$

However, in this case we can multiply $dg(x,y)$ by x, i.e.

$$dh(x,y) \equiv xdg(x,y)$$
$$= \left(3x^2y^2 + 2yx\right)dx + \left(2x^3y + x^2\right)dy.$$

Obviously $dh(x,y)$ again is an exact differential:

$$\underbrace{\frac{\partial}{\partial y}\left(3x^2y^2 + 2xy\right)\Big|_x}_{=6x^2y+2x} = \underbrace{\frac{\partial}{\partial x}\left(2x^3y + x^2\right)\Big|_y}_{=6x^2y+2x}.$$

Because of this the factor x is called integrating factor.

Notice that the above $df(x,y)$ is an exact differential, because the partial derivatives may be exchanged, i.e.

$$\frac{\partial}{\partial x}\frac{\partial f}{\partial y}\Big|_x\Big|_y = \frac{\partial}{\partial y}\frac{\partial f}{\partial x}\Big|_y\Big|_x$$

assuming continuity of the derivatives.

The special importance of exact differentials in thermodynamics is rooted in the following mathematical theorem: Let

$$dA(x,y) = Pdx + Qdy,$$

where P, Q, $\partial P/\partial y$, and $\partial Q/\partial x$ are single-valued and continuous in a simply- (or multiply-)connected region $\mathcal{R}$ bounded by a simple (or more) closed curve(s) $\mathcal{C}$. Then

$$\oint_C dA = \int_{\mathcal{R}} dx dy \left(\frac{\partial Q}{\partial x} \Big|_y - \frac{\partial P}{\partial y} \Big|_x \right).$$

This statement is called Green's theorem in the plane. A proof may be found in Spiegel (1971). We conclude immediately that if $dA(x, y)$ is an exact differential, and therefore

$$\frac{\partial Q}{\partial x} \Big|_y = \frac{\partial P}{\partial y} \Big|_x ,$$

we have

$$\oint_C dA = 0.$$

This means that if we divide a closed path C in the x-y-plane into two sections, i.e.

$$C = (x_1, y_1) \overset{\text{path I}}{\longrightarrow} (x_2, y_2) \overset{\text{path II}}{\longrightarrow} (x_1, y_1),$$

we find

$$\int_{1, pathI}^{2} dA + \int_{2, pathII}^{1} dA = 0$$

or

$$\int_{1, pathI}^{2} dA = \int_{1, pathII}^{2} dA.$$

Therefore the value of $A(x_2, y_2)$ does not depend on the path along which (x_2, y_2) is reached. Every function $A(x, y)$ possessing this property is called a state function. Thus, if $dA(x, y)$ is an exact differential then $A(x, y)$ is a state function and vice versa.

Example—Perpetual Motion Machine The physical significance of this is best explained using the internal energy E. Consider for simplicity a closed system containing a gas. We know from experience that the state of the gas is described completely if we know its temperature, T, and its volume, V. We want to study the change of E along a closed path C in the T-V-plane. Let us assume we find that

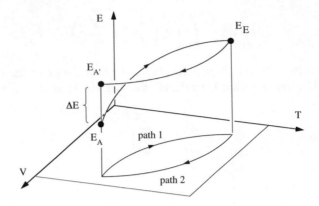

Fig. A.1 Hypothetical internal energy gain along a closed path in the T-V-plane

$$\oint_C dE = \Delta E \neq 0.$$

If $\Delta E > 0$ we may generate an arbitrary amount of energy simply by repeating the cyclic path in the T-V-plane (this situation is depicted in Fig. A.1). If $\Delta E < 0$ we reverse direction and again generate energy. A machine constructed on this principle is called a perpetual motion machine. However, no such device has been build thus far.

Example—dq Is Not an Exact Differential Let us study another instructive example. We consider a process involving volume change like the one we have discussed before (see p. 1). We want to show that

$$dq = dE + PdV$$

is not an exact differential. Using $E = E(T, V)$ we obtain

$$dq = \frac{\partial E}{\partial T}\Big|_V dT + \left(\frac{\partial E}{\partial V}\Big|_T + P\right)dV.$$

Exact differential would mean that

$$\underbrace{\frac{\partial}{\partial V}\frac{\partial E}{\partial T}\Big|_V\Big|_T}_{\overset{(*)}{=}\frac{\partial}{\partial T \partial V}\big|_T\big|_V} = \frac{\partial}{\partial T}\left(\frac{\partial E}{\partial V}\Big|_T + P\right)\Big|_V.$$

Here (*) holds, because dE is an exact differential (cf. above). Therefore we must have

$$\left.\frac{\partial P}{\partial T}\right|_V = 0.$$

This equation obviously cannot be correct, and therefore q is no state function.

Remark Because we have seen that S is a state function, we conclude that according to Eq. (1.47) $1/T$ is an integrating factor.

A.2 Three Useful Differential Relations

In the following we derive three useful differential relations. Consider $A = A(x, y)$ and $z = z(x, y)$. The differential of A is

$$dA = \left.\frac{\partial A}{\partial x}\right|_y dx + \left.\frac{\partial A}{\partial y}\right|_x dy,$$

and therefore

$$\left.\frac{\partial A}{\partial x}\right|_z = \left.\frac{\partial A}{\partial x}\right|_y \underbrace{\left.\frac{\partial x}{\partial x}\right|_z}_{=1} + \left.\frac{\partial A}{\partial y}\right|_x \left.\frac{\partial y}{\partial x}\right|_z$$

or

$$\left.\frac{\partial A}{\partial z}\right|_y = \left.\frac{\partial A}{\partial x}\right|_y \left.\frac{\partial x}{\partial z}\right|_y + \left.\frac{\partial A}{\partial y}\right|_x \underbrace{\left.\frac{\partial y}{\partial z}\right|_y}_{=0}.$$

Thus we find

$$\boxed{\left.\frac{\partial A}{\partial x}\right|_z = \left.\frac{\partial A}{\partial x}\right|_y + \left.\frac{\partial A}{\partial y}\right|_x \left.\frac{\partial y}{\partial x}\right|_z} \tag{A.1}$$

and

$$\boxed{\left.\frac{\partial A}{\partial z}\right|_y = \left.\frac{\partial A}{\partial x}\right|_y \left.\frac{\partial x}{\partial z}\right|_y.} \tag{A.2}$$

The third relation follows if we use $z = A$ in Eq. (A.1), i.e.

$$\underbrace{\left.\frac{\partial z}{\partial x}\right|_z}_{=0} = \left.\frac{\partial z}{\partial x}\right|_y + \left.\frac{\partial z}{\partial y}\right|_x \left.\frac{\partial y}{\partial x}\right|_z,$$

and therefore

$$\boxed{\left.\frac{\partial x}{\partial y}\right|_z = - \left.\frac{\partial x}{\partial z}\right|_y \left.\frac{\partial z}{\partial y}\right|_x,} \qquad (A.3)$$

where we have used

$$\left.\frac{\partial z}{\partial x}\right|_y = \frac{1}{\left.\frac{\partial x}{\partial z}\right|_y}$$

and

$$\left.\frac{\partial y}{\partial x}\right|_z = \frac{1}{\left.\frac{\partial x}{\partial y}\right|_z}.$$

A.3   Legendre Transformation

Consider

$$df = u\,dx + v\,dy$$

where

$$v = \left.\frac{\partial f}{\partial y}\right|_x. \qquad (A.4)$$

We define a new function g via

$$g = f - vy. \qquad (A.5)$$

Notice that g, computed for a certain y-value, is the intercept of the tangent of f at this y-value with the f-axis ($f(\ldots, y) = f'(\ldots, y)y + b$, where b is the intercept).

Next we compute dg, i.e.

$$dg = df - d(vy) = udx + vdy - vdy - ydv.$$

Therefore

$$dg = udx - ydv.$$

This tells us that g is a function of x and v, i.e. $g = g(x, v)$. The function $g(x, v)$ is called the Legendre transform of $f(x, y)$. It replaces the dependence on y by a dependence on v. The key to this replacement is the validity of $v = \partial f / \partial y \mid_x$.

Example—$f = p(x^2 + y^2)$ We consider the example

$$f(x, y) = p(x^2 + y^2), \tag{A.6}$$

where p is a parameter. We find

$$v = \frac{\partial f}{\partial y}\bigg|_x = 2py \quad \text{and thus} \quad y = \frac{v}{2p}. \tag{A.7}$$

Inserting this into Eq. (A.5) yields

$$g(x, v) - px^2 - \frac{1}{4p}v^2. \tag{A.8}$$

We can use this to illustrate an important point. Assume that the parameter p is changed, i.e. $p_{new} = p_{old} - \delta p$. If $\delta p > 0$ this means that

$$\delta f(x, y)\big|_{x,y} < 0, \tag{A.9}$$

where $\delta f = f(x, y; p_{new}) - f(x, y; p_{old})$. What happens to $g(x, v)$? The answer is

$$\begin{aligned} \delta g\big|_{x,v} &= (p - \delta p)x^2 - \frac{1}{4(p - \delta p)}v^2 - px^2 + \frac{1}{4p}v^2 \\ &\approx -\delta p x^2 - \frac{1}{4p}\left(1 + \frac{\delta p}{p}\right)v^2 + \frac{1}{4p}v^2 \\ &= -\left(x^2 + \frac{1}{4p^2}v^2\right)\delta p < 0. \end{aligned} \tag{A.10}$$

This means that the decrease of $f(x, y; p)$ at constant x and y is carried over to the Legendre transform $g(x, v)$ at constant x and v. Even though this is not a general proof, we can see easily from Eq. (A.5) in conjunction with (A.4) that a (local) shift of f at certain fixed variables x and y produces a corresponding shift of g at the attendant fixed values of x and v. This is very useful—as we shall see.

Appendix B
Grand-Canonical Monte Carlo: Methane on Graphite

"GCMC: adsorption of methane on graphite";
"units = Lennard-Jones units";
"temperature"; $T = 1.53$; Print ["$T =$", T];
"target bulk density"; ρ bulk = 0.05; Print ["ρ bulk =", ρ bulk];
"2nd virial coefficient";
B2 = NIntegrate $[-2\,\text{Pi}\,(\text{Exp}\,[-(4(1/s^\wedge 12 - 1/s^\wedge 6))/T] - 1)s^\wedge 2,$
$\{s, 0, \text{Infinity}\}]$;
"bulk pressure"; $P = T\rho$ bulk $(1 + \rho$ bulk B2$)$; Print ["P =", P];
"excess chemical potential"; μ ex = $2T\rho$ bulk B2; Print ["'μ ex =", μ ex];
"simulation box size (LxLxLz)"; $L = 6$; Lz = $2L$; $V = L^\wedge 2$ Lz;
$a = N[V\rho$ bulk Exp $[\mu\,\text{ex}/T]]$;
"cutoff radius"; rcut = 3.0;
"particle coordinates";
TLIST = Table $[\{\text{Random}\,[\text{Real}, \{1, L\}], \text{Random}\,[\text{Real}, \{1, L\}],$
Random $[\text{Real}, \{1, Lz\}]\}, \{i, 1, 2\}]$;
n = Length [TLIST];
"MC step counter"; mcsteps = 0;
"steps per MC-cycle"; maxmcsteps = 1000;
"cycle counter"; cycles = 0;
"total number of cycles"; maxcycles = 4000;
"initial values in density histogram"; nint = 100; ρ = Table $[0, \{i, 0, \text{nint}\}]$;
counter = 0;
While [cycles < maxcycles, cycles ++;
While [mcsteps < maxmcsteps cycles,
"particle insertion";
"1. random position";
$x = \{\text{Random}\,[\text{Real}, L], \text{Random}\,[\text{Real}, L], \text{Random}\,[\text{Real}, Lz]\}$;
"2. energy change";
$\Delta u = 0$;
Do $[y = \text{Extract}\,[\text{TLIST}, i] - x$;

© Springer Nature Switzerland AG 2022
R. Hentschke, *Thermodynamics*, Undergraduate Lecture Notes in Physics,
https://doi.org/10.1007/978-3-030-93879-6

$r = \text{Sqrt} \left[(y[[1]] - L \, \text{Round} \, [y[[1]]/L])^\wedge 2 + (y[[2]] - L \, \text{Round} \, [y[[2]]/L])^\wedge 2 + y[[3]]^\wedge 2 \right];$

If $[r < \text{rcut}, \Delta u \mathrel{+}= 4(r^\wedge(-12) - r^\wedge(-6)), \{\}], \{i, 1, n\}];$

$\Delta \, \text{usurf} = 17.908(0.4(1.034/x[[3]])^\wedge 10 - (1.034/x[[3]])^\wedge 4);$

"3. Metropolis";

If $\left[\text{Min} \left[1, \text{Check} \left[\frac{a}{n+1} \, \text{Exp} \, [-(\Delta u + \Delta \, \text{usurf})/T], 0 \right] \right] \geq \text{Random} \, [\,],$

$\{ \text{TLIST} = \text{Append} \, [\text{TLIST}, x]; n\mathrm{++}\}, \{\}];$

mcsteps ++;

"particle removal";

"1. random selection";

$p = \text{Random} \, [\text{Integer}, \{1, n\}];$

"2. energy change";

$\Delta u = 0;$

$\text{Do} \, [y = \text{Extract} \, [\text{TLIST}, i] - \text{Extract} \, [\text{TLIST}, p];$

$r = \text{Sqrt} \left[(y[[1]] - L \, \text{Round} \, [y[[1]]/L])^\wedge 2 + (y[[2]] - L \, \text{Round} \, [y[[2]]/L])^\wedge 2 + y[[3]]^\wedge 2 \right];$

If $[r > 0 \,\&\&\, r < \text{rcut}, \Delta u \mathrel{+}= 4(r^\wedge(-12) - r^\wedge(-6)), \{\}], \{i, 1, n\}];$

$\Delta \, \text{usurf} =$

$17.908(0.4(1.034/\text{Extract} \, [\text{TLIST}, p][[3]])^\wedge 10 -$

$(1.034/\text{Extract} \, [\text{TLIST}, p][[3]])^\wedge 4);$

"3. Metropolis"; If $\left[\text{Min} \left[1, \text{Check} \left[\frac{n}{a} \, \text{Exp} \, [(\Delta u + \Delta \, \text{usurf})/T], 0 \right] \right] \geq \text{Random} \, [\,],$

$\{ \text{TLIST} = \text{Delete} \, [\text{TLIST}, p]; n - - \}, \{\}];$

mcsteps ++];

"generate density profile normal to surface";

If [cycles > 5,

$\{ \text{Do} \, [\rho \, [[\text{Round} \, [\text{Extract} \, [\text{TLIST}, i][[3]]/(Lz/\text{nint})]]]\mathrm{++}, \{i, 1, \text{Length} \, [\text{TLIST}]\}];$

counter ++; "optional output: histogram";

If $[\text{False}, \{ \text{ListPlot} \, [\rho/(\text{counter} \, V/\text{Length} \, [\rho])] \}, \{\}];$

Print ["cycle", cycles, "of", maxcycles]}, {}]; "optional output: box";

If [False, {pts = Table [Point [Extract [TLIST, i]], $\{i, 1, \text{Length} \, [\text{TLIST}]\}];$

Show [Graphics3D [{PointSize [0.05], pts}]]}, {}]];

"complete density profile";

hist = {}; Do [hist = Append [hist, $\{iLz/\text{Length} \, [\rho], \rho[[i]]$

$/(\text{counter} \, V/(\text{nint} + 1))\}],$

$\{i, 1, \text{Length} \, [\rho]\}];$ ListPlot [hist, Joined $\rightarrow$ True, AxesLabel $\rightarrow$ {"z [LJ]",

"ρ [LJ]"},

PlotRange $\rightarrow \{0, 1\}$, PlotStyle $\rightarrow$ Black]

"box";

If [True, {pts = Table [Point [Extract [TLIST, i]], $\{i, 1, \text{Length} \, [\text{TLIST}]\}];$

Show [Graphics3D [{PointSize [0.05], pts}]]}]

Remark 1 *bulk* refers to the region far from the surface, where the system is homogeneous.

Remark 2 The program assumes that the bulk gas density is low. In particular it uses $\mu_{ex} \approx 2B_2(T)P$, where $B_2(T)$ is the second virial coefficient and P is the (bulk) gas pressure. This is obtain by integrating $\partial\mu/\partial P \,|_T = 1/\rho(P)$. $\rho(P)$ is obtained by inserting the expansion $\rho = c_1 P + c_2 P^2 + \ldots$ into the virial expansion of the pressure, $P = T\rho(1 + B_2\rho + \ldots)$ and comparing coefficients $(c_1 = T^{-1}$; $c_2 = -T^{-2}B_2, \ldots)$. Integration and subsequent subtraction of the ideal gas chemical potential yields $\mu_{ex} = \mu - \mu_{id} = 2B_2 P + \ldots$. The integral formula for the second virial coefficient can be found in every textbook on Statistical Mechanics.

Remark 3 The quantity $r = \text{Sqrt}[(y[[1]] - L\,\text{Round}[y[[1]]/L])^{\wedge}2 + (y[[2]] - L\,\text{Round}[y[[2]]/L])^{\wedge}2 + y[[3]]^{\wedge}2]$ is the *minimum image distance* between the particle x to be inserted or removed and another particle i in the system. The minimum image distance is the smallest distance within the set of all distances between x and i as well as i's periodic images (parallel to the surface). Subsequently the interactions are calculated only if $r < r_{cut}$—a suitable cutoff. In the present program all interactions of x with other particles or image particles are neglected if $r \geq r_{cut}$. r_{cut} must be large enough to justify the neglect of interactions. Simultaneously it must be small enough to avoid inclusion of interactions from one and the same particle more than once via its periodic images. Notice that the minimum image construction allows the actual particles to be anywhere in space—even outside the simulation box (cf. Frenkel and Smit 1996; Allen and Tildesley 1990).

Appendix C
Constants, Units, Tables

N_A	$6.02214\ldots \cdot 10^{23}$ mol^{-1}	Avogadro's number
R	$8.31447\ldots$ JK^{-1}mol^{-1}	Gas constant
hN_A	$3.99031\ldots \cdot 10^{-10}$ Jsmol^{-1}	h: Planck's constant ($\hbar = h/(2\pi)$)
$m_{amu}N_A$	10^{-3} kg	$m_{amu} = \frac{1}{12}m(^{12}C)$: Atomic mass constant
$F = eN_A$	$9.64853\ldots \cdot 10^4$ Cmol^{-1}	Faraday constant $= eN_A$; e: elementary charge
ε_o	$8.85418\ldots \cdot 10^{-12}$ Fm^{-1}	Electric constant
μ_o	$1.25663\ldots \cdot 10^{-6}$ NA^{-2}	Magnetic constant
c	$2.99792\ldots \cdot 10^8$ ms^{-1}	Vacuum speed of light ($c = (\varepsilon_o\mu_o)^{-1/2}$)
g	9.80665 ms^{-2}	Standard gravitational acceleration
G	$6.673\ldots \cdot 10^{-11}$ m^{3}kg^{-1}s^{-2}	Gravitational constant
1 bar	10^5 Pa	$1\,\text{Pa} = 1\,\text{Nm}^{-2}$
1 atm	101,325 Pa	
1 psi	703.0696 kgm^{-2}	
1 cmHg	1333.224 Pa	
1 Torr	133.322 Pa	
1 cal	4.1858 J	
1 eV	$1.60217\ldots \cdot 10^{-19}$ J $1.16045\ldots \cdot 10^4$ K	
1 kWh	$3.6 \cdot 10^6$ J	
0 °C	273.15 K	

© Springer Nature Switzerland AG 2022
R. Hentschke, *Thermodynamics*, Undergraduate Lecture Notes in Physics,
https://doi.org/10.1007/978-3-030-93879-6

Useful Tables:

HCP: Lide (2005)
HTTD: Lide and Kehiaian (1994)

Conversion Between Gaussian and SI-Units:

Quantity	Gaussian	SI
Speed of light	c	$(\mu_o \varepsilon_o)^{-1/2}$
Electric field	$\vec{E}$	$\sqrt{4\pi\varepsilon_o}\vec{E}$
Displacement	$\vec{D}$	$\sqrt{4\pi/\varepsilon_o}\vec{D}$
Charge	q	$q/\sqrt{4\pi\varepsilon_o}$
Magnetic induction	$\vec{B}$	$\sqrt{4\pi/\mu_o}\vec{B}$
Magnetic field	$\vec{H}$	$\sqrt{4\pi\mu_o}\vec{H}$
Magnetization	$\vec{M}$	$\sqrt{\mu_o/(4\pi)}\vec{M}$

References

Abramowitz, M., Stegun, I. (eds.): Handbook of Mathematical Functions. Dover, New York (1972)

Adair, G.S.: A theory of partial osmotic pressures and membrane equilibria, with special reference to the application of Dalton's law to haemoglobin solutions in the presence of salts. Proc. R. Soc. Ser. A **120**, 573 (1928)

Allen, G.: Reflexive catalysis, a possible mechanism of molecular duplication in prebiological evolution. Am. Nat. **91**, 65 (1957)

Allen, M.P., Tildesley, D.J.: Computer Simulation of Liquids. Clarendon Press, Oxford (1990)

Ashcroft, N.W., Mermin, N.D.: Solid State Physics. Saunders College Publishing, Philadelphia (1976)

Atkins, P.W.: Physical Chemistry. Oxford University Press, Oxford (1986)

Aydt, E.M., Hentschke, R.: Quantitative molecular dynamics simulation of high pressure adsorption isotherms of methane on graphite. Ber. Bunsenges. Phys. Chem. **101**, 79 (1997)

Bekenstein, J.: Black holes and entropy. Phys. Rev. D **7**, 2333 (1973)

Chandler, D.: Introduction to Modern Statistical Mechanics. Oxford, New York (1987)

de Gennes, P.-G.: Scaling Concepts in Polymer Physics. Cornell University Press, Ithaca (1988)

de Gennes, P.-G., Brochard-Wyart, F., Quéré, D.: Capillarity and Wetting Phenomena. Springer, New York (2004)

Dyson, F.J.: Existence of a phase-transition in a one-dimensional Ising ferromagnet. Commun. Math. Phys. **12**, 91 (1969)

Dyson, F.: Origins of Life. Cambridge University Press, Cambridge (1985)

Ebeling, W., Feistel, R.: Physik der Selbstorganisation und Evolution. Akademie-Verlag, Berlin (1986)

Eigen, M.: Viral quasispecies. Sci. Am. **269**, 42 (1993)

Eigen, M.: Steps Towards Life: A Perspective on Evolution. Oxford University Press, Oxford (1996)

Eigen, M., Winkler, R.: Laws of the Game: How the Principles of Nature Govern Chance. Princeton University Press, Princeton (1993)

Eigen, M., Gardiner, W., Schuster, P., Winkler-Oswatitsch, R.: The origin of genetic information. Sci. Am. **244**, 88 (1981)

Evans, D.F., Wennerström, H.: The Colloidal Domain. VCH, New York (1994)

Feigenbaum, M.: Universal behavior in nonlinear systems. Physica **7D**, 16 (1983)

Fermi, E.: Thermodynamics. Dover, New York (1956)

Feynman, R.P.: Statistical Mechanics. Addison Wesley, Reading MA (1972)

Feynman, R.P.: Feynman Lectures on Gravitation. Westview Press, Colorado (2003)

© Springer Nature Switzerland AG 2022   339
R. Hentschke, *Thermodynamics*, Undergraduate Lecture Notes in Physics,
https://doi.org/10.1007/978-3-030-93879-6

Fisher, M.E.: Renormalization group theory: its basis and formulation in statistical physics. Rev. Mod. Phys. **70**, 653 (1998)

Flory, P.J., Rehner, J.: Statistical mechanics of cross-linked polymer networks I. Rubberlike elasticity. J. Chem. Phys. **11**, 521 (1943)

Fowkes, F.M.: Attractive forces at interfaces. Ind. Eng. Chem. **56**, 40 (1964)

Frank, H.S.: Thermodynamics of a fluid substance in the electrostatic field. J. Chem. Phys. **23**, 2023 (1955)

Frenkel, D., Smit, B.: Understanding Molecular Simulation. Academic Press, California (1996)

Glansdorff, P., Prigogine, I.: Thermodynamic Theory of Structure, Stability and Fluctuations. Wiley-Interscience, New York (1971)

Goldberg, D.E.: Genetic Algorithms in Search, Optimization & Machine Learning. Addison-Wesley, Reading (1989)

Gould, H., Tobochnik, J.: An Introduction to Computer Simulation Methods. Addison-Wesley, Reading (1996)

Guggenheim, E.A.: On magnetic and electrostatic energy. Proc. R. Soc. Lond. **155A**, 49 (1936)

Guggenheim, E.A.: The thermodynamics of magnetization. Proc. R. Soc. Lond. **155A**, 70 (1936)

Guggenheim, E.A.: The principle of corresponding states. J. Chem. Phys. **13**, 253 (1945)

Guggenheim, E.A.: Thermodynamics—An Advanced Treatment for Chemists and Physicists. North-Holland, Amsterdam (1986)

Guse, C.: Properties of confined and unconfined water. Ph.D. thesis, Wuppertal University, 2011

Haldane, J.B.S.: The Causes of Evolution. Princeton University Press, Princeton (1990)

Hamer, W.J., Wu, Y.-C.: Osmotic coefficients and mean activity coefficients of uni-univalent electrolytes in water at 25 °C. J. Phys. Chem. Ref. Data **1**, 1047 (1972)

Haupt, A., Straub, J.: Evaluation of the isochoric heat capacity measurements at the critical isochore of SF6 performed during the German Spacelab Mission D-2. Phys. Rev. E **59**, 1795 (1999)

Hawking, S.: Black holes and thermodynamics. Phys. Rev. D **13**, 191 (1976)

Hendricks, R.C., et al.: Joule-Thomson inversion curves and related coefficients for several simple fluids. NASA Technical Note D-6807 (1972)

Hentschke, R.: Introductory Quantum Theory. Lecture Notes (2009). http://constanze.materials.uni-wuppertal.de

Hentschke, R., Hölbling, Ch.: A Short Course in General Relativity and Cosmology. Springer (2020)

Hentschke, R., Aydt, E.M., Fodi, B., Stöckelmann, E.: Molekulares Modellieren mit Kraftfeldern (2004). http://constanze.materials.uni-wuppertal.de

Herzfeld, J.: Entropically driven order in crowded solutions: from liquid crystals to cell biology. Acc. Chem. Res. **29**, 31 (1996)

Hill, T.L.: Statistical Mechanics. Dover, New York (1956)

Hill, T.L.: Statistical Thermodynamics. Dover, New York (1986)

Hirschfelder, J.O., Curtiss, C.F., Bird, R.B.: Molecular Theory of Gases and Liquids. Wiley, New York (1954)

Huang, K.: Statistical Mechanics. Wiley, New York (1963)

Israelachvili, J.: Intermolecular & Surface Forces. Academic Press, London (1992)

Kadanoff, L.P.: From Order to Chaos. World Scientific, Singapore (1993)

Kadanoff, L.P.: More is the same; phase transitions and mean field theories. J. Stat. Phys. **137**, 777 (2009)

Kaeble, D.H.: Dispersion-polar surface tension properties of organic solids. J. Adhes. **2**, 66 (1970)

Koenig, F.O.: The thermodynamics of the electric field with special reference to chemical equilibrium. J. Phys. Chem. **41**, 597 (1937)

Kondepudi, D., Prigogine, I.: Modern Thermodynamics. Wiley, New York (1998)

Koningsveld, R., Kleintjens, L.A.: Fluid phase equilibria. Acta Polym. **39**, 341–350 (1988)

Kramers, H.A., Wannier, G.H.: Statistics of the two-dimensional ferromagnet. Part I. Phys. Rev. **60**, 252 (1941)

Landau, L.D., Lifshitz, E.M., Pitaevskii, L.P.: Theory of Elasticity. Elsevier Science, Amsterdam (1986)

Levin, Y., Fisher, M.E.: Criticality in the hard-sphere ionic fluid. Phys. A **225**, 164 (1996)

Ley-Koo, M., Green, M.S.: Consequences of the renormalization group for the thermodynamics of fluids near the critical point. Phys. Rev. A **23**, 2650 (1981)

Lide, D.R.: Handbook of Chemistry and Physics. CRC Press, Boca Raton (2005)

Lide, D.R., Kehiaian, H.V.: Handbook of Thermophysical and Thermochemical Data. CRC Press, Boca Raton (1994)

Lifshitz, E.M., Landau, L.D., Pitaevskii, L.P.: Electrodynamics of Continuous Media. Elsevier Science, Amsterdam (2004)

Mather, J.C., et al.: A preliminary measurement of the cosmic microwave background spectrum by the COsmic Background Explorer (COBE) satellite. ApJ **354**, L37–L40 (1990)

Murad, S., Powles, J.G., Holtz, B.: Osmosis and reverse osmosis in solutions: Monte Carlo simulation and van der Waals one-fluid theory. Mol. Phys. **86**, 1473 (1995)

Nicolis, G., Progogine, I.: Self-Organization in Non-Equilibrium Systems. Wiley, New York (1977)

Nikitin, E.D.: The critical properties of thermally unstable substances: measurement methods, some results and correlations. High Temp. **36**, 305 (1998)

Nowak, M.A., Ohtsuki, H.: Prevolutionary dynamics and the origin of evolution. Proc. Natl. Acad. Sci. USA **105**14924 (2008)

Nüsslein-Volhard, Ch.: Coming to Life. How Genes Drive Development. Yale University Press, New Haven (2006)

Odijk, T.: Theory of lyotropic liquid crystals. Macromolecules **19**, 2313 (1986)

Onsager, L.: Reciprocal relations in irreversible processes. I. Phys. Rev. **37**, 405 (1931)

Onsager, L.: Electric moments of molecules in liquids. J. Am. Chem. Soc. **58**, 1486 (1936)

Onsager, L.: Crystal statistics. I. A two-dimensional model with an order-disorder transition. Phys. Rev. **65**, 117 (1944)

Oparin, A.I.: The Chemical Origin of Life. Thomas, Springfield (1964)

Owens, D., Wendt, R.: Estimation of the surface free energy of polymers. J. Appl. Polym. Sci. **13**, 1741 (1969)

Ozog, J.Z., Morrison, J.A.: Activity coefficients of acetone-chloroform solutions. J. Chem. Ed. **60**, 72 (1983)

Panagiotopoulos, A.Z.: Determination of phase coexistence properties of fluids by direct Monte Carlo simulation in a new ensemble. Mol. Phys. **61**, 813 (1987)

Panagiotopoulos, A.Z., Quirke, N., Stapleton, M., Tildesley, D.J.: Phase equilibria by simulation in the Gibbs ensemble. Mol. Phys. **63**, 527 (1988)

Pathria, R.K.: Statistical Mechanics. Pergamon Press, Oxford (1972)

Pauli, W.: Statistical Mechanics. Dover, New York (1972)

Pauli, W.: Thermodynamics and the Kinetic Theory of Gases. Dover, New York (1973)

Peierls, R.: On Ising's model of ferromagnetism. Proc. Camb. Philos. Soc. **32**, 477 (1936)

Pitzer, K.: Critical phenomena in ionic fluids. Acc. Chem. Res. **23**, 333 (1990)

Prigogine, I.: Etude Thermodynamique des Processus Irreversibles. Liège, Desoer (1947)

Rabel, W.: Einige Aspekte der Benetzungstheorie und ihre Anwendung auf die Untersuchung und Veränderung der Oberflächeneigenschaften von Polymeren. Farbe Lack **77**, 10 (1971)

Roe, R.-J., Zin, W.-C.: Determination of the polymer-polymer interaction parameter for the polystyrene-polybutadiene pair. Macromolecules **13**, 1221 (1980)

Rowlinson, J.S., Widom, B.: Molecular Theory of Capillarity. Oxford University Press, Oxford (1989)

Rubinstein, M., Colby, R.H.: Polymer Physics. Oxford University Press, Oxford (2003)

Salby, M.L.: Physics of the Atmosphere and Climate. Cambridge University Press, Cambridge (2012)

Schopf, J.W.: Fossil evidence of arcane life. Phil. Trans. R. Soc. B **361**, 869 (2006)

Schreiber, S., Hentschke, R.: Monte Carlo simulation of osmotic equilibria. J. Chem. Phys. **135**, 134106 (2011)

Sengers, J.V., Shanks, J.G.: Experimental critical exponent values for fluids. J. Stat. Phys. **137**, 857 (2009)

Shultz, A.R., Flory, P.J.: Phase equilibria in polymer-solvent systems. J. Am. Chem. Soc. **74**, 4760 (1952)

Specovious, J., Findenegg, G.H.: Physical adsorption of gases at high pressure: argon and methane onto graphitized carbon black. Ber. Bunsenges. Phys. Chem. **82**, 174 (1978)

Spiegel, M.R.: Advanced Mathematics. Schaum's Outline Series. McGraw-Hill, New York (1971)

Stanley, H.E.: Introduction to Phase Transitions and Critical Phenomena. Oxford University Press, Oxford (1971)

Strobl, G.: The Physics of Polymers. Springer, Berlin (1997)

Susskind, L., Lindsay, J.: An Introduction to Black Holes, Information and the String Theory Revolution. World Scientific, Singapore (2005)

Teller, E.: On the change of physical constants. Phys. Rev. **73**, 801 (1948)

Thompson, R.B., et al.: Origin of change in molecular-weight dependence for polymer surface tension. Phys. Rev. E **78**, 030801 (2008)

Treloar, L.R.G.: The elasticity and related properties of rubbers. Rep. Prog. Phys. **36**, 755 (1973)

Vilgis, T.A., Heinrich, G., Klüppel, M.: Reinforcement of Polymer Nano-Composites. Cambridge University Press, New York (2009)

Washburn, E.W.: The dynamics of capillary flow. Phys. Rev. **17**, 374 (1921)

Weinberg, S.: Cosmology. Oxford University Press, Oxford (2008)

Weiss, V.C.: Guggenheim's rule and the enthalpy of vaporization of simple and polar fluids, molten salts, and room temperature ionic liquids. J. Phys. Chem. B **114**, 9183 (2010)

Widom, B.: Equation of state in the neighborhood of the critical point. J. Chem. Phys. **43**, 3898 (1965)

Woese, C.R.: On the evolution of cells. PNAS **99**, 8742 (2002)

Wrana, C.: Introduction to Polymer Physics. LANXESS, Leverkusen (2009)

Yuan, Y., Lee, T.R.: Contact angle and wetting properties. In: Surface Science Techniques, pp. 3–34. Springer, Heidelberg (2013)

Zhi, F.L., Xian, L.S.: Creation of the Universe. World Scientific Publishing, Singapore (1989)

Index

© Springer Nature Switzerland AG 2022
R. Hentschke, *Thermodynamics*, Undergraduate Lecture Notes in Physics,
https://doi.org/10.1007/978-3-030-93879-6